Environmental Hazards

Assessing risk and reducing disaster

Third edition

Keith Smith

Routledge
Taylor & Francis Group

LONDON AND NEW YORK

First edition published 1991
by Routledge
11 New Fetter Lane, London EC4P 4EE

Simultaneously published in the USA and Canada
by Routledge
29 West 35th Street, New York, NY 10001

Second edition 1996
Third edition 2001

Reprinted 2002, 2003

Routledge is an imprint of the Taylor & Francis Group

Typeset in Garamond by
Florence Production Ltd, Stoodleigh, Devon
Printed and bound in Great Britain by
Biddles Ltd, Guildford and King's Lynn

British Library Cataloguing in Publication Data
A catalogue record for this book is available
from the British Library

Library of Congress Cataloguing in Publication Data
Smith, Keith
    Environmental hazards: assessing risk and reducing disaster/
    Keith Smith. – 3rd ed.
        p. cm. – (Routledge physical environment series)
        Includes bibliographical references (p.)
        1. Natural disasters. I. Title. II. Series.

GB5014.S6 2000
363.34–dc21   00–038251

ISBN 0–415–22463–2 (hbk)
ISBN 0–415–22464–0 (pbk)

Environmental Hazards

Disasters — both natural and human-made — continue to take their toll on human life and property worldwide. This substantially revised third edition maintains that existing knowledge about hazards and hazard mitigation must be better applied in order to reduce death and destruction, especially in developing countries. *Environmental Hazards* provides an objective and thoughtful survey of key findings from the natural and social sciences and integrates them to provide a comprehensive assessment of the threat from disasters and the policy responses which are required to achieve a safer world.

Environmental Hazards covers all the major rapid-onset events, whether natural, human or technological in origin, which directly harm humans, and what they value, on a community scale. The first half of this study comprises a broad overview of environmental risk which emphasises that all disasters arise from a combination of physical exposure and human vulnerability. It includes topics such as disaster databases, problems of monitoring the impact of disaster in both developed and developing countries together with the management and evaluation of risk. The second half provides a systematic treatment of environmental hazards. Different types of hazard create different losses in different countries. Extensive global case studies and policy responses are examined and the hazard reduction strategies that are relevant in each case are explained.

The revised third edition includes new material on disaster databases, El Niño events, sea-level rise and coastal flooding, global change and sustainability in mega-cities and elsewhere. This complicated subject is presented with a clarity which is reinforced with excellent summary tables and diagrams. A new colour plate section illustrates some disasters which have occurred since the last edition. References have also been fully updated and the author looks towards future problems in the twenty-first century. This important book provides a balanced and up-to-date approach to environmental hazards and is ideally suited to act as a course text.

Keith Smith is Emeritus Professor of Environmental Science at the University of Stirling.

A volume in the Routledge Physical Environment Series
Edited by Keith Richards
University of Cambridge

The Routledge Physical Environment series presents authoritative reviews of significant issues in physical geography and the environmental sciences. The series aims to become a complete text library, covering physical themes, specific environments, environmental change, policy and management, as well as developments in methodology, techniques and philosophy.

Other titles in the series:

Ice Age Earth
Late quaternary geology and climate
A. Dawson

The Geomorphology of Sand Dunes
N. Lancaster

Soils and Environment
S. Ellis and A. Mellor

Tropical Environments
The functioning and management of tropical ecosystems
M. Kellman and R. Tackaberry

Eco-hydrology
Ed.: Andrew J. Baird and Robert L. Wilby

Forthcoming:

Environmental Issues in the Mediterranean
J. Thornes and J. Wainwright

Mountain Geography
D. Funnell and R. Parish

Contents

Plates

Black and white

Colour

The colour plates appear between pp. 200–1.

Figures

Tables

Preface to the third edition

The first edition of *Environmental Hazards* was conceived at the start of the International Decade for Natural Disaster Reduction in 1990 and was rewritten some five years later. This edition appears at the end of the Decade and, perhaps more importantly, at the beginning of the twenty-first century. There is a unique opportunity, therefore, to review past developments in our understanding of environmental hazards and to anticipate what progress may be made as the new century begins to unfold. Such tasks are daunting – partly because of the current pace of global change – but also because the study of environmental hazards itself continues to broaden beyond the earlier concerns about site-specific risks to embrace complex, large-scale issues. Despite such difficulties, this new edition seeks to convey a flavour of the key initiatives, both theoretical and practical, that have already taken place and to provide a reasoned path which guides the reader through the growing maze of information towards a perspective on the future.

As in previous editions, the focus of *Environmental Hazards* remains on rapid-onset events that threaten human life and property on a scale which is sufficient to cause 'disasters'. In the interests of continuity, the basic structure and balance of the book has been retained but all the chapters have been revised and up-dated, some substantially so. Wherever possible, I have tried to sharpen the discussion in the text and to strengthen the case-study and illustrative material. These revisions have been aided by the helpful suggestions made by various colleagues and reviewers. No author can hope to capture more than a fraction of the hazards-related material now available but a wide range of references enables the reader to pursue chosen topics in greater detail.

Keith Smith
Braco, Perthshire
February 2000

Preface to the first edition

This book has been written primarily to provide an introductory text on environmental hazards for university and college students of geography, environmental science and related disciplines. It springs from my own experience in teaching such a course over several years and my specific inability to find a review of the field which matches my own priorities and prejudices. I hope, therefore, that this survey will prove useful as a basic source for appropriate intermediate to advanced undergraduate classes in British, North American and Antipodean institutions of higher education. If it encourages some students to pursue more advanced studies, or provides a means whereby other readers become more informed about hazardology, either as policy-makers or citizens, then I will be well satisfied. Without a wider appreciation of the factors underlying the designation by the United Nations of the 1990s as the International Decade for Natural Disaster Reduction (IDNDR), the important practical aims of the Decade to improve human safety and welfare are unlikely to be achieved.

The term 'environmental hazards' defies precise definition. Not everyone, therefore, will endorse either my choice of material or its treatment in terms of the balance between physical and social science concepts. In this book, the prime focus is on rapid-onset events, from either a natural or a technological origin, which directly threaten human life on a community scale through acute physical or chemical trauma. Such events are often associated with economic losses and some damage to ecosystems. Most disaster impact arises from 'natural' hazards and is mainly suffered by the poorest people in the world. Within this context, my intention, as expressed in the sub-title, has been to assess the threat posed by environmental hazards as a whole and to outline the actions which are needed to reduce the disaster potential.

The structure of the book reflects the need to distinguish between common principles and their application to individual case studies. Part I – the nature of hazard – seeks to show that, despite their diverse origins and differential impacts, environmental hazards create similar sorts of risks and disaster-reducing choices for people everywhere. Here the emphasis is on the identification and recognition of hazards, and their impact, together with the range of mitigating adjustments that humans can make. These loss-sharing and loss-reducing adjustments form a recurring theme throughout the book. In Part II – the experience and reduction of hazard – individual environmental threats are considered under five main generic headings (seismic hazards, mass movement hazards, atmospheric hazards, hydrologic hazards and technologic hazards). In this section the concern is for the assessment of specific hazards and the contribution which particular mitigation strategies either have made or may make to reducing the losses of life and property from that hazard.

Keith Smith
Braco, Perthshire
July 1990

Acknowledgements

This book could not have been completed without generous assistance from many sources. Over the years I have been grateful to the Drapers' Company of London for providing the opportunity to spend six months based at the University of Adelaide and to the Canadian High Commission for funding me to travel widely in that interesting country. The University of Stirling has supplied ongoing research facilities and has released me from other duties to permit several sabbatical leave visits to the USA, New Zealand and elsewhere in order to research environmental hazards.

This new edition continues to benefit from the informed comments of people much more expert in certain areas than I could ever be and I should like to acknowledge the painstaking work of Jeff Keaton (AGRA Earth and Environmental, Salt Lake City), Bill Murphy (University of Portsmouth), Mike Thomas (University of Stirling) and Roy Ward (formerly of the University of Hull). If I have ignored – or misunderstood – their suggestions, I can only apologise.

A number of individuals have kindly supplied unpublished information and helped with the diagrams and plates. I am particularly grateful to Darren Shaw for allowing me access to the CRED disaster database and for providing some key statistics and to Marjory Roy (formerly Meteorological Office, Edinburgh) for the wind speed data on which Figure 3.3 is based. Francis Urban (Economic Research Service of the US Department of Agriculture) and David Gilvear (University of Stirling) produced the data for Figures 12.6 and 12.7 respectively. Sue Pavin (International Federation of Red Cross and Red Crescent Societies) allowed me free access to that organisation's photo library in Geneva and her successor, Françoise Borst-Vermot, was equally efficient in supplying the slides from which many of the black and white plates are taken. My colleague Michael Thomas provided Colour Plate 3. Once again, the diagrams have been very ably prepared by Bill Jamieson and David Aitchison in the Department of Environmental Science at the University of Stirling.

The successful compilation of the book has depended heavily on the editorial advice received from Routledge's London office and, for this edition, Andrew Mould and Ann Michael have been invaluable in offering the right blend of encouragement and practical tips necessary to keep the project on schedule. Finally, my wife, Muriel, has not only provided the relaxed domestic environment necessary for writing but has also been actively involved in many aspects of the preparation of the book. No one else knows how much she has helped and this book continues to be dedicated to her in sincere, if inadequate, recognition of this long-standing support.

The author and publishers would like to thank the following learned societies, editors, publishers, organisations and individuals for permission to reprint, or reproduce in modified form, copyright material in various figures, tables and plates as indicated below. Every effort has been made to identify and acknowledge the original sources but, if there have been any accidental errors or omissions, we apologise to those concerned.

Learned societies

American Association for the Advancement of Science for Figure 3.1 and Table 1.1 from *Science* by C. Starr and for Table 13.1 from *Science* by C. Hohenemser *et al*.

American Geophysical Union for Table 6.7 from *EOS* by T.L. Holzer and Table 7.1 from the *Journal* by C. Newhall and S. Self.

American Meteorological Society, Boston, for Figures 9.2 and 9.3 from *Weather and Forecasting* by R.A. Pielke Jr and C.W. Landsea.

American Planning Association for Figure 11.11 from the *Journal* by E. David and J. Meyer.

International Mountain Society for Figure 8.9 from *Mountain Research and Development* by S.D. Oaks and L. Dexter.

Oceanography Society for Figure 9.7 from *Oceanography* by W.S. Broecker.

The Geographical Association for Figure 1.11 from *Geography* by M. Degg.

The Royal Academy of Engineering for Figure 2.3 from *Development at Risk?* by J. Twigg.

The Royal Society of London for Table 13.7 from *Risk: Analysis, Perception and Management* by D. Cox *et al*.

Publishers

Academic Press, Orlando, for Figure 7.7a from *Volcanic Activity and Human Ecology* by P.D. Sheets and D.K. Grayson (eds).

Australian Government Publishing Company, Canberra, for Figure 10.3 from *Bushfires in Australia* by R.H. Luke and A.G. McArthur.

Blackwell Publishers for Table 2.1 from *Disasters* by D.G. Sapir and C. Misson, Table 6.1 from *Disasters* by S. Parasuraman and Table 13.4 from *Professional Geographer* by S. Cutter and M. Ji.

Cambridge University Press for Figure 1.1 from *The Business of Risk* by P.G. Moore.

Elsevier Science for Table 14.1 from *Global Environmental Change* by R.J. Nicholls *et al*.

W.H. Freeman and Company, New York, for Figures 6.2, 6.3 and 6.5 from *Earthquakes* B.A. Bolt. © 1978, 1988, 1993 by W.H. Freeman and Company. Used with permission.

S. Karger AG, Basel, for Figure 4.1 from *Epidemiology of Natural Disasters* by J. Seaman, S. Leivesley and C. Hogg.

Kluwer Academic Publishers, Dordrecht, for Figure 6.10 from *Tsunamis: their Science and Engineering* by K. Iida and T. Iwasaki (eds).

MIT Press, Cambridge, Mass., for Figure 4.2 from *Reconstruction following Disaster* by J.E. Haas, R.W. Kates and M.J. Bowden.

National Academy Press, Washington, DC, for Figure 8.6 from *Confronting Natural Disasters* by G.W. Housner.

Osservatorio Vesuviano in co-operation with the United Nations IDNDR Secretariat for Table 3.4 from *STOP Disasters* by G. Wadge.

Oxford University Press, New York, for Figure 1.3 from *The Environment as Hazard* by I. Burton, R.W. Kates and G.F. White.

Pearson Education, Harlow, for Figure 11.6 from *Regions of Risk* by K. Hewitt.

Plenum Publishing Company for Table 13.5 from *Risk Analysis* by A.F. Fritzsche.

Reed Books, Australia, for Figure 10.4 from *Natural Disasters* by J.E. Butler.

Springer-Verlag, Heidelberg, for Tables 6.2 and 6.6 and Figures 7.2, 7.3 and 7.4 from *Geological Hazards* by B.A. Bolt, W.L. Horn, G.A. Macdonald and R.F. Scott and for Figures 5.3 and 7.6 from *Monitoring and Mitigation of Volcano Hazards* by R. Scarpa and R.I. Tilling (eds).

Thomas Telford Publishing, London, for Figures 6.6 and 14.1 and Table 1.3.

John Wiley and Sons, New York, for Figure 1.5 from *Climate Impact Assessment* by R.W. Kates, J.H. Ausubel and M. Berberian (eds) and Figure 9.6 from *Hurricanes* by R.A. Pielke Jr and R.A. Pielke Sr.

Organisations

Athens Center of Ekistics for Figure 8.1b from *Ekistics* by B.R. Armstrong.

California Seismic Safety Commission for Table 6.5 from *California at Risk* by W. Spangle and Associates Inc.

Centre for Resource and Environmental Studies, Canberra, for Figure 11.3 from *Flood Damage in the Richmond River Valley NSW* by D.I. Smith *et al.*

Illinois State Water Survey for Figure 11.8 from *The 1993 Flood on the Mississippi River in Illinois* by N.G. Bhowmik.

ISPAN Technical Support Center, Virginia, for Figure 11.1 from *Eastern Waters Study* by P. Rogers, P. Lydon and D. Seckler.

Met. Office, Bracknell, for Figures 10.1 and 14.2 from *Climate Change and its Impacts* by P. Martens *et al.* and R.J. Nicholls respectively.

United Nations Environment Programme, Nairobi, for Figure 2.4 from *Environmental Data Report.*

United States Department of Agriculture, Washington, DC, for Figure 12.6 originating in data from the Economic Research Service, USDA.

United States Geological Survey, Denver, Colorado, for Figure 11.9 from *Effects of Reservoirs on Flood Discharges on the Kansas and Missouri River Basins (Circular no. 1120E)* by C.A. Perry.

United States Geological Survey, Virginia, for Figure 7.7b from *The 1980 Eruptions of Mount St Helens* by C.D. Miller, D.R. Mullineaux and D.R. Crandell.

University of Toronto Department of Geography for Figure 1.4 and Table 1.2 from *The Hazardousness of a Place* by K. Hewitt and I. Burton.

University of Toronto Institute of Environmental Studies for Tables 3.1 and 3.3 from *Living with Risk: Environmental Risk Management in Canada* by I. Burton, C.D. Fowle and R.J. McCullough (eds).

World Bank, Washington, DC, for Table 11.1 from *Managing Natural Disasters and the Environment* by J. Brown and M. Muhsin.

Individuals

Professor R.G. Barry of University of Colorado for Figure 9.5.

Dr G. Berz of Munchener Ruch for Figures 2.7 and 9.4.

Dr K.R. Berryman of DSIR, Wellington, for Figure 6.9.

Dr W.S. Broecker of Columbia University, Palisades, New York, for Figure 9.7.

Professor R.J. Chorley of Cambridge University for Figure 9.5.

Dr D.R. Crandell of US Geological Survey, Denver, for Figure 7.7a.

Dr Jan de Vries, University of California, Berkeley, for Figure 1.5.

Dr D.R. Donald of US Department of Agriculture for Figure 12.3.

Dr D.J. Gilvear of Stirling University for Figure 12.7.

Dr T.S. Glickman of Resources for the Future, Washington, DC, for Figures 2.2 and 2.5.

Professor G.W. Housner of California Institute of Technology for Figure 6.1.

Professor R.W. Kates of Clark University for Figure 1.8.

Professor P.G. Moore of the London Graduate School of Business Studies for Figure 1.1.

Dr T. Omachi of the Infrastructure Development Institute, Japan, for Figure 12.8.

Professor D. Parker of Middlesex University, London, for Figure 11.5.

Dr D.E. Parker, Met. Office, and the editor of *Weather*, for Figure 12.5.

Dr R.A. Pielke Jr, National Center for Atmospheric Research, Boulder, Colorado, for Figure 2.8.

Dr D.J. Shaw of the Centre for Research on the Epidemiology of Disasters, University of Louvain, for Figure 2.1.

Professor P. Susman of Bucknell University, Louisburg, for Figure 2.10.

Dr J. Tomblin of the Department of Humanitarian Affairs, Geneva, for Figure 3.7.

Dr J. Whittow of Reading University for Figure 6.4.

Part I

The nature of hazard

1 Hazard in the environment

HAZARD IN CONTEXT

As we enter the twenty-first century, the earth supports a human population which, in general, is more numerous, healthier and wealthier than ever before. At the same time, there is an unprecedented awareness of risk in the environment together with a growing concern for the continuing death and destruction caused by 'natural' hazards. This paradox exists because natural hazards and human progress are rooted in the same ongoing processes of global change. As the world population grows and owns more material possessions, and as the built environment expands to accommodate such changes, greater numbers of people and property are put at risk from the forces of nature. These demographic and social trends also impose heavy burdens on precious natural assets, such as land and water. Many people in the poorest countries now have a fragile dependence on a degraded resource base which becomes progressively less able to withstand pressures from environmental forces. Human progress has also led to the emergence of 'man-made' threats. Environmental hazards are no longer limited to major geophysical events, such as earthquakes and floods, but include industrial explosions, major transport accidents and other technological threats. A growing recognition of hazard is encouraged because disasters make news. With the continuing improvement in global communications and widespread media reporting, the graphic results of hazards, both natural and man-made, feature repeatedly in newspapers and on television screens throughout the world.

What, then, is the reality? Is the world becoming a more dangerous place? Are natural hazards increasing? Why does human society appear more vulnerable to certain environmental processes? What is the added risk from the newer technological hazards? Is it possible to eliminate environmental hazards? If not, how can we define an acceptable level of risk? What is a disaster? What are the best means of mitigating disaster? Why, despite the investment in disaster reduction measures, do losses continue to rise?

Clear answers to all these questions remain elusive. Although a concern for risk can be traced back to the earliest recorded times (Covello and

Mumpower, 1985), broad-based research into natural hazards did not begin until almost the middle of the twentieth century. Until then, hazards had been viewed as isolated geophysical events, somehow divorced from society, which were to be tamed by engineering works, such as dams and levees. Gilbert White (1936, 1945) was the first person to question these attitudes by asserting that river control schemes were not necessarily the best – or the only – way to tackle flood problems in the USA. As a geographer, White's contribution was to introduce a social perspective. He cast natural hazards into a human ecological framework, operating at the interface of both natural and human systems, which allowed for other solutions than the 'structural' schemes as then implemented by civil engineers. Over the next 20–30 years, this pioneering work was extended by geographers associated with the 'Chicago School', whilst social scientists, such as sociologists, began their own interpretation of the role played by people in 'natural' hazards.

By the early 1970s, the study of natural hazards was highly fragmented. Scientists, such as geologists, hydrologists and civil engineers, continued with direct attempts to mitigate hazards through a strategy of environmental control based on the improved prediction of extreme natural events and the construction of physical works designed to resist them. More theoretical perspectives were split into two distinct camps (Mileti *et al.*, 1995). Geographers led a *hazard-based* approach predicated on the unifying concept of human ecology and the notion of mitigating losses by adding various human adjustments, such as better hazard perception and land use planning, to the existing use of physical control structures. Sociologists, on the other hand, adopted a more *disaster-based* view with an emphasis on understanding the role of collective human behaviour at times of community crisis and the need to improve preparedness for such mass emergencies.

During the 1970s extreme natural events became more prominent. For example, the prolonged Sahelian drought, the failures of the Peruvian anchovy harvest, the 1975–6 drought in north-west Europe and the severe North American winters of 1976–7 and 1977–8 exposed the vulnerability of many countries, including advanced nations, to climatic variability. Other significant events, such as the 1970 Bangladesh cyclone and the 1976 earthquake in China, prompted humanitarian concern for the apparent rapid rise in losses from natural hazards in the less developed countries. The decade also saw the publication of several important books, mainly from the North American research school inspired by White (White, 1974; White and Haas, 1975; Burton *et al.*, 1978, updated 1993).

The 1980s brought new developments. Emphasis was given to the relationships between under-development and hazard impact in the Third World, especially the extent to which socio-economic factors, such as economic dependency and a colonial legacy, exacerbates the effects of geophysical events. During this decade vulnerability to disaster, a characteristic of the poorest and most disadvantaged people, started to emerge as a key concept for social scientists. These trends were underpinned by reported statistics showing a rapid rise in losses from natural hazards, particularly in the developing world (Wijkman and Timberlake, 1984). In addition, following early signals of technological hazard in the 1970s – such as the Flixborough (UK) explosion in 1974, the release of dioxin at Seveso (Italy) in 1976 and the nuclear incident at Three Mile Island (USA) in 1979 – so-called 'man-made' hazards gained a new prominence. The year 1984 was a turning point with several major industrial accidents, including the release of methyl isocyanate at Bhopal (India). As a result, existing multidisciplinary research widened even further and earlier distinctions between 'natural' and 'man-made' hazards became ever more difficult to sustain.

By the late twentieth century, the study of environmental hazards had become complex and diverse. These features were reflected in a new generation of text-books (Bryant, 1991; Alexander, 1993; Cutter, 1993; Blaikie *et al.*, 1994; Hewitt, 1997). Alexander (1997) was critical of the fact that, despite this upsurge of natural hazards research, some of the basic concepts – such as vulnerability or disaster – still await agreed definitions, largely because academic protectionism has prevented a convergence of thought towards a truly integrated study of hazards. Natural and social scientists still retain differing views of the interaction between environment and society. Broadly speaking, natural scientists concentrate on hazards and adopt an *agent-specific model* whereby a geophysical event triggers a disaster, the severity of which depends on contributions from both environmental and social factors. Sociologists and others concentrate more on disasters and prefer a *response model* in which disasters are generated from within social systems themselves (Quarantelli, 1998).

Debate is inevitable in a field of study that embraces a wide range of disciplines and deals with both theory and practice. But, to policy makers charged with the task of reducing human suffering and the property destruction caused by environmental hazards, excessive argument can be confusing. It is important to note, therefore, that some important advances have recently been made. A more integrated approach to environmental hazards has been fostered by a focus on common methodologies, such as risk analysis and communication. Most countries no longer rely solely on the physical protection afforded by engineered structures but deploy a mix of anti-hazard measures, including better planning for response and recovery, emergency warnings and land use management. These are helpful steps but more needs to be done to develop the breadth of vision needed to deal with the large-scale disaster-related problems which now confront the world.

In terms of significant policy developments, a global programme to reduce the losses from natural hazards was adopted in December 1989 by the United Nations General Assembly proclaiming the 1990s as the International Decade for Natural Disaster Reduction (IDNDR). This initiative was stimulated by a belief that the losses from natural hazards were growing in many areas and were creating special problems for the poorest countries. The stated objective of the Decade was:

> to reduce through concerted international action, especially in developing countries, the loss of life, property damage, and social and economic disruption caused by natural disasters, such as earthquakes, windstorms, tsunamis, floods, landslides, volcanic eruptions, wildfires, grasshopper and locust infestations, drought and desertification and other calamities of natural origin.

Amongst other things, the IDNDR highlighted the fact that, whilst hazards may impact at the local scale, their consequences are increasingly due to worldwide factors, such as climate change, the rise of mega-cities and poverty. The losses from natural disasters create real doubts about the sustainability of further population growth and wealth creation in many countries. The challenge to researchers and practitioners alike within the hazards community is to blend their skills and adopt a wider perspective embracing global change in order to create a safer environment for all, now and in the future (Mileti *et al.*, 1999).

HAZARD AND RISK

Hazard is an inescapable part of life. Each day we all face some degree of personal risk, whether it is to life and limb in a road accident, to our possessions from theft or to our immediate surroundings from noise or other types of pollution. It is impossible to live in a totally risk-free environment. Fortunately, many of these threats are routine, that is, they pose chronic, rather than extreme, dangers and they do not lead to the comprehensive breakdown of social functions and structures associated with disasters.

Risk is sometimes taken as synonymous with hazard, but risk has the additional implication of the chance of a particular hazard actually occurring. *Hazard* is best viewed as a naturally occurring or human-induced process, or event, with the potential to create loss, that is, a general source of future danger. *Risk* is the actual exposure of something of human value to a hazard and is often regarded as the product of probability and loss. Thus, we may define hazard (or cause) as 'a potential threat to humans and their welfare' and risk (or consequence) as 'the probability of a hazard occurring and creating loss'. This distinction was illustrated by Okrent (1980) who considered two people crossing an ocean, one in a liner and the other in a rowing boat. The main hazard (deep water and large waves) is the same in both cases but the risk (probability of capsize and drowning) is very much greater for the person in the rowing boat. Thus, an earthquake hazard can exist in an uninhabited region but an earthquake risk can occur only in an area where people and their possessions exist.

Clearly, both hazard and risk can be increased and reduced by human actions. But, when large numbers of people are killed, injured or affected in some way, the event is termed a *disaster*. Unlike hazard and risk, a disaster is an actual happening, rather than a potential threat, so a disaster may be simply defined as 'the realisation of hazard'. Disasters are essentially social phenomena that occur when a community suffers an exceptional, non-routine level of stress and disruption. In many cases such crises will be triggered by extreme natural events or the failure of technological systems, although social factors within the community often influence the extent of loss. People, and what they value, are the essential point of reference for all disasters, but there is no universally agreed definition of the scale on which loss has to occur in order to qualify as a disaster (Quarantelli, 1998).

A more detailed disaster definition is:

> an event, concentrated in time and space, in which a community experiences severe danger and disruption of its essential functions, accompanied by widespread human, material or environmental losses, which often exceed the ability of the community to cope without external assistance.

This emphasises the social stress created by a disaster. Although no threshold or scale is given, it implies a major incident requiring the mobilisation of emergency services. In some highly localised incidents, such as a transport

Plate 1   The hazard vulnerability of a poor household in Bangladesh. The family includes several dependants living in a single room of fragile construction with few possessions. Such housing is unable to withstand the storm surge driven by a tropical cyclone. (Photo: International Federation)

accident, there may be more emergency service personnel than members of the public. For a more widespread disaster event, such as a hurricane, helpers will be outnumbered by victims and the effectiveness of the early response by the victims themselves will be a critical factor in mitigating loss.

In terms of decreasing severity, the following threats from environmental hazards can be recognised:

1   Hazards to people – death, injury, disease, stress.
2   Hazards to goods – property damage, economic loss.
3   Hazards to environment – loss of flora and fauna, pollution, loss of amenity.

Although the environment is clearly something that humans value, it is usually prioritised less by people when they are faced with immediate threats to their own life or possessions. For this reason, events which do not threaten people or their goods are excluded from this book, partly on the basis of available space but also because they represent a slightly different suite of hazards in terms of both source and impact. Just as hazard can be ranked, so the probability of an event can be placed on a theoretical scale from zero to certainty (0 to 1). The relationship between a hazard and its probability can then be used to determine the overall degree of risk, as shown in Figure 1.1. Whilst damage to goods and the environment can be costly in economic and social terms, a direct threat to life is the most serious risk faced by humans.

Given this framework, what is the risk from environmental hazards? In general terms, the high profile of disasters portrayed by the media is not matched by the actual incidence of deaths or damages. The main reason is that headline-seeking reports are, by definition, non-routine and arise relatively infrequently compared to other events within society. Table 1.1, abridged from Starr (1979), shows that many of the disasters that made headlines in the USA in the mid-twentieth century had an event frequency

Table 1.1 Major disasters that hit the headlines in the USA around the mid-twentieth century

Event type	Time period	Deaths per event		Frequency (events/year)
		Maximum	Average	
Air crashes	1965–1969	155	78	6.00
Earthquakes	1920–1970	180,000	25,000	0.50
Explosions	1950–1968	100	26	2.00
Major fires	1960–1968	322	35	0.67
Floods (tidal waves)	1887–1969	900,000	28,000	0.54
Hurricanes	1888–1969	11,000	1,105	0.41
Major rail crashes	1950–1966	79	30	1.00
Major marine accidents	1965–1969	300	61	6.00

Source: After Starr (1979)

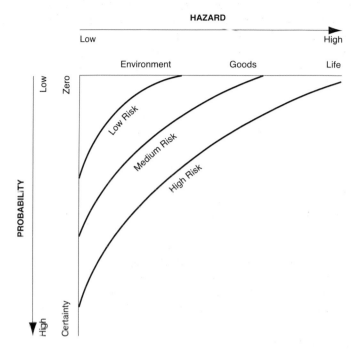

Figure 1.1   Theoretical relationships between the severity of environmental hazard, probability and risk. Hazards to human life are rated more highly than damage to economic goods or the environment.
Source: After Moore (1983)

of less than one per year. Another important point, noted by Sagan (1984), is that the premature deaths and injuries resulting from environmental hazards often involve acute bodily trauma and are treated primarily as *safety* issues. They are perceived by society quite differently from chronic human illnesses, which are viewed as *health* issues. Although the cumulative losses created by environmental hazards tend to be relatively low, they are often reported more prominently, and are often seen by the public as less acceptable, than those arising from other sources.

In the more developed countries (MDCs), average mortality from all causes is strongly dependent on age. The rate of death tends to be high during the first few years of life. It then drops sharply before rising steadily so that, at age 70 and beyond, the rate of death exceeds the infant mortality. This pattern reflects the importance of life-style factors and degenerative diseases in the western world, where some 90 per cent of all deaths are due to familiar medical disorders (heart disease, cancers, respiratory ailments). Tobacco consumption is a major factor. Worldwide about 3 million people die prematurely each year through smoking, many of whom are still in middle age and are losing, on average, some 20–25 years of life.

Consequently, *accidental* deaths from all causes usually constitute less than 3 per cent of overall mortality. Most of these deaths can be linked to common events, such as road accidents, thus leaving very few attributable to major disasters. For example, despite the fact that, between 1975 and 1994 natural hazards in the United States killed nearly 25,000 people (about 23 persons per week) and injured about 100,000 more (about 385 persons per month), only about one-quarter of the deaths and half the injuries resulted from recognised disasters (Mileti *et al.*, 1999). The majority of deaths stemmed from smaller but more frequent events (lightning strikes, car crashes in fog, and local landslides) and, according to Fritzsche (1992), a mere 0.01 per cent of the US population has died from severe natural disasters. Similarly, although every year natural hazards in the USA create about US$1 billion damage to public facilities, such as roads, water systems and buildings, these losses amount to only 0.5 per cent of the capital infrastructure owned by state and local governments. Disaster relief costs are, on average, less than 0.5 per cent of the total federal budget (Burby *et al.*, 1991).

For people in the 'less developed countries' (LDCs), the overall risk of disaster-related death is several times that in the richer nations. One estimate placed the risk of disaster-related death in the developing world as 12 times that in the industrialised countries (IFRCRCS, 1999). But, once again, the main causes are unlikely to be natural hazards. For example,

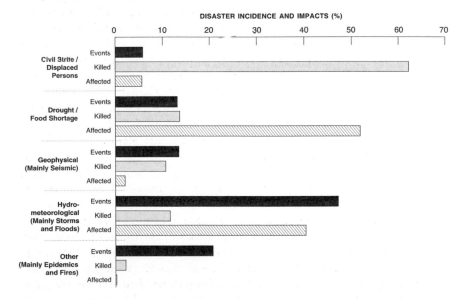

Figure 1.2   The percentage incidence of five disaster types over the 1964–89 period showing the proportion of people killed and adversely affected. Hydrometeorological disasters were the most frequent, civil strife killed the greatest number and drought affected the most people.
Source: Based on data in Heyman *et al.* (1991)

Heyman *et al.* (1991) showed that, in the period 1964–89, strife and displaced persons events accounted for well over half of all disaster deaths (Figure 1.2). Such events are largely confined to the Third World.

WHAT ARE ENVIRONMENTAL HAZARDS?

A precise definition of environmental hazards is difficult. Burton and Kates (1964a) defined *natural hazards* as 'those elements of the physical environment harmful to Man and caused by forces extraneous to him'. Natural hazards have also been seen as 'Acts of God'. These perspectives have not been helpful. They over-emphasise the 'surprise' factor in disaster when, in reality, it is now possible to delineate many hazard-prone areas and to recognise that common disasters, such as floods, are recurrent events at certain locations. In addition, because the 'Act of God' approach suggests – quite wrongly – that humans have no part to play in creating disasters, it also implies they have little hope of mitigating them.

In reality, most environmental hazards have both natural and human components. For example, flood problems may be exacerbated by fluctuations in climate, such as increased storm frequency, and also by human activities, such as land drainage and deforestation. The loss of life caused by a tropical cyclone will depend to some extent on storm severity but it can be greatly reduced by means of a warning message. The effects of a man-made nuclear accident will be influenced by the prevailing weather conditions controlling the downwind path and the rate of fallout from the radioactive plume. These interactions have led to the increasing recognition of hazards as *hybrid* events resulting from an overlap of environmental, technological and social processes (Jones, 1993). Other compound terms include 'quasi-natural' and 'na-tech' hazards. Thus, despite the convenience of the term, truly 'natural' hazards do not exist.

The human ecology perspective on natural hazards, illustrated in Figure 1.3, distinguishes between natural *events* and their interpretation as natural *hazards* (or resources). Since the Earth is a highly dynamic planet, most natural events show a wide range of variation through time in the use of energy and materials for environmental processes. The outer limits of this behaviour we call *extremes* and certain statistical measures, notably magnitude-frequency relationships, are used to describe such extremes. But extreme natural events are not considered disasters unless they cause large-scale death or damage to humans. The conflict of geophysical processes with people gives humans a central role in hazards and it is only by retaining a balance between resources and hazards that sustainable economic development can occur.

Many hazardous processes represent the extremes of a distribution of events that, in a slightly different context, would be regarded as a resource (Kates, 1971). There is often a fine line between environmental hazards and

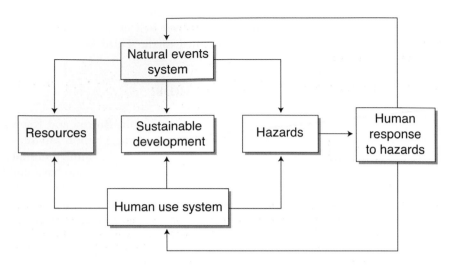

Figure 1.3   Environmental hazards exist at the interface between the natural events
and human use systems. Human responses to hazards can modify both
the natural events in, and the human use of, the environment.
Source: After Burton *et al.* (1993)

environmental resources, for example, between water out of control (flood
hazard) and water under control (reservoir resource). The atmosphere is
considered 'benign' when it produces holiday sunshine but 'hostile' when
it produces damaging storms. In reality, the environment is neither benign
nor hostile. It is 'neutral' and it is only human location, needs and perceptions which identify resources and hazards in the spectrum of natural events
(Burton *et al.*, 1993). Within this range, only a few high magnitude geophysical events create disaster.

Human sensitivity to environmental hazards represents a combination of
physical exposure, which reflects the range of potentially damaging events and
their statistical variability at a particular location, and *human vulnerability*,
which reflects the breadth of social and economic tolerance to such hazardous
events at the same site. In Figure 1.4 the shaded zone represents an acceptable range of variation for the magnitude of the physical variable, which
can be any environmental element relevant to human survival, such as rainfall. Most social and economic activities are geared to some expectation of
the 'average' conditions. As long as the variation of the environmental
element remains fairly close to this expected state, the element will be
perceived as mainly beneficial. However, when the variability exceeds some
threshold beyond the 'normal' band of tolerance, the same variable starts
to impose damage and becomes a hazard. Thus, very high or very low rainfall will be deemed to create a flood or a drought respectively. The exceedance
of a damage threshold enables two basic dimensions of a hazard to be identified: the hazard *magnitude* is determined by the peak deviation beyond the

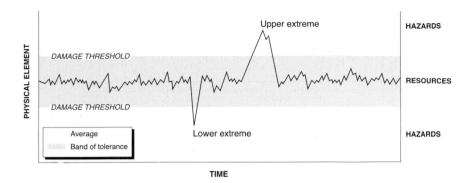

Figure 1.4 Sensitivity to environmental hazard expressed as a function of the variability of geophysical elements and the degree of socio-economic tolerance. Within the band of tolerance, events are perceived as resources; beyond the damage thresholds they are perceived as hazards. Source: Modified from Hewitt and Burton (1971)

threshold on the vertical scale and the hazard *duration* is determined by the length of time the threshold is exceeded on the horizontal scale.

It follows that the risk of disaster may vary through time with changes in physical exposure and human vulnerability in a given area. Some possibilities which give rise to increased risk are shown schematically in Figure 1.5. Case A represents a constant band of social tolerance and a constant variability of the geophysical factor but a decline in the mean value (perhaps a decrease in temperature). Case B represents a constant band of tolerance and constant mean but an increased variability (perhaps a trend to greater fluctuations in annual rainfall). Finally, in case C the physical variable does not change but the band of tolerance narrows and vulnerability increases (perhaps because population growth places more people at risk).

Human populations are most vulnerable on the margins of tolerance where small physical changes may create large socio-economic impacts, for example, the effects of rainfall variability on agriculture in semi-arid areas. However, the threshold may not always be a sharp boundary and it is unlikely that the relationships between event intensity and hazard impact will be linear once the damage threshold has been crossed. Figure 1.6 shows a schematic set of stage-damage curves used to assess the economic losses from flooding according to the depth of water entering individual houses. It can be seen that no damage occurs until the water rises above floor level but then losses rise sharply for comparatively small depth increments. For any given depth, actual losses are related to the economic status of the housing. Over a long period of time, frequent but unpredictable low-level variability around a very critical threshold, such as floor level, may well be more significant than the rare occurrence of more extreme events. For

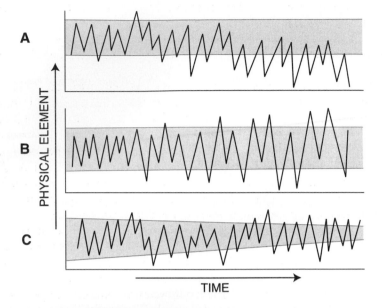

Figure 1.5 An illustration of changes in human sensitivity to environmental hazard due to variations in physical events and socio-economic tolerance. In each case the risk of disaster increases through time.
Source: After de Vries (1985)

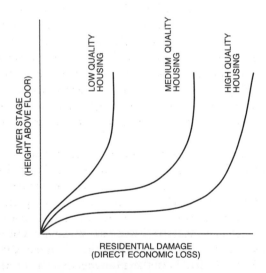

Figure 1.6 Schematic stage-damage curves showing the relationship between water depth and flooding in residential property. For a given stage, the absolute loss is greater as the quality of the housing stock increases.

example, 0°C is a critical weather threshold because water freezes at that temperature and more road accident deaths are likely to occur per hour in the marginal situation around 0°C than in much lower temperatures. But, in most mid-latitude winters, such a value could hardly be described as extreme.

The tendency towards admitting more human responsibility for 'natural' hazard is due to several factors. One factor has been the spread of human influence across the globe and the increased attention given to environmental pollution, which was described as a 'quasi-natural' hazard by Burton and Kates (1964a). Another factor has been the radical re-interpretation of natural hazard in the Third World (see Chapter 2). Most important of all, it is recognised that *technological hazards* constitute a new kind of threat and these are included in this book when the consequences are mainly due to accidents rather than to low-dose exposure to toxic substances.

Early concern about technological hazards was confined to the developed world but the risks now extend to less developed countries as technology transfer has grown. Perrow (1984) argued that technological disasters have been created by the spread of high-risk technologies which now incorporate systems so complex that it is impossible to anticipate all the potential interactions and failures. Because technology is dealing with ever more toxic substances, extending into ever more hostile environments and operating to increasing levels of performance, accidents can be regarded as inevitable or 'normal' in much the same way that natural hazards are a 'normal' part of the functioning of geophysical processes.

The term *environmental hazard* has the advantage of including a wide variety of hazard types ranging from 'natural' (geophysical) events, through 'technological' (man-made) events to 'social' (human behaviour) events (Figure 1.7). Some attempt can also be made to scale the range of hazards according to whether the impacts are intense and local, or diffuse and widespread within society. The extent to which hazards are *voluntary* or *involuntary* is particularly important. The degree of individual human responsibility for disaster increases greatly from the essentially accidental geophysical hazards (earthquake, tsunami) to the largely self-induced social hazards (smoking, mountaineering).

The great breadth of possible risk has led to some impossibly wide definitions of environmental hazard, for example, 'the threat potential posed to man or nature by events originating in, or transmitted by, the natural or built environment' (Kates, 1978). This definition can include long-term environmental deterioration (acidification of soils, build-up of atmospheric carbon dioxide) plus all the social hazards, both involuntary and communal (crime, terrorism, warfare), as well as voluntary and personal hazards (drug abuse, mountain climbing). These hazards have such different origins and impacts that a more focused definition is required.

In this book, environmental hazard will be mainly restricted to events which directly threaten human life and property on a community scale by

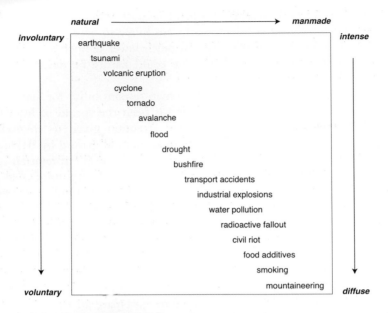

Figure 1.7   A general spectrum of environmental hazards from geophysical events
to human activities. Hazards which are increasingly man-made tend
to be more voluntary in terms of their acceptance and more diffuse in
terms of their impact.

means of acute physical or chemical trauma. This means an emphasis
on rapid-onset events, although there are notable exceptions, such as
drought. Acute bodily trauma, plus any related damage to property or the
environment, usually follows the sudden release of energy or materials in
concentrations which are greatly in excess of normal background levels.
Such releases may come from a natural source, such as a volcano, or from
a man-made source, such as a chemical factory.

Any manageable definition of environmental hazards will be both arbitrary and contentious but certain common features exist:

1   The origin of the damaging process or event is clear and produces
     characteristic threats to human life or well-being, for example, a flood
     causes death by drowning.
2   The warning time is normally short, that is, the hazards are often known
     as rapid-onset events. This means that they can be unexpected even
     though they occur within a known hazard zone, such as the floodplain
     of a small river basin.
3   Most of the direct losses, whether to life or property, are suffered fairly
     soon after the event, that is, within days or weeks.
4   The exposure to hazard, or assumed risk, is largely involuntary, normally
     due to the location of people in a hazardous area, for example, the

unplanned expansion of some Third World cities on to unstable hillsides.

5 The resulting disaster occurs with an intensity that justifies an emergency response, that is, the provision of specialist aid to the victims. The scale of response can vary from local to international.

The approach adopted here is not meant to imply that other environmental problems which are also influenced by human activity, such as deforestation, desertification, depletion of the stratospheric ozone layer, global warming and rising sea levels resulting from the enhanced greenhouse effect, are unimportant. But these are the so-called chronic or 'elusive' hazards. They comprise comparatively long-term, global issues, sometimes of obscure origin and consequence. Unlike environmental hazards, their effects are generally less concentrated in time and space. For example, soil erosion or industrial pollution rarely pose an immediate threat to human life on a community scale. Smets (1987) claimed that, apart from three industrial disasters involving the concentrated release of very toxic substances (mercury at Minamata, Japan in 1956; methyl isocyanate at Bhopal, India in 1984; and radioactive material at Chernobyl, former USSR in 1986), no instance of accidental pollution had so far *directly* caused more than fifty deaths anywhere in the world. Some choices in this book were arbitrary and the exclusion of pollution disasters is an obvious example. In this context it should be remembered that drought, which appears such a self-evident environmental hazard, is not a life-threatening event in the developed world. In view of the balance of previous work, and the availability of literature, the emphasis will be on the impact of geophysical and technological events and their management.

From this broad distinction between problems and hazards, it is possible to derive the following working definition of environmental hazards:

Extreme geophysical events, biological processes and major technological accidents, characterised by concentrated releases of energy or materials, which pose a largely unexpected threat to human life and can cause significant damage to goods and the environment.

Some links between rapid-onset and longer-term threats cannot be denied. For example, earthquakes are preceded by the slow build-up of stress in the Earth's crust. Equally, some of the longer-term trends towards environmental deterioration will exacerbate the damage potential of many hazard processes. Desertification and unwise land use contribute to the impact of the drought hazard. It is predicted that global warming during the next few decades will raise sea surface temperatures. In turn, this 'greenhouse effect' may increase the disaster threat from tropical cyclones. According to Emanuel (1987), the enhanced sea surface temperatures associated with a doubling of the present atmospheric concentrations of CO_2 could raise the maximum destructive potential of some of these storms by 60 per cent. Rising sea levels, also associated with global warming, will increase the catastrophe potential of storm surge hazards for low-lying coastal communities. The progressive warming of alpine areas will bring glacier retreat and an altitudinal rise in the permafrost zone which, in turn, is likely to release loose surface material in debris flows. The disaster impact of such environmental changes will be compounded in the LDCs by prolonged economic and social difficulties, some traceable back many decades to colonial times.

A TYPOLOGY OF HAZARD AND DISASTER

Most classifications of hazard have been dominated by geophysical processes (Table 1.2). It has also been usual to emphasise the impact of single elements, such as windspeed or rainfall, because this is relatively easy to do. In practice, the most severe hazards arise from compound or synergistic effects, as when wind combines with snow to produce a blizzard or when earthquakes set off landslides in steep terrain. The volcanic eruption of Mt St Helens, USA, in 1980 led to ashfalls, landslides, floods and wildfires. Alternatively, natural hazards can be divided into those of *endogenous* earth origin (such as earthquakes and volcanic hazards) and those of *exogenous* earth origin (such as floods, droughts and avalanches). Such physically bound classifications have limitations for disaster studies, although Hewitt and Burton (1971) itemised a variety of factors relating to damaging geophysical events which were not process-specific, including:

1 areal extent of damage zone;
2 intensity of impact at a point;
3 duration of impact at a point;
4 rate of onset of the event;
5 predictability of the event.

Hohenemser *et al.* (1983) have viewed technological hazards as a sequence of events, leading from human needs and wants to the selection of a particular technology through to harmful consequences. This chain of hazard evolution can be employed more widely, as illustrated for drought in Figure 1.8. The top line indicates seven stages of hazard development. The stages are identified generically at the top of each box and in terms of a sample development of hazard in the bottom. The stages are linked by causal pathways denoted by triangles. Six control stages are linked to pathways between hazard states by vertical arrows. Each is described generically as well as by specific control actions designed to eliminate or reduce the evolving hazard.

Rarely does a straightforward cause and effect situation apply. Hazards often consist of a complex chain of processes and impacts leading to disaster. For example, if we consider the fire and explosion caused by gas pipes

Table 1.2 Some potentially hazardous environmental elements or events

1 *Atmospheric*	
Single elements	*Combined elements/events*
Excess rainfall	Hurricanes
Freezing rain ('glaze')	'Glaze' storms
Hail	Thunderstorms
Heavy snowfalls	Blizzards
High windspeeds	Tornadoes
Extreme temperatures	Heat/cold stress

2 *Hydrologic*
Flood: freshwater from rivers/lakes/dambursts
Flood: coastal from marine storm surge/sea level rise
Wave action: coastal and lakeshore erosion
Drought: from rainfall deficit
Rapid glacier advance (surges)

3 *Geologic*
Mass-movement: landslides, avalanches, mudflows
Earthquake: ground shaking/tsunamis
Volcanic eruption: pyroclastic flows, ashfalls
Rapid sediment movement: severe erosion/siltation

4 *Biologic*
Severe epidemic in humans: Ebola fever, AIDS, malaria
Severe epidemics in plants
Severe epidemics in wild animals
Animal and plant invasions: locusts, grasshoppers, 'weeds'
Forest and grassland fires

5 *Technologic*
Transport accidents: air/train crashes, shipwrecks
Industrial explosions and fires
Accidental releases of toxic substances: dioxin, carbon monoxide
Nuclear power plant failures: release of radioactive materials
Collapse of public buildings or other major structures
Germ or nuclear warfare

Source: Modified after Hewitt and Burton (1971)

ruptured by the lateral spread of soil in an earthquake, such as that which struck San Francisco in 1906, the primary hazard is strong ground shaking, the secondary hazard is soil liquefaction and the tertiary hazard is fire and explosion. Hazards also produce a cascade of disaster impacts ranging from biophysical to economic. The eruption of Mt St Helens not only physically devastated an area of more than 500 km^2 but it also had an ordered sequence of effects on forestry (ranging through thrown trees, additional production costs and reduced income), and impacted widely through Washington State affecting recreation, construction, retailing and insurance. Certain disaster chains are so complex that they defy categorisation. This

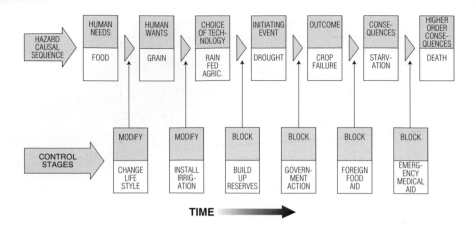

Figure 1.8   A schematic illustration of the causal chain of hazard development. The stages are expressed generically at the top of each box and in terms of a sample drought disaster in the lower segment. Six potential control phases, designed to reduce disaster, are linked to pathways between hazard stages by vertical arrows.
Source: Modified from Hohenemser *et al.* (1983)

is particularly so in the LDCs, where failures of government may compound the effects of natural processes by allowing a collapse of the economy or the outbreak of civil strife.

Although community loss is the major characteristic of disaster, some gains also arise. Therefore, it is necessary to categorise hazard impacts, not only into *gains* and *losses*, but also into other effects (Figure 1.9). *Direct* effects are those first-order consequences that occur immediately after an event, such as the deaths and economic loss caused by the throwing down of buildings in an earthquake. *Indirect* effects emerge later and may be much more difficult to attribute directly to the event. These include factors such as mental illness resulting from shock, bereavement and relocation from the area. *Tangible* effects are those to which it is possible to assign reasonably reliable monetary values, such as the replacement of damaged property. *Intangible* effects, although real, cannot be satisfactorily assessed in monetary terms. These distinctions are not entirely static. The loss of human life, for example, has proved notoriously difficult to assess financially in the past but, as methods of assessment improve, it is becoming a more tangible effect of disaster.

Direct losses are the most visible consequence of disasters caused by the immediate damage which is created, such as building collapse. They are comparatively easy to measure but they are not always the most significant outcome. *Direct gains* represent the benefits which may flow to surviving residents in the area after a disaster. These can include various forms of aid and even some longer-term enhancement of the environment, for example, fertile deposits from volcanic eruptions or river floods. On the Icelandic

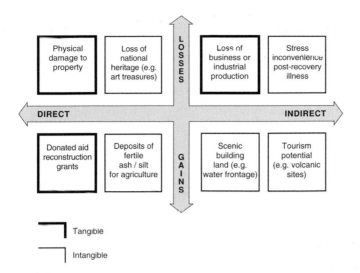

Figure 1.9   The potential impact of environmental hazards in terms of losses and gains, both direct and indirect, with an indication of some tangible and intangible effects.

island of Heimaey, the volcanic ash cleared from the town of Vestmannaeyjar was used as foundation material to extend the runway at the island's airport and to level out an old lava flow for a new settlement to replace the houses destroyed during the eruption in 1973.

Indirect losses arise mainly through the second-order consequences of disaster, such as the disruption of economic and social activities in a community. For example, an 'inverse multiplier' effect may occur whereby property values fall, consumers save rather than spend, business becomes less profitable and unemployment rises. Ill-health effects often outlast the direct losses (Ahearn and Cohen, 1984). Whilst the relief and rescue phase of disaster response may bring out a strong community spirit amongst the healthy survivors, psychological stress is known to affect the victims of disaster directly and also to have an indirect influence on many others such as family and rescue workers. The symptoms can include shock, anxiety, stress and apathy and are expressed through sleep disturbance, belligerence and alcohol abuse. After some disasters, attitudes of blame, resentment and hostility occur and many pre-existing social problems are exacerbated by evacuation and relocation.

Direct gains are comparatively rare and are usually confined to a distinct sub-set of the population affected by disaster. For example, some victims may, through a variety of means, be able to obtain more than their fair share of disaster aid whilst those with practical skills, for example, in the construction trade, may obtain very well-paid employment due to a demand for such workers in the restoration phase following the event.

Indirect gains are less well understood. They represent the very long-term benefits enjoyed by a community as a result of its hazard-prone location. Very little systematic research has been undertaken, for example, into the balance between the continuing advantages of a riverside site (flat building land, good communications, water supply and amenity) compared with the occasional losses suffered during periodic flooding.

FROM HAZARD TO DISASTER

Physical exposure to hazard

In simple terms, hazard exposure is created by people and property being in the wrong place at the wrong time. In the worst-case outcomes both the time and the spatial dimensions are usually highly concentrated in order to produce large-scale losses, as shown below.

Time is important in creating the element of 'surprise' which is often taken as a keynote of disaster. There is a major mismatch between the timescales of extreme geophysical events and those for human societies. Planet Earth is at least 4.6 billion years old and many of the natural processes influencing its continuing evolution function on timescales of centuries or millennia. By comparison, individual humans exist for a few decades only and societies can change significantly within even shorter spans. Therefore, a volcano which erupts violently every thousand years or so will be operating routinely according to ongoing tectonic processes although the event will be anything but routine for a population which may have settled the slopes in the last fifty years.

The time dimension is crucial in understanding magnitude–frequency relationships. The total amount of energy and materials released by geophysical events of a given size (or duration) is a product of their magnitude (or duration) times their frequency of occurrence. When the magnitude of the event is plotted against its frequency, it usually exhibits the type of relationship shown in Figure 1.10(a). The *recurrence interval* (or return period) is the time which, on average, elapses between two events that equal, or exceed, a particular magnitude. A plot of recurrence intervals versus associated magnitudes usually produces a group of points that approximates a straight line on semi-logarithmic paper (Figure 1.10(b)).

As already stated, the potential for disaster is concentrated in a small number of large, relatively rare events. For example, the total energy released in an earthquake creating a great disaster might well be as much as 30,000 times greater than the energy of a small disaster with few fatalities. A handful of very large earthquakes account for most of the total seismic energy released worldwide, and the five largest events recorded in the twentieth century were responsible for over half of all the earthquake-related deaths. The physical magnitude of most natural hazards can be measured

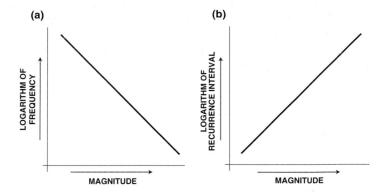

Figure 1.10 The general relationship between magnitude, frequency and recurrence interval for natural events. A few high-magnitude, but rare, events are responsible for most disasters but it is difficult to define a threshold beyond which geophysical events become disastrous.

objectively on various scientific scales, for example, earthquakes – Richter scale; tornadoes – Fujita scale; hurricanes – Saffir–Simpson scale. However, event size alone is a poor guide to disaster impact, mainly because of the great differences in the degree of physical exposure and human vulnerability of the communities at risk.

Space also leads to a concentration of disasters. This is partly explained by the fact that some hazardous geophysical processes are restricted to certain geographical regions of the world. Figure 1.11 shows the global distribution of five key natural hazards: earthquakes, volcanic eruptions, tsunamis and severe storms (tropical and extra-tropical). Within this general distribution, the most unsafe physical settings – which cannot be shown on maps at this scale – are steep, tectonically active mountains with high rainfall inputs and extensive tracts of low land near coasts. Areal concentrations of population greatly compound the risk. For example, more than half the world's population now lives within 60 km of the ocean, often in low-lying tracts of land subject to river and coastal flooding. The greatest potential for disaster exists in the most populous cities. Degg (1992) has indicated that over three-quarters of these urban centres, which together accommodate 10 per cent of the global population, are exposed to at least one natural hazard. The greatest concern is for the fastest growing cities, all of which are located in the Third World.

For multiple natural hazards to occur, it is necessary for a region to have many active geophysical processes. Alexander (1987a) showed that as much as 70 per cent of Italy is at risk from earthquakes, landslides, floods and avalanches. For a developed nation, Japan also suffers severely from environmental hazards. The country lies on a tectonic plate boundary, contains nearly thirty active volcanoes and is at risk from major earthquakes, such

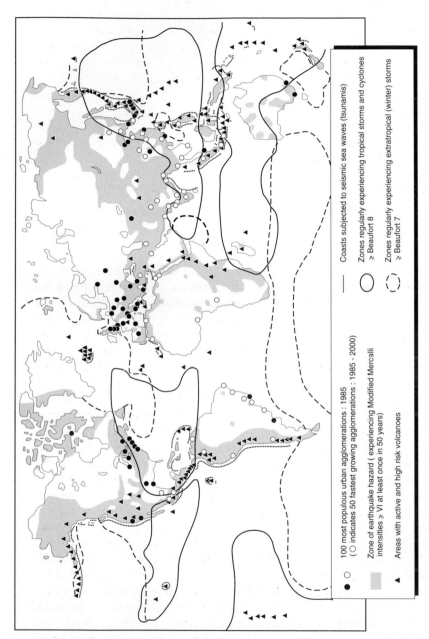

Figure 1.11 The location of the world's largest 100 urban areas (1985) in relation to earthquake zones, volcanoes, tsunami-affected coasts and windstorm hazards.
Source: After Degg (1992)

Legend:

○ 100 most populous urban agglomerations : 1985
(○ indicates 50 fastest growing agglomerations : 1985 - 2000)

Zone of earthquake hazard (experiencing Modified Mercalli intensities ⩾ VI at least once in 50 years)

▲ Areas with active and high risk volcanoes

⋯⋯ Coasts subjected to seismic sea waves (tsunamis)

◯ Zones regularly experiencing tropical storms and cyclones ⩾ Beaufort 8

◌ Zones regularly experiencing extratropical (winter) storms ⩾ Beaufort 7

as the Kobe disaster of 1995. In an average year, three or four tropical cyclones come ashore whilst the northern islands are subject to severe winters with heavy snowfalls. Economic and social factors are also implicated. The population pressure is intense on the comparatively few areas of flat, habitable land. About 45 per cent of the total population is concentrated in the three metropolitan zones of Tokyo, Osaka and Nagoya at densities which exceed 5,000 people per km^2. Economic growth has greatly increased the exposed risk through environmental pollution and rapid social changes.

Great spatial differences exist in the population and in the wealth exposed to hazard, not least between the LDCs and the MDCs. For example, India has about 16 per cent of the world's population but has less than 2 per cent of the world's income. The equivalent figures for the USA are 5 per cent and 36 per cent respectively. Although this 'two worlds' picture is frequently adopted for convenience in this book, it remains a crude and unsatisfactory division. Within individual countries, regions, towns and streets, there are highly local variations in hazard exposure and in human vulnerability which are difficult to capture through available statistics.

Human vulnerability to hazard

The concept of *vulnerability*, like risk and hazard, indicates a possible future state. It implies a measure of risk combined with a relative inability to cope with the resulting stress. Timmerman (1981) defined vulnerability at the society or community scale as 'the degree to which a system, or part of a system, may react adversely to the occurrence of a hazardous event'. Most attitudes to reduce system-scale vulnerability can be seen as expressions of either resilience or reliability. Anderson (2000) showed how the concept of human vulnerability has been refined through time, although there is still no fully acceptable, and discipline-free, definition available.

Resilience is a measure of the rate of recovery from a stressful experience, reflecting the social capacity to absorb and recover from the occurrence of a hazardous event. Traditionally, resilience has been the main weapon against hazard in the LDCs where disaster is often accepted as a 'normal' part of life. In this situation, group coping strategies are important. For example, nomadic herdsmen in semi-arid areas have tended to accumulate cattle during years with good pasture as an insurance against drought. Urban resilience can also be developed in the MDCs, as shown by the rapid recovery of the Los Angeles electricity supply following the Northridge earthquake in 1994 (Table 1.3). *Reliability*, on the other hand, reflects the frequency with which protective devices against hazard fail. This approach is applicable to the MDCs, where technology and engineering design have provided what is perceived to be a high degree of reliability for most urban services. But extreme stress, for example from an earthquake, can easily disrupt road networks, electric power lines or water systems. When such systems fail, there is frequently no alternative source of supply.

Table 1.3 Restoration of power supplies in Los Angeles following the Northridge earthquake in 1994

Time	Number of people without power
Initially	2,000,000
By dusk	1,100,000
After 24 hours	725,000
After 3 days	7,500
After 10 days	Almost all power restored

Source: After Institution of Civil Engineers (1995)

This largely western-based, society-level view of vulnerability has been challenged by Blaikie *et al.* (1994) and others on the grounds that it is people – rather than disembodied systems – which have to deal with disaster. For vulnerable people, access to resources at either a household or an individual scale is often the most critical factor in either achieving a secure livelihood or recovering effectively from disaster. Those households with direct access to capital, land, tools and equipment, and which also have able-bodied members with specialist skills, are the most resilient when disaster strikes. Access to information and the availability of a social network which can mobilise support from outside the household are important too. The poor, and those otherwise disadvantaged by age, gender, ethnicity or health status, are comparatively defenceless against disaster and are also least likely to have a voice in influencing decisions about community responses to future threats. Vulnerability exists in wealthy countries too. For example, 20 per cent of the US population have been defined as suffering from some disability: a proportion likely to increase as the population ages. In general, elderly people are less mobile and capable of undertaking damage-reducing measures, such as emergency evacuation, than the population as a whole.

About 25 per cent of the world's population live in areas at risk from natural disasters. Most of these people are in the LDCs, where vulnerability arising from poverty, discrimination and lack of political representation hampers the development process (Anderson, 1995). The poorest people often have little choice but to locate in unsafe settings, whether this is urban shanties or degraded rural environments. In terms of loss of life and relative economic impact, disasters hit hardest where poverty-stricken people are concentrated. Despite the fact that, in most LDCs, rural inhabitants still outnumber those in urban areas, there are now more urban dwellers in the Third World than in Europe, North America and Japan combined (Cairncross *et al.*, 1990). In urban squatter settlements, population densities may reach 150,000 per km², perhaps ten times the level in established areas and orders of magnitude greater than the densities in rich areas (Davis, 1978). Many buildings are erected on steep slopes or flood-prone land, exposed to strong winds and landslides without suitable materials or construction skills. In highly populated rural areas, population densities can exceed 1,000 per km^2 and life is a recurrent struggle to secure cultivable land. Many people are landless and disadvantaged by land tenure systems which deny them access to the means to support themselves. Some farmers in the LDCs can pay one-third of their crop income to absentee landlords and disasters often exacerbate the poverty gap, as when food prices increase during droughts and grain merchants make large profits.

In the LDCs these broad and complex socio-economic problems combine with insecure physical environments to create a high degree of vulnerability. Environmental degradation and development decisions contribute increasingly to disaster impact (Kreimer and Munasinghe, 1991). Common problems include the basic organisational structure, which embraces everything from poor roads and untrained civil servants to the total lack of welfare programmes. This results in inadequate housing and health provision combined with a low nutritional status. Risk varies according to occupation, social class, ethnicity, caste, age and gender. For example, in the Bangladesh cyclone disaster of 1970 more than half the deaths were suffered by children under age ten, who comprised only one-third of the population, whereas males between 15 and 45 years escaped comparatively lightly (Sommer and Mosely, 1972). The very young and the very old are especially at risk. In some LDCs as many as 50 per cent of the population are under 15 years of age and highly vulnerable to economic exploitation and disease. Older people, especially widows, face difficulties in maintaining their livelihood after disaster. Similarly, people with on-going disabilities or chronic malnourishment suffer more from water-related diseases common in floods, such as dysentery. As a result of such factors, the scale of disaster impact is often a function of human vulnerability rather than of the physical magnitude of the event.

2   Dimensions of disaster

AUDITING DISASTER

There is no agreed definition of 'disaster'; therefore, the reporting and archiving of disasters is subjective and inconsistent. In addition, some of the raw information will be inaccurate and prone to bias. Although official disaster reports are issued by many agencies, ranging from governments to insurance companies, the news media are an important – but often unreliable – source of information when databases are compiled. Finally, there are pitfalls in the interpretation and analysis of disaster databases themselves.

In general, the media tend to over-emphasise rapid-onset events, especially if they are unusual, at the expense of more pervasive disasters (Greenberg et al., 1989; Wrathall, 1988). The prominence given on television news to disasters is determined by journalistic criteria, such as timeliness, immediate human interest and – above all – the visual impact of film reports, rather than the degree of social impact. According to Garner and Huff (1997), media reporting is similar for all types of disaster and is marred by an excessive concentration on the emergency phase immediately after the event, especially if it can be illustrated by dramatic images of helpless victims. In contrast, the media rarely seek to educate the public about risk or provide balanced, in-depth reporting of responses designed to mitigate future disasters. Another bias is created by an undue emphasis on events close to home. For example, Adams (1986) studied the geographical coverage of natural disasters by American television and found that the severity of the events accounted for comparatively little of the differences in reporting. Following an analysis of thirty-five major disasters (each causing at least 300 fatalities) in various parts of the world, it was concluded that the world was prioritised with a western perspective whereby the death of one western European equalled three eastern Europeans equalled nine Latin Americans equalled eleven Middle Easterners equalled twelve Asians. A more subtle media filter arises when there is a dependence on advertising revenue, which promotes a tendency to concentrate on the prosperous target markets of the media sponsors. This could lead to the under-reporting of disaster impacts on poorer social groups and geographical areas, a bias which was noted in the case of the 1994 Northridge earthquake (Rodrigue and Rovai, 1995).

Apart from bias in the initial reporting of events, many disasters create genuine problems of classification and may be recorded under more than one heading. For example, cyclones and floods often occur together; landslides can be associated with earthquakes and volcanic eruptions as well as with storms. To avoid double-counting, it is important to record such compound events once only according to the major cause of impact. Counting the loss of life is not easy, especially in the LDCs. For example, which total should be used when, as commonly happens, a wide range of deaths is given or simply reported as 'in the thousands'? What should be done about the persons reported missing or those who die later from secondary effects, such as famines and epidemics? Above all, any disaster audit limited to deaths, injury and damages fails to capture the intangible impacts. Many people suffer increased hardship after disaster, including the loss of a relative, destruction of the family home, malnutrition, loss of employment, debt and even forced migration. Of all these losses, homelessness is the only one for which reasonably consistent statistics exist.

The subsequent archiving of events can itself be a difficult task. Disaster databases are usually compiled on best-estimates of specified impact criteria, such as death, injury and economic loss, which breach predefined thresholds. Thus, events responsible for at least 100 deaths *or* at least 100 people injured *or* at least US$1 million damage have traditionally been archived. This method has basic weaknesses. For example, the lack of reliable social statistics in many LDCs often precludes a precise total of the people killed by an event. What constitutes an 'injury' has never been properly defined. Economic loss is difficult to assess and about 70 per cent of all disaster reports lack reliable information on financial loss or infrastructural damage. The inevitable result of missing or unreliable data is that major discrepancies can exist in disaster counting, even for developed countries. Over a 33-year period, one agency recorded more than twice the number of meteorological- and hydrological-related deaths than another organisation in Japan (Mitchell, 1989).

Disaster statistics compiled for individual nation states are then assembled into global datasets but direct international comparisons based on such information can be misleading. One problem is that such comparisons disguise large differences in the relative disaster impact between (and even within) the LDCs and the MDCs. For example, a US$1 million loss would be caused by a much lower magnitude and higher frequency event in, say, California compared to Bangladesh. At the same time, California would be more likely to have the resources to recover from such an event. National statistics also fail to capture the impact of disaster on the most vulnerable groups, such as poor people and ethnic minorities. Small, isolated communities are especially at risk. The loss of ten able-bodied men from a remote fishing village housing 200 people would be far more devastating for the survival of that community than the death of 100 men in a large city. In other words, even when the raw statistics are reliable, the impact of a disaster on any area should always be placed in the context of local population numbers, the nature and scale of economic functions and the financial resources available in both the public and private sector. This rarely happens.

Despite these problems, global datasets of disaster have been compiled by several organisations. Those used in this book are:

1   *The NHRAIC database* The Natural Hazards Research and Applications Information Center (NHRAIC) at Boulder, Colorado, compiled a 35-year archive of natural disasters for the period 1947–81 (Thompson, 1982). Drought was excluded, a common omission because of the difficulties of defining drought and distinguishing its effects on food shortage from the endemic hunger that exists in many LDCs. Major events were restricted to a threshold of at least 100 deaths or injuries, or damage of at least US$1 million. This latter figure was subsequently raised for price inflation.

2   *The RFF database* Resources for the Future (RFF) in Washington, DC, assembled a database from 1945–86 covering both natural and industrial disasters (Glickman *et al.*, 1992). This archive also excluded drought and industrial accidents were limited to those in which a hazardous material caught fire, exploded or was released in a toxic cloud. The recording threshold was set at a relatively low minimum of twenty-five fatalities for natural disasters compared with five for industrial accidents. The difference in weighting was justified on the grounds that the public judges industrial accidents more severely than the death toll alone would suggest.

3   *The CRED database* The Centre for Research on the Epidemiology of Disasters (CRED), at the University of Louvain, Belgium, has developed a record, called EM-DAT, covering both natural and technological disasters from 1900 to the present (Sapir and Misson, 1992). Since 1995 this archive has replaced the database previously maintained by the US Office of Foreign Disaster Assistance (OFDA). CRED information is obtained from UN bodies, NGOs, reinsurance companies and the press. For inclusion, a disaster must have killed ten or more persons, or affected at least 100 people, although an appeal for international assistance or a government disaster declaration will take precedence over the first two criteria. In order for displaced persons, drought and famine to register, at least 2,000 people have to be affected. The individual disaster types are shown in Table 2.1. The CRED database is now regarded as the primary reference because it is more comprehensive than others and also because it is used by the disaster agencies themselves.

CRED data can be manipulated to identify so-called 'significant' natural disasters for which the more stringent criteria are:

Table 2.1 List of disaster types recorded in EM-DAT

Accident	Earthquake	Insect infestation
Avalanche	Epidemic	Landslide
Chemical accident	Famine	Storm
Civil strife	Fire	Tsunami
Cyclone	Flood	Typhoon
Displaced persons	Heat/cold wave	Volcano
Drought	Hurricane	

Source: After Sapir and Misson (1992)

1   number of deaths per event – 100 or more;
2   significant damage – 1 per cent or more of the total annual national gross domestic product (GDP);
3   affected people – 1 per cent or more of the total national population.

These 'significant' natural disasters comprise a sub-set of global events which impact severely on the countries involved. In particular, the relative measures adopted for damage and affected people indicate more accurately than absolute values the effect of disaster on countries with weak economies and small populations. They are, therefore, a better measure of impacts in many LDCs. In general, technological disasters do not create such large-scale national impacts. Therefore, when comparisons are made in this book, for example between the number of 'significant' natural disasters and the number of technological disasters from the CRED database, it is important to realise that the latter category uses the much lower recording thresholds adopted by CRED.

Finally, the analysis of any disaster database poses special difficulties. In particular, it is important to specify the data period when making any statement about disaster impact. Although the CRED archive starts in 1900, the recording of natural disasters worldwide did not become reasonably comprehensive until the creation of OFDA in 1964. Since then, as shown in Figure 2.1, there has been a steep rise in the number of reported disasters from an annual average of fewer than 50 to around 250 per year in the 1980s and 1990s. Considerable debate exists about the extent to which this rise represents a real increase in disaster impact as opposed to the increased efficiency in disaster reporting which has taken place during recent decades.

The short-term sampling of databases can lead to unrepresentative conclusions because of the concentration in time of a few relatively rare, but high magnitude, events. Thus, it is likely that the 1970 Bangladesh cyclone and the 1976 Tangshan (China) earthquake together accounted for about 95 per cent of all disaster-related deaths during the 1970s, as indicated in Figure 2.2. Indeed, the RFF database compiled by Glickman *et al.* (1992) shows that just three natural disasters and two industrial accidents caused most of the deaths between the 1940s and the 1980s. In addition, the

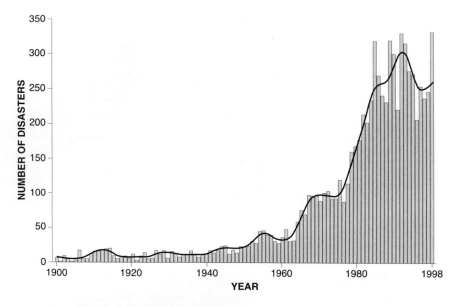

Figure 2.1   The annual total of natural disasters archived in the CRED database, 1900–98. Increases in the number of recorded events occur after 1964 (when OFDA was created) and after 1973 (when CRED was created). Part of the overall rise since the 1960s is due to improvements in communications and reporting efficiency.
Source: After CRED

nature of disaster impact is highly dependent on the type of event. For example, earthquakes tend to kill and injure more people than floods, but floods affect greater numbers and create more homelessness. Therefore, reliable conclusions from disaster archives are best drawn over reasonably extended and representative sampling periods whilst recognising that, over several decades, factors such as news reporting conditions and human vulnerability may have changed sufficiently to undermine attempts to standardise the record. Ideally, in order for disaster statistics to be used for reliable time-trend analysis of hazard impacts, the death totals should be standardised for the number of people at risk (for example, to compensate for population growth) and the damage totals should be standardised for price inflation (to compensate for changes in monetary value). These refinements are rarely introduced.

PATTERNS OF DISASTER

The best available estimates indicate that, in the past 1,000 years, about 15 million people have died as a result of at least 100,000 natural disasters

(a)

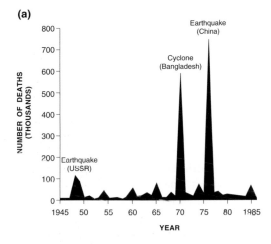

(b)

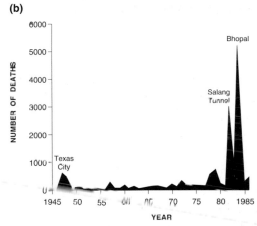

Figure 2.2 The mortality from (a) natural disasters and (b) major industrial accidents worldwide, 1945–86. The Salang tunnel disaster occurred when a fuel truck crashed in Afghanistan and the Texas City event was due to the explosion of ammonium nitrate fertilizer. The ten worst natural disasters and the ten worst industrial accidents accounted for 66 per cent and 63 per cent respectively of all lives lost.
Source: After Glickman *et al.* (1992)

(Munich Re, 1999). In the second half of the twentieth century, about 250 great natural disasters were reported. By the late 1980s and 1990s, the recorded level of events was approximately four times higher than in the 1950s, mainly due to an increase in climate-related disasters. In the 1970s and 1980s over twenty countries suffered individual natural disasters that killed more than 10,000 people and seven countries lost more than 100,000 lives in one event. According to Clarke and Munasinghe (1995), over

3 million disaster-related deaths were recorded between the mid-1980s and the mid-1990s. Less than 1 per cent of this total was due to technological disasters. More than 1 billion people were adversely affected by natural disasters and there was much material damage, including more than US$100 billion in 1991 and 1992. Elo (1994) estimated that in 1992 alone the world economy lost more money from natural disasters in the LDCs (US$62 billion) than it spent on development aid (US$60 billion).

These losses were created by a consistent and recurring pattern of natural agencies. Table 2.2 shows that over different sampling periods, both 35 years in length, two independently compiled databases illustrate a very similar occurrence of hazard types. In both cases, floods are the most common cause of disaster, accounting for about one-third of all recorded events. If tropical cyclones are added, the cumulative total exceeds 50 per cent. It should be noted that floods and tropical cyclones are important because, apart from killing people, they destroy large amounts of property and are responsible for much homelessness. The omission of drought from the NHRAIC database complicates any further direct comparison but it can be seen that earthquakes are also a frequent source of disaster, accounting for 10–15 per cent of events.

A similar consistency exists in the geographical pattern of hazard impact. There is a truism that, in disaster, the poor lose their lives while the rich lose their money. This is because over 90 per cent of disaster-related deaths occur among the two-thirds of the world's population who live in the LDCs and about three-quarters of all the economic damage is confined to the

Table 2.2 Global totals of the main types of natural disaster compiled for two different databases over contrasting 35-year sampling periods

NHRAIC data 1947–81			CRED data 1964–98		
Agent	Number	Per cent	Agent	Number	Per cent
Flood	343	32	Flood	456	33
Tropical cyclone	211	20	Tropical cyclone	298	21
Earthquake	161	15	Drought	205	15
Tornado	127	12	Earthquake	133	10
Snowstorm	40	4	Storm	115	8
Thunderstorm	36	3	Landslide	65	5
Landslide	29	3	Heatwave	29	2
Rainstorm	29	3	Coldwave	28	2
Heatwave	22	2	Volcano	23	2
Volcano	18	2	Avalanche	10	1
Coldwave	17	2	Tsunami	10	1
Avalanche	12	1	Forest fire	6	0
Tsunami	10	1	Insect infestation	2	0
Total	1,055	100	Total	1,380	100

Sources: Adapted from Thompson (1982) and CRED database

MDCs. Such a distribution suggests important relationships between disasters and wealth. Thus, Asia suffers greatly from natural disasters because of its large population, many of whom live in poverty and are concentrated in dense clusters in tectonically active zones or near low-lying coasts subject to cyclones and floods. Technological disasters tend to be concentrated in areas that are industrialising quickly without adequate health and safety regulation. Table 2.3 shows 'significant' natural disasters and technological disasters recorded over the 35-year period 1964–98 according to continental location. Asia experienced over half of all 'significant' natural disasters and, together with Africa, the Caribbean and Central America, accounted for over 80 per cent of the incidence. Asia was also the continent with most technological disasters. In turn, this global pattern is due to the repetitive occurrence of disaster in certain countries. Table 2.4, using the same 35-year database, reveals that the same group of nations appears in a ranking of the five most hazardous countries worldwide according to both natural and technological agents. Four of the five countries are in Asia, with India and China being especially exposed to both types of disaster. The presence of the USA on both lists may be partly a function of higher disaster reporting standards in that country.

There is plenty of evidence that natural disasters consistently claim most lives in the poorest countries, whether the impact is calculated absolutely

Plate 2 Homelessness is one of the few intangible consequences of disaster for which data can be systematically collected. In June 1990 a major earthquake (M = 7.3) struck the Gilan and Zanjan provinces of north-western Iran. In the most affected area, 60–90 per cent of the houses collapsed with 40,000 dead and 60,000 injured; 500,000 people were made homeless. (Photo: International Federation)

Table 2.3 The number of 'significant' natural disasters and of technological disasters recorded by continental area, 1964–98

Continent	Natural disasters		Technological disasters	
	Total	Per cent	Total	Per cent
Asia	743	53	1,472	46
Europe	33	3	625	19
Africa	250	18	410	13
Caribbean and				
Central America	145	10	178	6
North America	32	2	274	8
South America	124	9	227	7
Australia and Oceania	76	5	36	1
Total	1,403	100	3,222	100

Source: CRED database

Table 2.4 Ranked list of the five countries recording the highest number of 'significant' natural disasters and technological disasters, 1964–98

Natural disasters		Technological disasters	
India	153	India	344
China	106	China	233
Bangladesh	76	USA	227
Philippines	74	Philippines	175
USA	31	Bangladesh	91
Totals	440		1,070

Source: CRED database

or relatively according to the total population at risk. Table 2.5 ranks all the countries which experienced an annual average of at least 1,000 disaster-related deaths over the 1947–81 and 1964–98 periods. The top four countries in the earlier period, all in Asia, also appear in the second period. This pattern persists in the incidence of disaster-related deaths per million population (Table 2.6) where the top six countries in 1947–81 also appear in the 1964–98 list. The major difference between the two periods is the emergence of African countries in the 1964–98 CRED lists, a feature which reflects the inclusion of drought disasters in the CRED database and the effects of the prolonged drought in the Sahel during the 1970s and 1980s. Despite such variations, it is significant that two countries – Bangladesh and Iran – feature throughout both tables. The broad inverse relationship between disaster-related deaths and national wealth can also be validated for individual hazards over short periods of time. For example, Japan, the

Table 2.5 Ranked list of countries with an average of more than 1,000 deaths per year from natural disasters, 1947–81 and 1964–98

1947–81 period		*1964–98 period*	
Country	*Deaths per year*	*Country*	*Deaths per year*
China	10,983	India	46,042
Bangladesh	10,494	Ethiopia	34,370
India	3,213	Bangladesh	15,934
Iran	1,750	China	9,214
Korea, South	1,135	Sudan	4,334
Pakistan	1,059	Mozambique	3,191
		Iran	2,476
		Peru	2,078
		Philippines	1,250

Source: After Thompson (1982) and CRED database

Table 2.6 Ranked list of countries with more than 1,000 deaths per million of population from natural disasters, 1947–81 and 1964–98

1947–81 period		*1964–98 period*	
Country	*Deaths per million*	*Country*	*Deaths per million*
Bangladesh	3,958	Ethiopia	20,167
Guatemala	3,174	Mozambique	5,915
Nicaragua	2,590	Sudan	5,361
Honduras	1,995	Bangladesh	4,470
Iran	1,539	Peru	2,933
Peru	1,309	Honduras	2,816
New Guinea	1,283	Nicaragua	2,716
Haiti	1,189	Somalia	2,397
Korea, South	1,021	Guatemala	2,258
		India	1,641
		Iran	1,316

Source: After Thompson (1982) and CRED database

Philippines and Bangladesh are all Asian countries exposed to tropical cyclones but they differ greatly in economic strength. Table 2.7 shows that the loss of life from cyclones in each country has little to do with the incidence of such storms but a great deal to do with the link between low income and vulnerability to environmental hazard.

Most of the absolute economic loss from environmental hazard occurs in the developed world, especially North America. Rubin *et al.* (1986) claimed that over 70 per cent of the cost of US natural disasters was associated with flooding and hurricanes and it was estimated that the federal outlay on relief represented less than one-quarter of the total costs involved. But, if

Table 2.7 The incidence of tropical cyclones 1964–98 and number of people killed in selected high-, middle- and low-income countries

Economy	Cyclone events	Deaths	Deaths per event
Japan – high-income	48	2,215	46
Philippines – middle-income	149	22,063	148
Bangladesh – low-income	42	508,516	12,107

Source: CRED database

economic damage is considered as a proportion of national wealth, disasters bear most heavily on the smallest and poorest countries. According to Porfiriev (1992) natural disasters have taxed the economy of the former Soviet Union three to four times more than the USA. In the MDCs major events rarely cost more than 0.1 per cent of GDP but the economic impact on the poorest countries can be twenty to thirty times greater (Zupka, 1988). Some regions are especially vulnerable. For example, Otero and Martí (1995) cited the following costs of specific natural disasters on the yearly GDP of the relevant Central American country: 1985 Mexico City earthquake – 3 per cent; 1986 San Salvador earthquake – 24 per cent; 1988 Nicaragua hurricane – 40 per cent; and the 1987 Ecuador earthquake – 10 per cent. Small island countries in the Caribbean and the Pacific Oceans which depend on a narrow range of primary products have suffered damage from hurricanes equivalent to 15 per cent of their GDP. After cyclone Ofa in 1990 the microstate of Niue in the South Pacific faced a bill amounting to 40 per cent of GDP simply to repair the damage to government-owned buildings (Twigg, 1998). The recurrent effect of typhoons and drought on the GDP of Fiji is shown in Figure 2.3.

Underdevelopment leads to a lack of effective disaster planning because the costs of preparedness are high (Funaro-Curtis, 1982). For example, although earthquakes were not unknown in the Nicaraguan capital of Managua, the country had neither a relief organisation nor a contingency plan in place before the disaster of 1972. This event rendered 70 per cent of the population homeless and destroyed at least 10 per cent of the nation's industrial capacity. Within such countries, repeated disaster strikes accumulate to prolong economic and social problems. In April 1992 the eruption of the Cerro Negro volcano blanketed an extensive area of Nicaragua after a decade-long recession (Economic Commission for Latin America and the Caribbean, 1992a). The physical damage was estimated at US$19 million, creating losses in productive capacity of about 0.3 per cent, but the indirect effects were much greater. Large tracts of farmland had to be rehabilitated by deep-level ploughing and natural drainage channels were clogged with ashfall, causing flood problems in the rainy season. The main effect was on 150,000 people living high on the volcano's slopes and engaged in low-yield agriculture which did not meet their basic needs. Since they

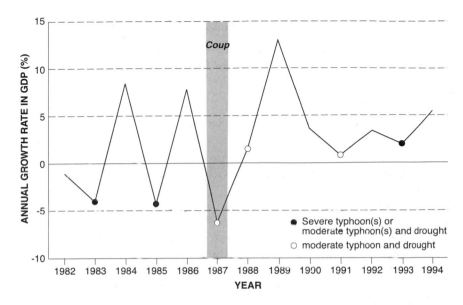

Figure 2.3   The adverse effect of typhoons and drought on the annual gross domestic
product (GDP) of Fiji, 1982–94. It can be seen that the most severe
loss occurred in 1987 when moderate natural events coincided with
civil disorder.
Source; After Twigg (1998)

were unable to find paid employment, they engaged in environmentally
harmful wood-cutting activities supplemented with emergency aid. This
event was followed in September 1992 by a tsunami which killed at least
116 people and affected twenty-six towns along the Pacific coast (Economic
Commission for Latin America and the Caribbean, 1992b). Again the direct
costs were relatively low, at US$25 million, but the event reduced GDP
by about 0.4 per cent. The poor suffered most heavily and over 20,000
small-scale fishermen and merchants lost their means of income.

DISASTER TRENDS

Just as the scale of hazard impact can often be attributed to human vulnerability, so some apparent trends in disaster are due to socio-economic factors
rather than the frequency and magnitude of geophysical processes. In all
cases, the statistical evidence must be interpreted cautiously. For example,
ongoing improvements in event monitoring and global communications
tend to produce a progressive, but artificial, increase in the number of
reported disasters. Figure 2.4 shows how the 120-year time-series (1860–
1980) for volcanoes reported as active per year throughout the world shows

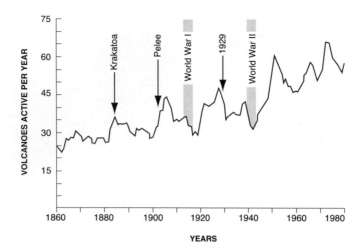

Figure 2.4   Variations in the number of volcanoes reported as active per year
throughout the world from 1860 to 1980 (three-year running average)
in relation to selected key events, including the world economic recession in 1929.
Source: After UNEP (1987)

external influences rather than a true picture of volcanic eruptions. First,
the long-term upward trend is a measure of improved global monitoring
rather than a real increase in volcanic activity. Second, it can be seen that
a preoccupation with wars or economic depression may have decreased
reporting efficiency, while major volcanic events have enhanced awareness
and increased reporting for a period of several years after the event.

Bearing these limitations in mind, recent decades have provided evidence
of increased disaster impact. Figure 2.5 shows the general upward global
trend during 1945–86 in deaths from natural hazards and industrial
accidents for groups of at least twenty-five and five people respectively
(Glickman *et al.*, 1992). Given the low mortality threshold for this database, the graph is a sensitive illustration of change and it is likely that
some of the upward trend is due to the improved reporting of small-scale
events in remote areas. The apparent decline in the 1980s, especially in
industrial accidents, may reflect organisational and regulatory changes for
improved safety, although data up to 1991 tended to confirm the rise
(Glickman *et al.*, 1993). Figure 2.6 employs the CRED database and reveals
an upward trend in 'significant' disasters over a 30-year period. Whilst the
number of events claiming at least 100 deaths had more than doubled by
the early 1990s, disasters creating economic damage equivalent to 1 per
cent or more of national GDP had risen well over four-fold.

Rising economic losses can be validated by insurance data. Figure 2.7
shows the incidence of 'great natural catastrophes' recorded by the Munich

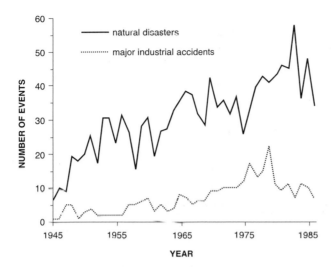

Figure 2.5 Trends in the number of natural disasters and major industrial accidents recorded worldwide 1945–86 by the RFF database. The mortality threshold is set at 25 or more for natural disasters and at 5 or more for industrial accidents.
Source: After Glickman *et al.* (1992)

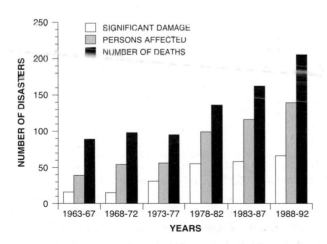

Figure 2.6 The trend in 'significant' disasters worldwide between 1963 and 1992 based on the CRED database showing the number of disasters which individually breach the thresholds for deaths, economic loss and persons affected.
Source: After IFRCRCS (1994)

(a)

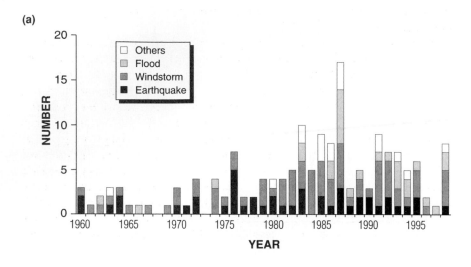

(b)

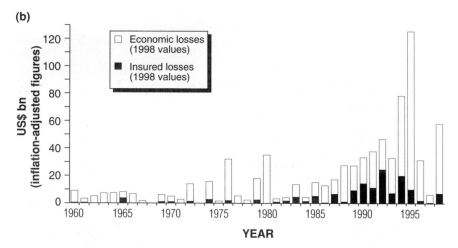

Figure 2.7   The trend in disasters according to insurance data: (a) shows the number
of 'great natural catastrophes' recorded 1960–98 and (b) shows the
economic losses and the insured losses which occurred over the same
period (at 1995 US$ values).
Source: After Berz (1999)

Reinsurance Company 1960–98 and the associated material losses. Over
this period the number of such events has risen by a factor of three. After
adjusting for inflation, the economic losses have increased nine times and
the insured losses have increased fifteen times (Berz, 1999). In view of
improved hazard awareness within the insurance industry, these values may
over-estimate the real position but the trend is clear. Before 1987 only one
event cost the insurance industry more than US$1 billion. Since 1987 no

fewer than twenty-three such disasters have occurred, twenty-one of which
have been recorded since 1990.

These trends have not been evenly distributed across the globe. Broadly
speaking, there has been a tendency during much of the twentieth century
for rising economic losses in the MDCs to be offset by a fall in fatalities,
at least for some weather-related hazards. If the USA is taken as an example,
with mortality statistics normalised by the population at risk and loss data
adjusted for inflation, deaths have followed a progressive decline for both
tornadoes and hurricanes (Riebsame *et al.*, 1986). Despite the growing coastal
population exposed to hurricanes, improved weather forecasting combined
with safer building regulations and pre-planned evacuation measures have
created greater safety along many shorelines. However, the economic losses
from hurricanes are still rising (Figure 2.8) and any recent levelling off is
more a reflection of temporarily reduced hurricane activity than of a genuine
decline in hazard (Gross, 1991). Pielke (1997) has stressed that such trends

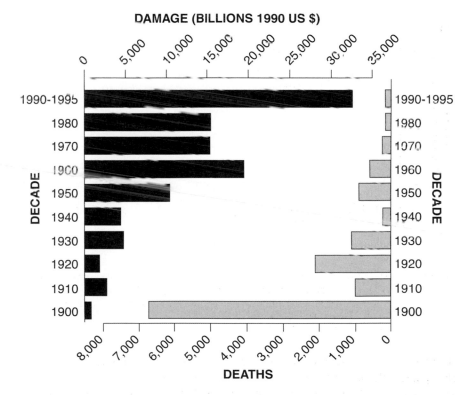

Figure 2.8   The property damage (at 1990 US$ values) and the loss of life in the
continental United States due to tropical cyclones in each decade of
the twentieth century. There is a broadly inverse relationship between
damages and deaths.
Source: After Pielke and Pielke (2000)

are due to societal changes, rather than to the incidence of hurricanes, and that the USA remains exposed to great potential loss from future storms (see Chapter 9).

Despite the welcome fall in fatalities for some hazards in many of the wealthier countries, the world trend is towards more disaster-related deaths and damages driven mainly by increased vulnerability in the LDCs. There are several reasons, outlined below, why disaster impact is growing, even if the frequency of extreme geophysical events is unchanged and despite the many positive steps being taken to reduce disasters.

Population growth

The world population total reached an estimated 6.25 billion in the year 2000. This means that the number of people exposed to hazard is increasing, especially because some 90 per cent of the population growth is taking place in the LDCs. In these countries, human vulnerability is already high through dense concentrations of population in unsafe physical settings. Continued population growth outstrips the ability of governments to invest in education and other social support services and creates further competition for land resources. In the very poorest countries, the human use of natural resources has created a problem of food security and fragile livelihoods. Only a quarter of the people in Africa have access to safe drinking water, millions lack sufficient food supplies to maintain active lives and drought can lead to widespread famine. But in countries where families survive only through supplying labour and the oldest members depend on support from the young, the pressure for large families persists. Conversely, the demographic trend in the MDCs is creating a rise in the elderly population, who sometimes need specialist support in disaster. For example, in the UK about 70 per cent of the adults categorised as disabled are aged 60 years or over.

Land pressure

Rural land pressure adds to the problem: perhaps as many as 850 million people live in areas suffering severe environmental degradation. In many LDCs more than 80 per cent of the population is dependent on agriculture but many are denied an unequal access to land resources. Poverty forces the adoption of unsustainable land use practices and countries with a legacy of deforestation, soil erosion and over-cultivation find their environment more vulnerable to environmental hazard, especially floods and droughts. The 'modernisation' of agriculture can bring problems. In the tropics large areas have been given over to capital-intensive plantation agriculture, often accompanied by reservoir construction for water supplies. This displaces farmers from their land, and the controlled supply of irrigation water for the monocultural cash crops reduces the seasonal flooding necessary for flood-retreat agriculture. Low-lying coasts have been made more vulnerable to storm surge by the clearance of mangrove forests, often for fish farming and salt production in the LDCs and for tourist development in the MDCs. Inland, the drainage of wetlands leads to a loss of common property resources such as fisheries and forests. As dietary habits change, the traditional crops are likely to be replaced, with a consequent potential loss of biodiversity and genetic resources. The impoverishment of the agricultural base results in large shifts of population from rural areas to urban centres.

Urbanisation

Rural–urban migration, driven by local land pressure and global economic forces, is concentrating people into badly built and over-crowded cities. Some 20–30 million of the world's poorest people move each year from rural to urban areas. Already some of these cities, exposed to destructive earthquakes and other hazards, have between one-third and two-thirds of their population in squatter settlements. According to Davis (1978), rural–urban migration can cause squatter settlements to double in size every 5–7 years, about twice as fast as the overall growth rate for Third World cities. The rural migrants form the poorest urban dwellers. In Turkey, for example, most of the rural–urban migration has been to cities with high seismic risk (Torry, 1980). Apart from being on unsafe sites, these shanties have poor water supplies and sanitation. Coupled with poor diets, this results in inadequate nutrition and endemic disease. There has also been considerable growth in coastal cities, many of which are exposed to hurricanes. Even in the MDCs, coastal cities in seismically active areas, such as the west coast of the USA and Japan, have been built on loosely compacted sediments or landfill sites which are likely to perform poorly in earthquakes.

Inequality

Disaster vulnerability is closely associated with the economic gap between rich and poor, which is growing in many areas. Over 20 million people cannot afford a nutritionally adequate diet. In Asia and the Middle East, about one-third of the population live in poverty, a proportion which rises to nearly half in sub-Saharan Africa. Globally, it has been estimated that some 20 per cent of the population control 80 per cent of the wealth. National disparities continue to increase, thereby exacerbating vulnerability. For example, in Chile the wealthiest 20 per cent of the population expanded their control of national income from around 50 per cent to 60 per cent between 1978 and 1985. Over the same period, the income share of the poorest 40 per cent of the population fell from 15 per cent to 10 per cent (IFRCRCS, 1994).

Climate change

Global warming is likely to bring significant changes in the world's climate over the coming decades. The expected temperature change will be greater and faster than that seen over the past 10,000 years. The probable physical consequences range from the more frequent inundation of some low-lying coasts, especially where natural ecosystems such as salt marsh or mangroves have been removed, to increased riverflow from snowmelt in alpine areas. It is probable that the most significant effects will be experienced in countries highly dependent on natural resource use and will influence activities such as agricultural development, forestry, wetland reclamation and river management. Future shifts in disease patterns may newly threaten animal and human populations. The overall result is likely to widen the gap between developing and developed nations because the impacts will be most severe on ecosystems already under stress and for countries which have few spare resources for adapting to, or mitigating, climate change.

Political change

The richer countries appear to be reducing their commitment to internal welfare and to the international community. For example, in many western countries, health spending per person has declined since 1980 and the role of the welfare state has been deliberately reduced. Over eastern Europe and in the former Soviet Union, the collapse of communism has removed the influence of the state with respect to health care, education and social provision. State paternalism has been replaced with an unregulated scramble towards free-market ideals in which the weakest members of society are ill-equipped to compete. At the same time many of the MDCs fail to contribute adequately to development aid. During recent decades the volume of development aid has declined, resulting in greater vulnerability to disaster as aid agencies are left to fill the welfare role vacated by governments.

Economic growth

Rising vulnerability to hazard is not simply the prerogative of the LDCs. Economic growth in the wealthy countries has increased the exposure to catastrophic property damage. Along with the growing complexity and cost of the physical plant responsible for the world's industrial output, capital development has ensured that each hazard will encounter an increasing amount of property unless steps are taken to reduce the risks within cities and on industrial sites. Partly in response to the growing shortage of building land, some of the growth has occurred in areas subject to natural hazards, whilst man-made hazards such as toxic chemicals and the use of nuclear power have added to the loss potential. The availability of increased leisure time has led to the construction of many second homes built in potentially dangerous locations, such as mountain and sea-shore environments.

Technological innovation

The rising technology of the rich countries is normally seen as helping to prevent disaster through better forecasting systems and safer construction techniques. However, the more a society becomes dependent on advanced technology, the greater is the potential for disaster if the technology fails. New high-rise buildings, large dams, building construction on man-made islands in coastal areas, the proliferation of nuclear reactors, the reliance on mobile homes for low-cost housing, more extensive transportation (especially air travel) are all examples of trends which create additional vulnerability to hazard. In the LDCs the introduction of even low-level technology, such as the building of a new road through mountainous terrain, may increase landslides through the logging of steep slopes, and some building innovations, such as 'modern' concrete houses, may be unable to withstand earthquakes (Coburn *et al.*, 1984). The disastrous release of a toxic chemical at Bhopal, India, was directly related to modernisation efforts which introduced a complex and poorly managed industrial production system into a society unable to cope with it.

Social expectations

Hazard vulnerability may be increased because of rising social expectations, particularly in the MDCs. People have become much more mobile in recent years and expect to be transported around the world in the minimum elapsed time, irrespective of adverse environmental conditions such as severe weather. The same security of service is expected by consumers from most weather-dependent enterprises such as energy supply or water supply. Frequently the drive for greater competition in commerce and industry has resulted in reduced manning and smaller operating margins. In turn, these apparent improvements allow less scope for an effective corporate response to environmental hazard.

Global interdependence

The functioning of the world economy works against the LDCs by reinforcing hazard vulnerability. Most of the Third World's export earnings come from primary commodities for which market prices have either fallen over several decades or remain highly unstable. The LDCs have little opportunity to process and market what they produce and are dependent on the importation from the industrialised nations of manufactured goods which are often highly priced or tied to aid packages. The progressive impoverishment of the small-scale farmer, combined with a foreign debt burden that may be many times the normal annual export earnings, takes resources away from long-term development in a process that has been described as a transfusion of blood from the sick to the healthy. The cycle is reinforced when natural disaster destroys local products and undermines incentives for investment. Major disasters, such as the Sahelian drought, disrupt not only local economies, but can also bring shortages in neighbouring regions, create floods of international refugees and stimulate aid programmes to the extent that the repercussions of environmental hazard are truly global. Figure 2.9 illustrates how the effects of a major disaster can extend from the victims in the immediate hazard zone to reach the world through the media and appeals for aid.

PARADIGMS OF HAZARD

The interpretation of hazard, in terms of both causes and cures, has passed through several phases. The earliest responses, which were to flood hazard, attempted to exert limited control over the forces of nature and levees were constructed on the floodplains of the Nile and Tigris–Euphrates rivers some 3–4,000 years ago. Gradually, the disadvantages of hydraulic engineering began to emerge; Marsh (1864) demonstrated how such human actions were adversely changing the environment. This fundamentally ecological view-point was rediscovered in the USA in the 1920s by writers such as Barrows (1923) and was absorbed into the 'behavioural' approach to hazards advo-cated by the influential North American school of geographers led by Gilbert White. Since about 1975 a split has been apparent between this view of hazard and the more recent 'structural' interpretation, which typically derives from anthropologists and development workers with field experience in the Third World.

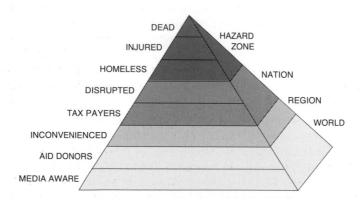

Figure 2.9   A disaster impact pyramid showing the spread of the event from the death of a small number of victims in the hazard zone to an aware-ness by the global population communicated by the mass media.

The behavioural paradigm

Modern environmental engineering was pioneered in the USA during the 1930s by a generation newly aware of the perils of soil erosion and floods. In 1936 the US Congress passed an important Flood Control Act which designated the Army Corps of Engineers as the federal agency responsible for large-scale watershed management. This organisation commenced an ambitious programme of engineering works to control flood waters and protect floodplain property, of which the Tennessee Valley Authority scheme became one of the best-known examples. The approach which began here characterised attitudes to environmental hazards for the next fifty years.

Although soil and water conservation objectives lay behind the civil engineering, flood control works were constructed on the premise that geophysical extremes are the cause of disaster. Since the blame was assumed to lie with nature, it appeared logical that the prediction of natural events and their control, through a strengthening of the built environment, would provide an effective cure. Such goals appeared to be both attainable and desirable in countries like the United States during the 1930s and 1940s because of the confidence associated with the rapid growth in relevant scientific fields (meteorology, hydrology), demands for greater development of natural resources and the availability of capital for major engineering projects. White (1945) moderated this view by urging a wider approach using non-structural methods integrated with flood defence works into a comprehensive scheme of floodplain management. The exacerbating role played by the victims themselves was also recognised. In the MDCs 'behavioural' faults were attributed to the poor perception of hazards by the flood control authorities and the flood victims which allowed settlement on floodplains to expand. Within the LDCs it was felt that disasters were compounded by even less rational behaviour, such as deforestation or the over-grazing of land, by so-called 'folk' societies. It was further believed that the universal consequence of disaster was a disruption of 'normal' life reflected in a breakdown of economic production and a failure of the social system. This lack of order was seen as a temporary interruption of stability in the MDCs but regarded mainly as a function of the inherent lack of a stable, western-style approach in much of the developing world.

Based on this rather paternalistic diagnosis, a solution was sought in the power of applied science and technology. A 'technical fix' approach was often advocated in the more industrialised countries. In the fulness of time, it was thought that the transfer of the appropriate technology to the LDCs, as part of an overall modernisation process, would eventually solve their problems too. Inevitably, the emphasis on high technology led to a rather authoritarian organisational pattern. Only government-backed institutions possessed the financial resources and technical expertise needed to apply science on the scale deemed necessary and had the power to re-impose order and rationality after a major disaster. The United Nations, in particular, sprouted a number of agencies with interests in disaster mitigation.

This paradigm has been characterised by three thrusts (Hewitt, 1983):

1   An emphasis on field monitoring and the scientific explanation of geophysical processes. This was aimed at the modelling and prediction of damaging events and has employed advanced technical tools such as remote sensing and telemetry.
2   Despite an acknowledgement of the role of human perception and behaviour, an underlying commitment to physical and managerial control was also present. Often the aim has been to contain nature through environmental engineering works, such as flood embankments or avalanche sheds.
3   Another clear strand has been the formulation of disaster plans and emergency responses. This role has often been given to the armed forces, mainly on the assumption that only a military-style organisation can function properly in a disaster area, but it is convenient that this role also underpins the notion of the state re-imposing order on a devastated community.

It has since been recognised that this paradigm, which still represents the dominant view from government, is an essentially western interpretation of environmental hazard which is rooted in materialism. It has been described by its critics as an optimistic, deterministic evaluation which reflects undue faith in technology and capitalism. This view has also been criticised because it emphasises the role of individual choice in hazards, either from the decision maker or the victim, at the expense of wider social and economic forces.

The structural paradigm

Since the mid-1970s an alternative interpretation has been gaining ground (Emel and Peet, 1989). This view can be termed structuralist in approach because it is less hazard-specific and emphasises the constraints which are placed on individual action by more powerful institutional forces. There is little doubt that an alternative philosophy was sought because the earlier approach had not proved wholly successful in reducing disaster impact, especially in the LDCs. Indeed, the structuralist view has been largely developed by social scientists with first-hand experience in the Third World (Waddell, 1983).

The structuralist view forges a link between environmental disasters and the underdevelopment and economic dependency of the Third World. It originates in the belief that disasters in the LDCs arise more from the workings of the global economy, from the spread of capitalism and the marginalisation of poor people than from the effects of geophysical events. Consequently, the proponents of this view argue for a clearer distinction between what they see as the geophysical 'triggers' of natural disaster and the ongoing economic, social and political problems of the LDCs. It is a radical interpretation of disaster which envisages solutions based on the redistribution of wealth and power in society to provide access to resources rather than on the application of science and technology to control nature.

The alternative view challenges the behavioural paradigm at several key points:

1 It asserts that environmental disasters are not primarily dependent on physical processes. In the LDCs especially, it is argued that growing poverty has created greater vulnerability for the population either as a rural proletariat (dispossessed of land and compelled to grow cash crops rather than subsistence food) or as an urban proletariat (forced into shanty towns in the most dangerous built-up areas). Thus, the severity of disaster impact is related more to the scale of human exploitation than to the stresses imposed by nature.

2 It queries the assumption that disasters are such unusual phenomena in a Third World context. This point stems from the fact that many disasters occur in areas experiencing rapid environmental and social change. It is argued that physical processes creating cyclones or droughts have existed for much longer than human activities, such as urbanisation, and more truly reflect the fundamental nature of such regions. That disasters regularly recur in poor countries is increasingly linked to the fact that the available responses are severely limited by a lack of resources

3 It asserts that disaster victims are not to blame for their own misfor tunes. They do not necessarily lack adequate perception or engage in irrational, hazard-inducing behaviour, but – especially in the Third World – they have little choice but to locate in unsafe settings where daily survival is the main objective. They lack the time to prepare for emergency action and the resources to recover from disaster. Again, rural over-population and the attendant rise of hazard-prone primate cities are interpreted as a symptom of capitalism, which is seen as a root cause of environmental disaster.

4 Given a belief that disasters in the Third World are characteristic rather than accidental, it follows that disaster mitigation depends on structural change taking place in society. But the structuralist view rejects the modernisation theory and its triumphalist optimism, as expressed by the behavioural school (Blaikie *et al.*, 1994). Even long-term 'development' can be counter-productive. Maskrey (1989) cited the example of Peru, which has become more hazard-prone with the post-colonial shift of population from mountain communities to high-risk urban centres. This is especially the case in the capital, Lima, which is located in an active seismic zone where houses of Spanish design with heavy roofs are crowded with low-income families.

5 For sustainable development and disaster reduction to occur together, the structuralist view stresses a reliance on local knowledge rather than

imported technology. In rural areas, for example, successful development means 'bottom-up' strategies that start at community level, such as the establishment of cooperatives to provide seed banks, crop insurance and credit for tools and other assets lost in disasters. Such measures would help to stabilise the rural base and halt the migration to unsafe urban environments. Conversely, technical aid, particularly beyond the emergency relief phases, is perceived as increasing vulnerability by making a short-term problem semi-permanent through additional dependency.

In summary, the alternative view is based on the theory that under-development is not a temporary state but is an ongoing process of Third World impoverishment perpetuated by technological dependency and unequal trading arrangements between rich and poor nations (Susman *et al.*, 1983). Within the LDCs, it leads to the process of marginalisation, as depicted in Figure 2.10. Disaster simply reinforces the growing gap between rich and poor. Even in 'normal' times the poorest sections of society are pressured to over-use the land and, when disaster strikes, the conventional responses merely accelerate the continued under-development and marginalisation. Thus, the differential degree of damage between various parts of the city caused by the Guatamala City earthquake in 1976 led to the designation of this event as a 'classquake'.

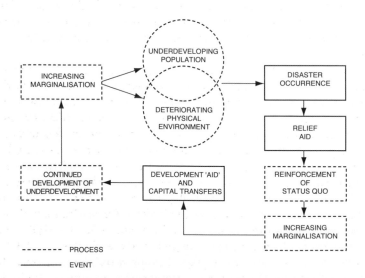

Figure 2.10   The process of marginalisation typical of the LDCs and the relationship to disaster. Increasing vulnerability is the result of a population suffering from underdevelopment and a deteriorating physical environment. Relief and development aid simply reinforce the state of under-development that creates the disaster potential in the first place. Source: After Susman *et al.* (1983)

TOWARDS A BALANCED VIEW OF DISASTER

From the standpoint of practical policy making, there is merit in achieving a compromise between these two paradigms. Such a compromise has to be based on a recognition of different causes and consequences of disaster between developed and developing countries because there is little point in replacing a wholly environmental and technocratic perspective with an equally polarised view based on political and economic determinism.

The behavioural school can be criticised for not embracing the work of some anthropologists and sociologists in the Third World and for not addressing the role of institutional factors and global forces, such as poverty and technical change, in increasing hazard vulnerability (Torry, 1979). It emphasises adjustment to hazard but, because it is underpinned by ecological theory, it can perhaps be criticised most severely for failing to consider *maladjustment* to hazard expressed through the adverse ecological effects that arise from the self-destructive union that many hazard-prone societies have with their environment. In particular, the behaviouralist view relies on the quantification of risks, the erection of physical structures and the investment of public funds to 'design-out' hazards from the environment. In some cases, the 'trade-off' costs of such disaster reduction strategies for the environment have been under-estimated. On the other hand, the behavioural approach has always been pragmatic and willing to adopt existing knowledge to mitigate disaster. It can be defended on the grounds that it provides a framework for practical hazard mitigation, especially as there is now more awareness of the need for economic and environmental sustainability in disaster reduction.

The structuralist school draws theoretical support from political economy. This approach relies more on qualitative risk assessment and queries both the cost-effectiveness and the intangible trade-off between ecological and aesthetic costs of large-scale engineering schemes. It can be criticised for rather stridently expressed views which, at worst, simply call for social revolution and attempt to deny the success of any devices – such as flood banks or forecasting and warning systems – in mitigating disaster. The 'green' lobby within the structuralist school sometimes presents the side effects of hazard reduction as worse than the original problem. Most of the research inspired by this paradigm has focused on the Third World, although there is now an emerging concern for disadvantaged minorities in the MDCs, as reflected in the growing body of material on gender issues (Fothergill, 1996; Fordham, 1998). However, the real strength of the structuralist approach lies in the way it is refining the concept of poverty to help protect the lives and livelihoods of the most vulnerable members of the LDCs through the advocacy of community-based disaster mitigation strategies which are assimilated into traditional, local activities.

A balanced interpretation of disaster allows that the structural paradigm has relevance outside the Third World and that science and technology have a role to play beyond the MDCs. All individuals are constrained to some extent by the social and institutional circumstances in which they find themselves. In turn, institutions are moulded by macro-scale economic and political processes, sometimes operating at an international level. At the same time, there will always be a need for some degree of physical protection and reliance on modern technology at locations highly vulnerable to the loss of life and property.

Attempts have been made to find a balance between the behavioural and the structural approaches based on a greater recognition of relevant local factors (Mitchell, 1990; Palm, 1990). For example, Penning-Rowsell (1987) showed how British flood studies in the 1970s were dominated by work in North America that stressed behavioural-type issues of floodplain encroachment, poor hazard perception and the limited use of measures other than engineering controls. But, in England and Wales, land use control is in the hands of the local planning authority and the scope for individual action is limited. There is little doubt that institutional inertia and vested interests, at national and local levels, have to be recognised in all environmental hazard appraisal. Political control can be an important aspect, as can the funding arrangements, which frequently ensure the selection of one type of hazard adjustment rather than another. Professional bias exists. For example, the traditional dominance of civil engineers in the agencies charged with flood reduction has, not surprisingly, led to an emphasis on engineering structures as a solution to flood problems. Other forces in society, such as landowners or the voice of the environmental movement, all help to influence attitudes.

The challenge in the post-industrial age is to adopt a culture based more on 'living with hazards' than on confronting them in a purely physical way. In many cases, this could be achieved by integrating disaster-reducing strategies with more rational risk-taking and better land use management. There is already a trend away from site-specific engineering to a consideration of larger issues that reflect the importance of human vulnerability and the need for more sustainable development. This means protecting the poorest members of society whilst making the best use of natural assets at present and ensuring that resources are available to meet the needs of future generations.

3   Risk assessment and disaster management

THE NATURE OF RISK

Risk is an integral part of life. Indeed, the Chinese word for risk, *weij-ji*, combines the characters meaning 'opportunity' and 'danger' to imply that uncertainty always involves some balance between profit and loss. Since risk cannot be eliminated, the only option is to manage it. *Risk management* means reducing the threats posed by known hazards, whilst simultaneously accepting unmanageable risks, and maximising any related benefits. Achieving optimum safety is a broad task involving economics, legal standards and available technology in what are often controversial value judgements. There are great difficulties in deciding what is an acceptable level of risk, who benefits from risk management, who pays and what constitutes success or failure in risk reduction policy.

Risk assessment is the first step designed to find out what the problems are. It involves evaluating the significance of a risk, either quantitatively or qualitatively. In practice, quantitative risk assessment has not been attempted for many environmental hazards but, following Fournier d'Albe (1979), can be conceptualised as:

$$\text{Risk} = \frac{\text{Hazard (probability)} \times \text{Loss (expected)}}{\text{Preparedness (loss mitigation)}}$$

Even when risks have been quantified by statistical or other methods, uncertainties usually attach to the estimate obtained and its interpretation when drawing up spending priorities for limited resources. Quantitative risk assessment is a process accessible only to a technically well-informed minority. At best it may limit public debate about hazard reduction and, at worst, it can be used to justify the conclusion that the practitioner wishes. Risks also need to be assessed in a more qualitative way. As Keeney (1995) stated, the sound evaluation of risk requires both good science and good judgement.

Neither risk assessment nor risk management can be divorced from value judgements and choices which, in turn, are conditioned by individual beliefs and circumstances. Many people make decisions and take actions regarding hazards based on their personal perception of risk rather than on some externally derived measure of threat. Thus, *risk perception* has to be regarded as a valid component of risk management alongside more scientific assessments. Distinctions are frequently drawn between *objective* and *perceived* risks. This is because individuals perceive risks intuitively and often quite differently from the results obtained by more objective assessment models based on analogies with financial cost-benefit analysis (Starr and Whipple, 1980). Resolving the conflict between the outcome of technical risk analysis and more subjective risk perception is a major factor in many hazard management strategies.

The type and degree of perceived risk varies greatly, even between individuals of the same age and sex, according to personal factors such as location, occupation and life-style. Cross-national variations also exist (Rohrmann, 1994). It is common to classify risks into two main categories:

1   *Involuntary risks* These are risks which are not knowingly or willingly undertaken. They often relate to rare events with a catastrophic potential impact. The risk may be unknown to the exposed person. If the risk is perceived, it may be seen as inevitable or uncontrollable, as in the case of earthquakes. Most of the hazards considered in this book fall into this category and represent the risks resulting from a person living in a hazard-prone environment.
2   *Voluntary risks* These are risks which are more willingly accepted by people through their own actions. Such risks are likely to be more common, have less catastrophe-potential and be more susceptible to control. Unlike involuntary risks, they are rated more directly by individuals according to their own judgements and life-style. The greater scope for control over voluntary risks is seen in either modifications of individual behaviour (stopping smoking or ceasing participation in a dangerous sport) or some form of government action (the introduction of safety legislation or pollution control). Man-made hazards, including risks from technology, are usually placed in this group.

This division between risk categories is often less clear than it appears. For example, while cigarette smoking and mountain climbing are obvious cases of voluntary risk-taking activities, the same cannot be so firmly stated for some other hazardous activities. Driving a car may be essential to get to work and the alternative to working in a dangerous chemical factory may be unemployment. In other words, a risk is generally less voluntary than another risk if its avoidance is connected with a greater personal sacrifice on the part of the risk-bearer. As already stated, natural hazards are usually seen as involuntary, although the inundation of some active floodplains is sufficiently frequent and well-publicised to cast doubts on

Plate 3 The expansion of cities in the tropical parts of Australia has led to an increased exposure to hazard. Much of Townsville, northern Queensland, has been built on a low-lying coast from which mangrove forest has been removed and is at risk from storm surge associated with tropical cyclones and from flooding by the Ross river. (Photo: K. Smith)

this. Some floodplain dwellers may elect to buy a home which is cheaper than an equivalent property in a safer part of town and then feel less need for expenditure on house maintenance. Viewed in this light, the locational decision is both voluntary and economically rational. Voluntary risk takers are sometimes seen as more personally identifiable than 'statistical' risk takers because they may be grouped in a specific hazard-prone location. But, even if the overall probability of risk can be estimated fairly accurately, as in the case of a large population of cigarette smokers, it is still not possible to know exactly which individuals will die from tobacco-related diseases.

Despite this, people have been shown to react differently to voluntary risks compared to risks which are imposed by some outside body. In a pioneering study of public attitudes towards various technologies, Starr (1969) attempted a correlation between the risk of death to an individual, expressed as the probability of death per hour of exposure to a particular hazardous activity (P_f), and the assumed social benefit of that activity converted into a dollar equivalent. From Figure 3.1 it can be seen that there was a major difference between voluntary and involuntary risk perception, with people willing to accept voluntary risks approximately 1,000 times greater than involuntary risks. For example, voluntary risks such as driving, flying and smoking were apparently accepted even though they produced a risk of death of one in 100,000 or more per person per year

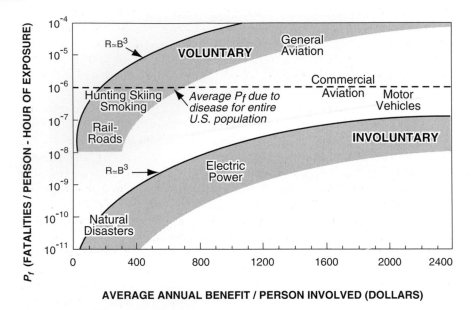

Figure 3.1   Risk (P_f) plotted relative to benefit and grouped for various kinds of voluntary and involuntary human activities which involve exposure to hazard. The diagram also shows the approximate third-power relationship between risks and benefits. The average risk of death from disease is indicated for comparison.
Source: After Starr (1969). Reprinted with permission from *Science* 165: 1332–8. © 1969 AAAS

while the involuntary risks exposed people to a risk of about one in 10 million or less per person per year. Interestingly, the acceptability of risk from a technology appeared to be approximately equal to the third power of the benefits, either real or imagined; that is, the technologies with the greater dangers have the greater assumed benefits. However, other workers have suggested that people are not always willing to make such trade-offs between risks and benefits and that perceived dangers influence risk attitudes more strongly than perceived benefits. This interpretation is most likely in connection with so-called 'dread' hazards, for example, nuclear power.

Given that the provision of absolute safety is impossible, there is great sense in trying to determine the level of risk that is acceptable for any activity or situation. *Acceptable risk* seems a sensible concept but it is a much misunderstood term. For example, it does not necessarily describe either the level of risk that people are happy with or the lowest risk possible. Fischhoff *et al.* (1981) concluded that the term has limited applicability because it really describes the least unacceptable option, and the risk associated with this option is not truly 'acceptable' in any absolute sense.

One must always specify acceptable risk as acceptable *to whom*. Even actual behaviour does not necessarily reflect the optimum choice. For example, in the case of a consumer buying a car, the mere act of purchase need not imply that the product is safe *enough*, just that the trade-off with other forms of transport is the best available. Therefore, 'risk tolerance' might be a better term than 'acceptable risk'.

We still lack understanding of informed consent about risk taking and the information that is needed to enable any individual to make rational hazard-related decisions. One difficulty is the constantly changing social view of risks. For example, after a major disaster, there are usually calls for community protection to a much higher level of future safety. The 1993 floods in the upper Mississippi river basin had an estimated return period of more than one in 200 years, yet some people who were flooded asserted that this event should now be regarded as an unacceptable risk. Such arguments ignore the benefits derived by those communities from their floodplain location over the previous 100 years or so, when few flood losses occurred, and the high cost to the taxpayer of providing protection against a flood of the 1993 magnitude. Although growing safety concerns will continue to raise the level of acceptable risk, the level will also depend on the prosperity of a country, since a poor nation does not have the resources to tackle many public hazards. In summary, there is no fully objective, value free approach to risk decisions and, since there is often uncertainty about the best way to manage many hazards, quantitative analysis is best viewed as a partial, rather than an absolute, function.

RISK ASSESSMENT

According to Kates and Kasperson (1983), risk assessment comprises three distinct steps:

1 The identification of local hazards likely to result in disasters, that is, what hazardous events may occur?
2 The estimation of the risks of such events, that is, what is the probability of each event?
3 The evaluation of the social consequences of the derived risk, that is, what is the likely loss created by each event?

However, for sound risk *management* to occur, there should be a fourth step which addresses the need to take post-audits of all risk assessment exercises. In practice, very few studies have followed risk assessment through to ask what happened after the assessment was undertaken and to discover whether any protective action taken was effective. The general lack of such feedback is one of the most serious deficiencies in the reduction of environmental hazards at the present time.

The statistical analysis of risk is based on theories of probability. When analysis is undertaken, risk (*R*) is taken as some product of probability (*p*) and loss (*L*):

$$R = p \times L$$

If every event resulted in the same consequences, it would be necessary only to calculate the frequency of occurrences. But most environmental hazards, such as severe storms, have highly variable impacts and some assessment of these consequences, as well as their probability, is required.

A detailed illustration has been provided by Krewski *et al.* (1982). From experience, it is known that *n* different mutually exclusive events $E_1 \ldots E_n$ may occur. These events might be a series of damaging floods or urban landslides but it should be appreciated that the method does depend on the availability of a good database accumulated over a period of time. Thus, the method is less satisfactory for rare natural events, such as large magnitude earthquakes, or for some technological hazards, such as the release of radionuclides from nuclear facilities.

From historical data, it can be determined that event E_j will occur with probability p_j and cause a loss equivalent to L_j, where subscript *j* represents any of the individual numbers $1 \ldots n$ and $L_1 \ldots L_n$ are measured in the same units, for example, £ sterling or lives lost. It is assumed that all the possible events can be identified in advance. Therefore, $p_1 + p_2 \ldots p_n = 1$.

After arranging the *n* events in order of increasing loss ($L_1 < \ldots < L_n$), the cumulative probability for an individual event can be calculated as $P_j = p_j + \ldots p_n$. This specifies the probability of the occurrence of an event for which the loss is as great as or greater than L_j, as shown in Table 3.1. If we can categorise all possible events in terms of the property loss (expressed in £ sterling), it may be possible to produce a risk analysis along the following lines:

Property loss (£)	Probability (p)	Cumulative probability of exceedance (P)
0	0.950	1.000
10,000	0.030	0.050
50,000	0.015	0.020
100,000	0.005	0.005

This theoretical example shows that there is a 95 per cent chance of no property loss and only a 2 per cent chance of a property loss of £50,000 or greater.

In some circumstances, it may be necessary or desirable to produce a summary measure of risk (*R*). This can be done by calculating the *total probable loss*:

$$R = p_1L_1 \ldots + p_nL_n$$

Table 3.1 Basic elements of quantitative risk analysis

Event	Probability	Loss[a]	Cumulative probability
E_1	p_1	L_1	$P_1 = p_1 + \ldots + p_n = 1$
E_j	p_j	L_j	$P_j = p_j + \ldots + p_n$
E_n	p_n	L_n	$P_n = p_n$

Source: After Krewski *et al.* (1982)

Note
a Arranged in increasing order ($L_1 \leq \ldots \leq L_n$).

In this example, R would be £1,550. Alternatively, the *maximum loss* could be calculated. This is a rather extreme summary which ignores the probability of occurrence and rakes the risk to be equal to the maximum loss which, in this case, would be £100,000. Because of the skewed distribution, another way would be to take a given *percentile loss*, for example 98 per cent level of loss.

The same methodology can be applied when damaging events cause loss of life. For the above example, an appropriate tabulation might be:

Number of deaths	Probability	Cumulative probability
0	0.990	1.000
1	0.006	0.010
2	0.003	0.004
3	0.001	0.001

This procedure is not always followed. In the earth sciences, for example, the traditional practice has been to make the majority of risk assessments solely on the basis of the probability of an event, rather than on the probability of an event causing a certain loss. This approach reduces some of the uncertainty but, arguably, is incomplete. For many threats, especially the newer technological hazards, the historical database may be inadequate to support a reliable statistical assessment of risk. In these cases *event tree techniques* are used. According to Vesely (1984), event tree technique is a process of inductive logic that can be applied whenever a known chain of events must occur to produce disaster. A simple example might be an industrial plant which has two safety systems designed to protect against the failure of a pipe carrying a toxic gas: a primary System A (relief valve) and a secondary System B (safety seal). After the initiating event (perhaps an increase in gas pressure within the pipe), several accident sequences are possible. The various possible combinations of system success and failure, with hypothetical probabilities attached, are shown in the event tree (Figure 3.2). This approach has also been used in seismic and geological risk assessment.

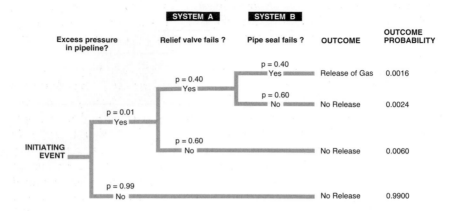

Figure 3.2   A probabilistic event tree for a hypothetical gas pipeline accident. The performance of safety systems A and B determines the outcome probability of the potentially disastrous initiating event.
Source: Diagram refined courtesy of Dr J.R. Keaton

EXTREME EVENT ANALYSIS

Because probability analysis is the most common method of resolving uncertainty, a knowledge of the magnitude and frequency of damaging events is a vital element in hazard management. For example, given adequate records, statistical methods can be used to show that floods of certain sizes may, on average, be expected annually, every ten years, every 100 years and so on. Such information will guide the engineer charged with designing structures like dams or levees. The *design event* is an important concept in engineering risk assessment because it specifies the magnitude of event which a structure is built to withstand during its lifetime. In the case of flood assessment for dam design, for example, there may be more than one design event. One might be the *spillway design flood*, that is, the flood which a dam will be called upon to pass safely without over-topping – and thereby endangering – the whole structure. Another design event might be the *reservoir design flood*, that is, the flood against which the reservoir will be able to provide downstream protection through storage. The reservoir design flood is usually smaller than the spillway design flood.

To some extent, the problem of engineering design is an economic one. The trick is to avoid the excess costs associated with either under- or over-design. Thus, the initial cost of building a bridge capable of passing the 5-year flood may be relatively small but the long-term costs of restoring some structural damage at an average 10- or 20-year interval could be much larger. Conversely, a bridge built to pass the 200-year flood at the same site might also fail to be cost-effective. Apart from these engineering applications, a knowledge of flood frequency is necessary for the deployment of other hazard management strategies such as floodplain zoning and flood insurance.

It is important to stress that extreme event analysis is only as valid as the data available. Probability methods work best for clearly defined hazards with a recurrence interval which conforms to human memory, for example, with floods rather than earthquakes. Sampling error ensures that attempts to extrapolate the 100-year flood from a database 30 years long are likely to have a large margin of error. This places serious limitations on the ability to predict rare events. Most hazardous events also tend to be random in time rather than regular in occurrence. This means that the 100-year flood has a probability of 1:100 in any year and has an *average* return period of one century: in practice such a flood could occur next year, not for 200 years, or be exceeded several times in the next 100 years.

Most extreme event analysis is concerned with the distribution of relevant annual maximum or minimum values at a given site. For example, annual maximum wind gusts recorded at Tiree in western Scotland over a 59-year period from 1927 to 1985 are available to determine the potential for windstorm damage. These events can be given a rank, *m*, starting with *m* = 1 for the highest value, *m* = 2 for the next highest and so on in descending order. The return period or recurrence interval *Tr* (in years) can then be computed from:

$$Tr \text{ (years)} = (n + 1)/m$$

where *m* = event ranking and *n* = number of events in the period of record. The percentage probability for each event may then be obtained from:

$$P \text{ (per cent)} = 100/Tr$$

The annual frequency (*AF*), which is a useful concept because it is at the annual level that probabilistic parameters are combined for all-hazards-at-a-site calculations, can be given by:

$$1/Tr \text{ (years)} = AF$$

Figure 3.3 shows the Tiree wind gusts plotted linearly and return periods plotted logarithmically so that the data fall on a straight line. In some datasets the presence of major outliers, or 'wild' values, creates problems of interpretation. By means of limited extrapolation it is possible to estimate the return period corresponding to a desired speed, or the speed which has a desired return period, slightly beyond the available record. However, extrapolation beyond the period of data introduces uncertainty into the estimate of extreme events.

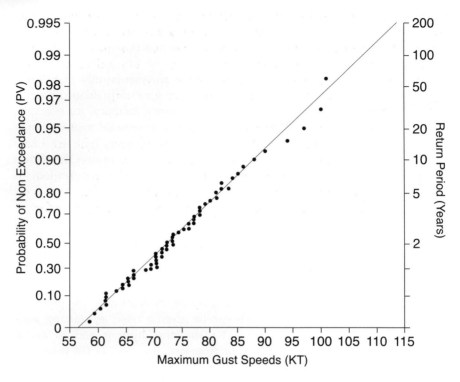

Figure 3.3   Annual maximum wind gusts (in knots) at Tiree, western Scotland, 1927–85, plotted in terms of probability and return period.

For some hazards, extreme events can be estimated from indirect evidence. In the case of floods, an absence of peak flow measurement can sometimes be compensated for by geomorphological or sedimentological information which permits reconstruction of the maximum height reached by flood waters. For earthquakes, the Modified Mercalli scale (Appendix 1), which employs structural damage as a means of estimating earthquake intensity and disaster impact, can be used as a rough guide to the maximum earthquake magnitude. But these are not fully satisfactory substitutes for the direct measurement of extreme events as part of a quality-controlled database.

Another weakness of extreme event analysis lies in the limited representativeness of data in both time and space. When past records are used to predict the future, there is an underlying assumption that there will be no change in the factors causing the extreme events. This assumption, known as *stationarity*, clearly ignores the possibility of environmental change. Sometimes these changes occur naturally over very long periods. For example, in China, where all major earthquakes over the last 2,000 years have been documented, there have been relatively active and quiet phases, often persisting for hundreds of years. In recent decades the increasing human interference with the environment has affected the stationarity of some datasets. For example, deforestation of a drainage basin is likely to increase the frequency of floods and mudslides. Growing concentrations of green-house gases in the atmosphere are predicted to increase the frequency of heatwaves and may well influence other hazardous aspects of weather and climate.

As a result, the frequency of extreme events can vary with changes in the mean and standard deviation of records which are not part of a stationary time-series, for example, as a result of climate change (Parker *et al.*, 1986). Figure 3.4 illustrates the situation when the mean remains constant over a period but the variability, expressed by the standard deviation, increases. For any normally distributed variable, such as air temperature, the effect is

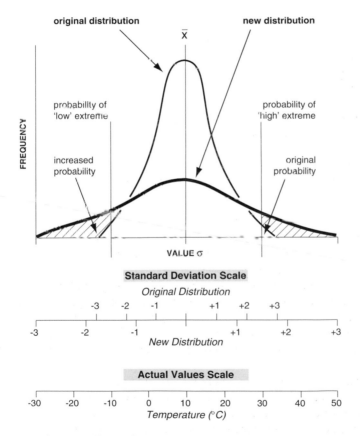

Figure 3.4   The effects of a change to increased variability on the occurrence of extreme events. Both the upper and lower hazard impact thresholds are breached more frequently as a result of the increased standard deviation, although the mean value remains constant.
Source: Adapted from Parker *et al.* (1986)

to increase the frequency of both 'high' and 'low' extreme events relative to the thresholds which define the relevant social band of tolerance. Although the mean remains the same, the impact thresholds are breached more frequently with an increase in the standard deviation from 3 to 5. On the other hand, Figure 3.5 shows the consequences of an increase in the mean value with constant variability, as expected with global warming. In this case the frequency of 'high' extremes relative to the threshold rises, whilst the incidence of 'low' extremes falls. This effect would be reversed with a lower mean value.

Statistical methods require several key qualities from the database before any analysis can be deemed reliable. Apart from the fact that each event should be drawn from the same homogeneous population in order to ensure stationarity, each event should also be independent, should ideally follow a normal distribution curve and should fit some underlying frequency distribution. For assessment purposes, it is always helpful if the data, when plotted, approximate to a straight line. Some data, such as that for floods, may be regionalised in order to make the best use of the information (Smith and Ward, 1998). For example, regionalisation allows the derivation of flood frequency curves for any site within a defined area with similar physical characteristics even though the individual site may lack flow records. Although most statistical techniques attempt to model hazardous events as

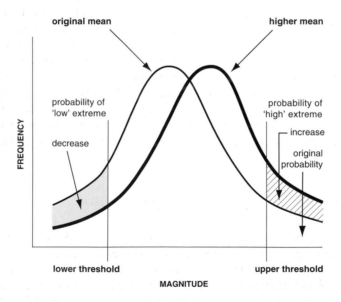

Figure 3.5   The effects of a change to an increased mean value on the distribution of extreme events. In this case the shift to a higher mean value results in a greater frequency of hazard impact from 'high' extreme events with a corresponding reduction in 'low' magnitude events.
Source: After Parker *et al.* (1986)

random over time, not all geophysical events are truly random. Thus, an extreme earthquake follows the accumulation and subsequent release of pressure in the Earth's crust and is not likely to be repeated in a given locality until sufficient stress has built up again.

Even when the data meet the underlying assumptions, problems may still exist (Wigley, 1983). Daily rainfall data, for example, have a skewed, rather than a normal, statistical distribution. Autocorrelation may exist between the primary data and hazard impact data whilst the predictor variables may themselves be inter-correlated. Where non-linear relationships exist, probability functions become extremely sensitive to changes in the mean value (Wigley, 1985). A shift in the mean value of only one standard deviation would cause an extreme event expected once in twenty years to become five times more frequent. Similarly, the return period for the one-in-a 100 year event would fall to only 11 years, an increase in probability of nine times. The full implications become apparent in the context of two successive extreme events, which have great significance in disaster studies because of their cumulative impact. Thus, a community might be able to mobilise sufficient resources to recover from a single hurricane but be unable to withstand the effect of a similar event the following year. If it is assumed that extreme events occur at random with probability p, the probability of two extremes in consecutive years will be p^2, which is far more sensitive to changes in either the mean or the variability. For example, let us assume that, in a given area, hurricanes occur with an annual probability of $p = 0.1$, which means that a single hurricane has one chance in ten of occurring in any given year. Therefore, for the occurrence of a single hurricane the return period is 10 years and the return period for two hurricanes in successive years is 100 years. If p becomes 0.2, then the return period for a single event is halved to 5 years but the recurrence interval for two consecutive hurricane seasons is reduced by a factor of four to only 25 years. In other words, a doubling of the annual probability significantly increases the chances that an individual will experience the event during an average lifetime.

RISK PERCEPTION AND COMMUNICATION

It is not possible to maintain a clear distinction between 'objective statistical' and 'subjective perceived' risk for hazard management decisions. This is because 'objective' risk estimation itself involves value judgements, such as the definition of hazard events and the sampling frame, or dataset, chosen for analysis. The model of decision taking most widely employed in the hazards field is centred on individual choice and relies on the concept of *bounded rationality*. Simon (1956) argued that perception is a filter through which the decision maker views the 'objective' environment and its hazards. All the potential responses to hazard are not necessarily known or reliably assessed. Faced with a complex situation and an imperfect knowledge base, the decision maker is forced to simplify and tends to aim for a satisfactory – rather than a maximum – outcome. Kates (1962) stressed that such choices are based on individual knowledge (the so-called 'prison of experience'). It is this individual experience which ensures that hazard victims and hazard managers respond to environmental risk in different ways according to their own personal 'prison'.

Risk perception is very different from risk assessment. Risk assessment is a specialised scientific function usually carried out by trained experts on behalf of others. It seeks to exclude all emotive aspects associated with personal preferences in order to produce repeatable results applicable at the group level. By contrast, risk perception has no single, replicable outcome because risk means different things to different people. All individual perceptions of risk are regarded as equally valid and, for any given threat, each individual chooses – within the available limits – his or her own response. Whyte and Burton (1982) noted that, for the lay risk perceiver, the consequences of a threat often assume much greater significance than the probability, so that the standard risk analysis formula can be rewritten:

$$R = p \times L^x$$

where x is a power that depends on a number of factors but whose numerical value is always greater than 1.

The basic differences between risk assessment and risk perception, summarised in Table 3.2, can lead to conflict in hazard management. Technical experts often assess infrequent hazards that take many lives at one time as equal to regular hazards that, over an equivalent period, take a similar number of lives just one at a time. Conversely, most lay perceivers, including the media, give greater weight to hazards that take many lives at one time, that is, to major disasters. Experts might well rate voluntary and involuntary risks equally whereas non-experts would show greater concern for involuntary risks.

Table 3.2 Some major differences between risk assessment and risk perception

Phase of analysis	Risk assessment processes	Risk perception processes
Risk identification	Event monitoring Statistical inference	Individual intuition Personal awareness
Risk estimation	Magnitude/frequency Economic costs	Personal experience Intangible losses
Risk evaluation	Cost–benefit analysis Community policy	Personality factors Individual action

Some risk analysts regard perceptions as invalid because they arise from subjective influences. But, to the lay person, perceptions are the only relevant view because they incorporate the expert's analysis together with individual judgement based on experience, social context and other factors. The fact that this view is less 'scientific' does not render it invalid. Problems arise in hazard management when risk analysts expect their conclusions to be accepted because they are 'objective' whilst lay people reject such interpretations precisely because they ignore individual concerns and fears. In practice, it has to be recognised that lay people are not making direct judgements about resource allocation for risk management at the community level, although they may influence such decisions indirectly via the political process. When the lay perspective of hazard is exaggerated well beyond the results of objective analysis, and where the lay person has little information about the costs and benefits of risk reduction, it would be wasteful to spend large amounts of money to achieve only a negligible improvement in safety.

Given this conflict between technical assessors and the public, there is a need for better communication between the two groups, especially bearing in mind the adversarial context of many debates about environmental hazards. From a practical standpoint, such communication should aim to enable the lay perceiver to converge better with the objective view. Slovic (1986) has specified many of the difficulties faced by risk communicators in presenting to a lay audience complex technical material that is clouded by uncertainty. For example, strong initial beliefs are hard to modify whilst people lacking firm opinions can be greatly influenced by the way in which the risks are expressed.

Risk perception may be studied in two basic ways. The *revealed preferences* approach observes how people behave and takes this as a reflection of public perception by assuming that, through trial and error, society has arrived at an 'acceptable' balance between the risks and benefits associated with any activity. This approach has been used to formulate 'laws' of observed behaviour, such as those illustrated in Figure 3.1. The *expressed preferences* method uses questionnaire surveys to ask a sample of people to express verbally what their preferences are. This method permits the gathering of more specific information but respondents may not always act in the way they suppose when actually faced with a hazardous situation. With either method, there are difficulties in sampling opinions in a way that gives a reliable view of the public that can be used for risk management purposes.

Hazard perception is influenced by many interrelated factors including past experiences, present attitudes, personality and values, together with future expectations. The dominant influence comes from past experience in that those with direct personal knowledge of previous hazard events have more accurate views regarding the probability of future occurrences. Direct experience is also the most powerful incentive to hazard mitigation, as illustrated by the rush to enact various hazard-reducing measures after the 1971 earthquake at San Fernando, California. Following this event, 46 per cent of residents in San Fernando and nearby Sylmar took personal steps to mitigate future seismic hazards, but this dropped to 24 per cent for the rest of the San Fernando valley and fell to only 11 per cent for the Los Angeles basin as a whole (Meltsner, 1978). When direct personal experience of disaster is lacking, as it is for most people, individuals learn about hazards from many indirect sources, including the media. Television is a powerful source of information shaping hazard perception, and it has already been shown in Chapter 2 that a considerable in-built bias exists in television news reporting. Through such influences, hazard perception is likely to be moulded quite differently from more objective risk analysis outcomes.

There are many other reasons why lay people perceive hazards differently from technical experts, including geographical location and aspects of personality. For example, early work on floods revealed that rural dwellers, such as farmers, often reveal hazard perceptions which are closer to objectively derived estimates than those of urban dwellers. These group perceptions may well be influenced by social or cultural factors. The influence of personality is exercised mainly through the so-called 'locus of control'. This classifies people according to the extent that they believe hazardous events are dependent on fate (external control) or within their own responsibility (internal control). In order to reduce the stress associated with uncertainty, hazard perceivers tend to adopt certain recognisable models of risk perception with which they are more comfortable. These can be grouped into three basic types, all of which conflict in some way with more objective risk analysis:

1   *Determinate perception* Many lay people find it difficult to accept the random element of hazardous events and therefore seek to view their occurrence in a more ordered fashion. A determinate perspective recognises that hazards exist but seeks to place extreme events in some pattern, perhaps associated with regular intervals or even coming in a repeating cycle. For some events, such as certain types of earthquake, this need not be an inaccurate perspective but it does not fit the temporally random pattern associated with most threats.

2   *Dissonant perception* The most negative form of perception is dissonance or threat denial. Like determinate perception, it can take several forms. For example, past events can be viewed as freaks, and therefore unlikely to be repeated; or the existence of past events can even be dismissed as not happening. Dissonant perception is often associated with people who have much material wealth and are at risk from a major disaster, such as those living adjacent to the San Andreas fault in California. On the other hand, it has to be recognised that continuing vague threats

of earthquakes are much more difficult for the public to come to terms with psychologically than short-term warnings of other hazards. In this case threat denial may be an attempt to conceptualise reality in a way that makes the extended risk from earthquake more comfortable on a day-to-day basis.

3   *Probabilistic perception* This type of perception is the most sophisticated in that it accepts that disasters will occur and also perceives that many events are random. It also accords best with the views of those charged with making resource allocation decisions about risks. But, in some cases, the acceptance of risk is often combined with a need to transfer the responsibility for dealing with the hazard to a higher authority, which may range from the government to God. Indeed, the probabilistic view has sometimes led to the fatalistic 'Acts of God' syndrome whereby individuals feel no personal responsibility for hazard response and wish to avoid expenditure on risk reduction.

Attempts have been made to study the social amplification of risk whereby relatively minor threats elicit a disproportionately strong degree of public concern (Kasperson *et al.*, 1988). Some of the factors thought to increase or reduce public risk perception are listed in Table 3.3. Risks are taken more seriously if the impact is likely to be life-threatening, immediate and direct. This means that an earthquake would normally be rated as more serious than a drought. The nature of the potential victim can be significant since risk perception is not restricted to purely personal concerns. Awareness is heightened if children are specifically at risk or if the victims form a readily identifiable group of people. Thus, any threats to a school party would tend to be amplified. Level of knowledge can be an important element, particularly when it is related to the level of belief in the sources of hazard information. This is a factor in the perception of complex technological risks, especially if a lack of scientific understanding is combined with a disbelief of opinions expressed by technical experts. Age is also a factor; a study by Fischer *et al.* (1991) found that students emphasise risks to the environment whilst older people emphasise health and safety issues.

As technological hazards become more prominent, it is likely that the public will start to view many environmental hazards as avoidable events which are capable of some human control. The media have a relevant role here. This is because over-dramatic, victim-orientated journalism not only leads to an over-emphasis on some risks at the expense of others but also implies that little can be done to reduce risks. More weight needs to be given to the common hazards. For example, in New Zealand the death toll on the roads every six months exceeds the total loss of life due to earthquakes throughout the recorded history of that country.

Table 3.3 Twelve factors influencing public risk perception with some examples of relative safety judgements

Factors tending to increase risk perception	Factors tending to decrease risk perception
Involuntary hazard (radioactive fallout)	Voluntary hazard (mountaineering)
Immediate impact (wildfire)	Delayed impact (drought)
Direct impact (earthquake)	Indirect impact (drought)
Dreaded hazard (cancer)	Common hazard (road accident)
Many fatalities per event (air crash)	Few fatalities per event (car crash)
Deaths grouped in space/time (avalanche)	Deaths random in space/time (drought)
Identifiable victims (chemical plant workers)	Statistical victims (cigarette smokers)
Processes not well understood (nuclear accident)	Processes well understood (snowstorm)
Uncontrollable hazard (tropical cyclone)	Controllable hazard (ice on highways)
Unfamiliar hazard (tsunami)	Familiar hazard (river flood)
Lack of belief in authority (private industrialist)	Belief in authority (university scientist)
Much media attention (nuclear plant)	Little media attention (chemical plant)

Source: Modified after Whyte and Burton (1982)

DISASTER MANAGEMENT

It is clear that risk assessment and risk perception have to be combined in the attempts made by governments and others to reduce environmental hazards. It is unfortunate, therefore, that the quantitative and qualitative approaches to risk management are sometimes polarised by adherents of the 'behavioural' and 'structuralist' schools respectively (see Chapter 2). The fundamental risk management dilemma, as expressed by Zeckhauser and Shepard (1984), is: Where should we spend whose money to undertake what programmes to save which lives with what probability? According to Somers (1995), this dilemma is resolved through the following groups of policy options available to risk managers:

1   *Educational* Public information programmes can be used to raise hazard awareness so that people can take voluntary action to reduce risk.
2   *Economic* Financial instruments, such as subsidies, tax credits and fines, can be used by governments to encourage compliance with hazard reduction policies.
3   *Regulatory* The relevant authority enforces compliance with safety requirements through the force of law and the threat of prosecution for non-compliance.

In practice, the acceptance of risk by the public depends crucially on the degree of confidence placed in the organisation charged with its management. These authorities operate on a variety of spatial scales. For example, continental-scale EC Directives on industrial safety are incorporated by the British government into codes of practice, which are then adopted by national bodies such as the Health and Safety Executive, and are enforced by local authorities (Royal Society, 1992). In turn, these bodies implement a variety of regulatory instruments to manage risks, which range from the corrective (structural) type of measures to the more preventative (non-structural) measures detailed later in this book. The differing scales and responsibilities of these institutions can create problems. For example, flood management bodies typically have responsibility for large geographical areas and tax bases, despite the fact that most of the hazard impacts are local. Agencies funded out of general taxation have often proved ineffective in protecting floodplains from further hazard-prone settlement because they have exerted little control over local 'rent-seeking' processes of land development.

All communities pursue a variety of social and economic goals, and public safety has to take its place with other competing demands. If it is accepted that hazard reduction is not cost-free, and that the total amount of money to be spent on improved safety is not fixed in some arbitrary way, then it follows that investment in risk management must be evaluated. Like many other commodities, safety is something that can be bought. In a totally rational world, resources would not be allocated to mitigate hazard beyond the point where the cost of any extra reduction just equals the benefit. Techniques such as cost-benefit analysis and risk-benefit analysis allow some comparison of different strategies to be made, without which very different amounts of money could be spent to save lives at the decision margins (Marin, 1986). In general, the amount of risk-related government spending, even in the MDCs, appears to be small, as shown by Zimmerman (1990) for the USA. In the UK direct public spending on health and safety regulations is fairly constant at about 0.1 per cent of total central government expenditure (Royal Society, 1992). Even then, some of the investment in increased safety is likely to be traded-off against other values, as when spending on flood defence works leads to greater property values and economic risk on floodplains, the so-called 'levee effect' (see Chapter 11).

Safeguarding human life is a key priority and, although there may be ethical objections and technical difficulties to putting a price on the activity, it has to be done. The conventional approach has been based on the notion of *human capital* and makes a financial assessment on the basis of an individual's lost future earning capacity. This places a zero value on those people who, for whatever reason, are unable to work. The preferred way is to ask how much people would be willing to pay in order to achieve a slight reduction in their chance of a premature death (Jones-Lee *et al.*, 1985). *Willingness to pay* is a better valuation tool because it measures risk aversion, that is, the value people place directly on reducing the risk of death and injury, rather than some more abstract concept based on long-term economic output. Willingness to pay can be assessed by questionnaires which ask the respondents to estimate either the levels of compensation required for assuming an increased risk or the premium they would pay for a specified reduction in risk. These studies have found that the valuation of risk should include some allowance for the pain from and degree of aversion to the potential form of death (for example, high values for death by cancer), and that willingness to pay tends to decline after middle age as the risk of mortality from natural causes increases.

Effective risk management depends on the implementation of a sequential series of actions. The individual stages often overlap but it is crucial that they operate as a closed loop because a major aim of hazard management is to learn from experience and feedback. As shown in Figure 3.6 the process has pre- and post-disaster phases:

1 *Pre-disaster protection* This covers a wide range of activities, starting with risk assessment involving the accumulation of data and the preparation of loss estimates. This leads to the use of various measures, such as the construction of engineering works, insurance and land use planning, aimed at the long-term, cost-effective, mitigation of hazard. The preparedness phase, which reflects the degree to which a community is alert immediately before the disaster strike, covers shorter-term emergency planning, hazard warning and evacuation procedures plus the stock-piling of supplies. Much attention has been given to improving preparedness in recent years.

2 *Post-disaster recovery* The relief period covers the first few 'golden' hours or days after the disaster impact. After the initial rescue of survivors, it is concerned with the distribution of basic supplies (food, water, clothing, shelter, medical care) to ensure no further loss of life. The rehabilitation phase involves the following few weeks or months during which the priority is to enable the area to start to function again. A

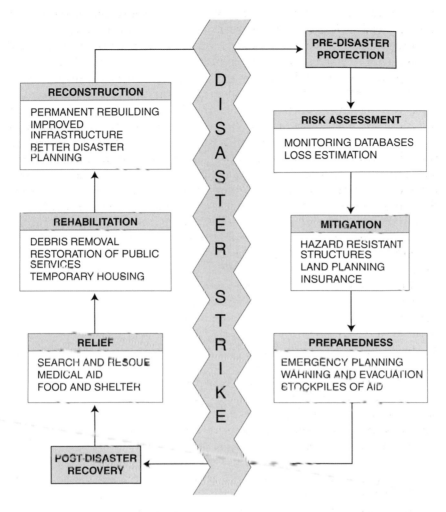

Figure 3.6   The reduction of risk through pre-disaster protection and post-disaster recovery activities. The time-scales needed for the various activities shown range from hours to decades.

common and expensive priority is the removal of disaster debris, such as building rubble blocking roads or food spoiled due to power failure (Solis *et al.*, 1996). Reconstruction is a much longer-term activity which attempts to return an area to 'normality' after severe devastation. If possible, improved disaster planning should occur at this stage, for example the construction of a new flood control reservoir.

Without adequate feedback, risk management is unlikely to be fully effective, and Figure 3.7 illustrates similar disaster processes in a way that highlights the learning element. A closing of the disaster mitigation cycle through the education of people, both victims and managers at all levels, is essential. At the community level, local officials and hazard-zone occupants need to understand the capabilities and the limitations of hazard mitigation. This can be done through the use of brochures, maps, videos and more formal seminars, workshops and training exercises aimed at improving disaster response. At the world level, international organisations and relief agencies require greater technical support in disaster management and need to pool their resources and experience in order to achieve the global aims of the IDNDR.

An effective, integrated approach to disaster reduction is rarely achieved. It is often difficult to quantify the combined risks from multiple hazards, especially those created by low frequency/high magnitude events. The risks may also be spread very unevenly between different communities and social groups. Estimating the costs of mitigation is also problematic, not least for the purpose of saving lives. As a result, the money spent can vary greatly for the same statistical degree of risk. When funds are allocated, institutional weakness, lack of technical expertise and the poor enforcement of legislation weaken the effectiveness of disaster reduction strategies. In the poorest countries the lack of financial and human resources means that pre-disaster

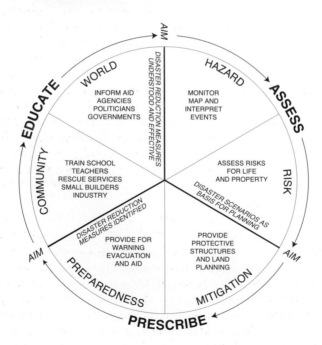

Figure 3.7   The risk management cycle showing the sequence of assessment, response and education which is essential for successful disaster reduction.
Source: Based on an original diagram by Dr John Tomblin, DHA, Geneva

planning and preparedness are inadequate. After major events the country may be dependent on external aid and the whole concept of a return to 'normality' after a disaster strike is inappropriate. Such disadvantage leads to reduced aspirations about the level of risk reduction which can be achieved (Sokolowska and Tyszka, 1995). Even in the advanced nations, the dominant culture is often based on a reactive, emergency response to disaster strikes rather than on a more proactive strategy that prevents disaster victims in the first place.

INFORMATION TECHNOLOGY

In recent years an improved understanding of 'natural hazards' has led to a greater emphasis on more anticipatory forms of risk management, that is, preparedness. This has been helped by new developments in information technology, including Geographical Information Systems (GIS), Global Positioning Systems (GPS) and remote sensing. From the late 1970s, the increased computing capacity provided by microprocessor-based devices has offered fresh opportunities in project planning and real-time decision making in emergencies to a range of organisations. These advances have been supplemented by advances in remote sensing which also have wide applications in disaster management. By the early 1990s, relatively powerful and networked desk-top computer systems were an integral part of disaster management operations, especially in the MDCs (Stephenson and Anderson, 1997). For example, Drabek (1991) indicated a fairly wide use of PC-based decision-support systems in the USA, especially during the emergency phase of disasters when damage assessment, route designation for evacuation and the availability of shelters are critical issues. Although networked computer systems are liable to suffer from power failure during disasters, the increasing reliability of portable radio-based transmission systems can allow communication even when the ground-based infrastructure is destroyed. Battery-operated notebook computers can be easily carried into remote or devastated areas where GPS technology permits almost instant location-fixing and vehicle tracking. Satellite-based telemetry can then be deployed for imaging and field survey work.

Geographical Information Systems provide a key resource for disaster mitigation with many local governments routinely holding archives of contours, rivers, geology, soils, highways, census data, phone listings and areas subject to flooding, or other hazards, for their area. GIS is now used on PCs at an affordable cost to aid all aspects of disaster management, including land zoning decisions, the warning of residents and the routeing of emergency vehicles. For example, Emmi and Horton (1993) presented a GIS-based method for estimating the earthquake risk for both property and casualties which can be applied to disaster planning and land zoning in large communities, whilst Mejía-Navarro and Garcia (1996) demonstrated a GIS suitable for assessing a range of geological hazards backed up with a decision-support system for planning purposes. Dymon (1999) described how GIS models were used to calculate the height of the potential storm surge before hurricane 'Fran' reached the North Carolina coast in 1996. Emergency managers used the information to identify the areas to be evacuated whilst, after the storm, detailed data on residential locations was used to verify insurance claims. Potential vulnerability to disaster, expressed by the location of the poorest groups, the elderly and women-headed households, can also be captured in a GIS in order to promote better emergency responses (Morrow, 1999). GIS-based systems work best for those hazards that can be mapped at a suitable scale. For example, the greatest success has been achieved with the monitoring and forecasting of meteorological and flood hazards in the MDCs which, in turn, has led to improved warning and evacuation systems.

The various forms of *remote sensing* also provide great potential for risk management, especially in terms of hazard forecasting. Table 3.4 gives a summary of the applications of remote sensing in disaster management (Wadge, 1993). For example, hurricanes can be tracked with the aid of geostationary satellites (for example, Meteosat) which provide global cover between 50°N and 50°S at half-hourly intervals. Such repetitiveness enables fairly close monitoring of storm movements towards landfall. Coastal radar can then take over to provide continuous cover, thus illustrating the effectiveness of different types of remote sensing used in combination. A similar situation arises with tornadoes which, in the USA, can be tracked by the National Severe Storms Forecast Center by geostationary satellites in combination with Doppler radar (Ray and Burgess, 1979). These sophisticated instruments can also detect the rotation of a tornado and its ground speed of movement. As a result, Doppler radar has increased the probability of detecting damaging tornadoes from 69 per cent to more than 94 per cent and false alarms have reduced from 73 per cent to 24 per cent. Less success has been achieved with some other hazards, although satellites can provide cost-effective, global coverage of volcanic activity which is unavailable by other means. Similarly, large-scale drought monitoring is possible using changes in soil moisture, surface waters and vegetation which are indicated through changes in surface albedo (Teng, 1990). In addition, the topographic information necessary for hazard mapping can be provided by both SPOT and Fuyo-1 satellites which have optical stereo-imaging for this purpose.

There are limitations. For example, most remotely sensed imagery needs filtering and correction. Even then, many sensors are unable to penetrate a cloud cover. Those satellites, such as SPOT and ERS-1, which are at a sufficiently low orbit (a few hundred kilometres) to give high spatial resolution (a few tens of metres in size) typically image each area only every few days. Generally, satellite images are at a regional scale and do not have resolutions below 120 m on the ground, so small features can be missed. In this situation, ground truthing assumes great significance. Perhaps most important of all, remote sensing has so far been of little help with the mitigation of earthquake disasters.

Table 3.4 The use of remotely sensed imagery in hazard management

Hazard	Remote sensor	Nature of data	Uses
Storms	Geostationary satellites (5) (e.g. Meteosat)	Global, 5 km resolution, every half hour: gives cloud, water vapour	Storm tracking, weather forecasting
	Polar-orbiting satellites (2) (e.g. NOAA)	Global, 1 km resolution, every 6 hours: gives cloud, temperatures	Storm tracking
	Ground-based VLF (e.g. SFERIC service)	Global, time and position of lightning	
Floods	Landsat (SPOT, NOAA)	Near-infrared discrimination of land/water	Flood extent mapping
	Satellite radar (e.g. ERS-1)	Water content from backscatter for soil/snow	Runoff/snowmelt models
	Ground-based radar	Intensity of rainfall	Weather forecasting/ runoff models
Earthquakes	Satellite/airborne radar (e.g. ERS-1)	Interferometric mapping of surface deformation	Prediction of earth- quakes?
	Differential GPS	Point monitoring of surface deformation	Prediction of earth- quakes?
	Landsat/SPOT/ Fuyo-1	Detection of topographic evidence for earlier faults and offsets	Estimate of earth- quake recurrence
Volcanic eruptions	NOAA/TOMS	Eruption plume height, motion and SO_2	Aircraft warning, eruption monitoring
	Landsat (thematic mapper)	Size and temperature of emitted radiation	Eruption precursor/ monitoring
	Satellite/airborne radar	Deformation of volcano's surface	Eruption precursor/ monitoring
Drought/pests	Meteosat & NOAA	Cloud temperatures and vegetation indices	African storm warnings, drought monitoring and pest migration prediction

Table 3.4 continued

Hazard	Remote sensor	Nature of data	Uses
Fires	NOAA (thermal infrared)	Night-time thermal emissive anomalies give temperature and size of fires	Wildfire monitoring
Landslides	SPOT	Topography from stereopairs	Landslide inventory and susceptibility mapping
	Landsat	Spectral character of landslide ground	

Source: After Wadge (1993)

4 Adjustment to hazard

Accepting and sharing the loss

RANGE OF ADJUSTMENTS

Most environmental hazards which threaten human populations produce some response designed to mitigate the risk. The range of human adjustment to hazard can be organised into three groups, although the most effective adjustments often involve a combination of measures drawn from more than one group. Also, to be effective, all responses have to be in place well before the hazard event. The options are:

1   *Modify the loss burden* The most limited responses seek to spread the financial burden as widely as possible beyond the immediate victims through relief and insurance programmes These schemes are essentially loss-sharing, rather than loss-reducing, measures. But, increasingly, disaster aid and insurance measures are being used to encourage future loss reduction.
2   *Modify the hazard event* This involves reducing losses by adjusting damaging events to people. Where possible, hazard-resistant design and engineered structures are deployed to safeguard lives and property from selected phenomena in certain high-risk settings. However, many natural hazards are insufficiently well understood, or manageable, to be physically suppressed through environmental control.
3   *Modify human vulnerability* This is the largest group of responses and includes all measures designed to reduce hazard losses by adjusting people to damaging events. It is achieved mainly through preparedness programmes and hazard avoidance strategies, such as the installation of forecasting and warning schemes and longer-term planning for more appropriate land use allocations.

CHOICE OF ADJUSTMENTS

Comparatively little is known about how individuals and organisations choose between these alternatives. One of the main factors influencing choice is the level of understanding about hazard processes and the extent to which technology can be deployed against them. Clearly, some of the poorer nations will be limited to a sub-set of options on financial grounds. All adjustments are not equally appropriate for all types of hazard. For example, a relative lack of understanding of the physics of the Earth's crust means that credible forecasting and warning schemes are not yet available for most tectonic hazards. Consequently, earthquake hazard adjustment relies mainly on hazard-resistant design and community preparedness. In contrast, most atmospheric threats can be forecast with varying degrees of reliability and warning schemes are a major tool in reducing the threat from these hazards.

Hazard reduction has to take its place alongside all the everyday problems requiring solution. In general, natural hazards are not rated highly by residents or political decision makers compared with other social problems such as inflation, unemployment and crime. Radical approaches to risk, such as relocating people away from hazardous areas, are rarely practical or acceptable. Therefore, compromises have to be reached, often on the basis of cost–benefit assessments. The selection of one option may affect the likely uptake of other adjustments. For example, the construction of river engineering works may appear to offer such a high degree of physical protection from floods that other options, such as insurance, appear less necessary. Choices are also influenced by recurring debates about the balance between public and private risks and who should bear the cost of mitigating hazards.

Different hazards operate in different landscape settings. Some, such as floods, are found in relatively well-defined topographic settings and floodplains can be identified and mapped more accurately than areas prone to earthquakes or tropical cyclones. This has important implications for the effectiveness of land use zoning as a hazard adjustment measure. Other types of information, such as the extent of the database of hazard events, will vary greatly between different hazards and different areas. Community settings differ too. Many people are unaware that they live in a hazardous location and individual perceptions tend to underestimate low-probability events relative to the more frequent hazards. Before any choice is made, local authorities and people in the community must be above the hazard perception threshold. That is, they must be aware of the problem before they can act.

At the individual level, three factors are known to influence the choice of adjustment:

1 *Experience* Individuals with no experience of hazard are less likely to accept information, or recognise the seriousness of the threat, and take appropriate action.
2 *Material wealth* Individuals with capital and resources, which also includes access to information and technology, have an ability to choose between a wider range of adjustments than poorer people.

3 *Personality* This is a factor in that an individual's sense of his/her ability to control events, and his/her confidence in the future, will influence decisions.

Working on the premise that floodplains are for floods rather than for people, it follows that the best response of all is the total avoidance of hazardous land. This has clearly not been possible in the past. It is even less likely in the future because of the growing pressure on land for development and the positive attractions of certain hazard zones, such as flat sites on floodplains or outstanding scenic views in many coastal or landslide areas. Once development has occurred, it is comparatively rare that permanent evacuation becomes a viable response. This is especially true for cities because, even after severe disasters, there is usually sufficient infrastructure remaining to encourage rebuilding on the same site, although there may be some detailed relocation of facilities.

When urban relocation does occur, the distance involved is usually small. For that reason alone, the exercise may not be completely successful. The city of Scupi, Yugoslavia, was damaged by an earthquake in AD 581. It was relocated a few kilometres away as Skopje and was badly damaged again in 1963. The relocation of whole communities is most likely in the case of small, island settlements overwhelmed by natural disaster. Even then, it may not be permanent. The entire population of 264 people was evacuated from the island of Tristan da Cunha in the remote South Atlantic following a volcanic eruption in 1961. The islanders were relocated in Britain but, within two years, most had elected to return to their original home.

ACCEPTING THE LOSS

On the basis that the best policy decisions are made when government intervenes least, there is a theoretical case for allowing individuals to assume whatever environmental risk they wish. In this free-choice situation, people would be allowed to locate anywhere as long as they were prepared to accept the consequences of such a decision. This 'no-action response' would allow the purchase of a property on a floodplain, perhaps at a lower cost than a similar house in a less hazardous setting, as long as the purchaser took the risk of being flooded. In theory, this passive attitude gives people the responsibility of living with their own decisions. In practice, the effects are rarely acceptable. Lack of information and capital, rather than calculated choice, encourages so many people to locate in hazardous locations that large sections of communities, often disadvantaged in other ways, are placed at risk. When disaster strikes, it is difficult for a responsible government to ignore the consequences. However, a passive approach may be justified on cost–benefit grounds where the scale of risk is small or where the lack of financial and technical resources may preclude more positive action.

SHARING THE LOSS

Disaster aid

Disaster aid is the inevitable outcome of humanitarian concern following a serious disaster. It is part of the general flow of aid from rich to poor countries which, in turn, increasingly blurs the distinction between hazard mitigation and the support of broader development motives. Aid will always remain a necessary response to major disasters but it can never fully redress the profound economic and social disparities around the world which are responsible for creating so much hazard vulnerability. In addition, aid is not ideal as a long-term measure for disaster reduction. It may even be counter-productive if victims come to rely on such external support after a protracted series of damaging events or if it encourages settlement in high-risk areas in the expectation that disaster victims will be compensated at no direct cost to themselves. The efficiency of disaster aid is closely linked to the state of pre-disaster preparation (see Chapter 5).

Disaster aid flows to victims via governments, charitable non-governmental organisations (NGOs) and private donors. Some aid may originate in the same area as the disaster but large-scale responses involve international support. As many governments in the MDCs become less willing to take welfare responsibility for the poor and the vulnerable, the role of NGOs such as the International Federation of Red Cross and Red Crescent Societies (IFRCRCS), Oxfam and the church agencies has increased. The NGOs are funded by governments and disaster appeals. The success of appeals is dependent on the nature of the event, and sudden-onset disasters, such as earthquakes, tend to attract more donations than slow-onset disasters, such as droughts. Another factor is media coverage, which is not always helpful. For example, early reports of disaster impact are often inaccurate due to a confused emergency situation and may not be corrected before the issue is dropped in favour of new stories. As indicated in Chapter 2, there is a distinct lack of more reflective reporting which puts disasters in longer-term perspective and which indicates any lessons which may have been learned. In general, the media are guilty of generating recurrent images of the helpless victims of disaster without reporting the successes of disaster mitigation. This emphasis can lead to 'donor fatigue' and the lessening of support for disaster appeals.

Disaster aid is used for the three purposes of relief, rehabilitation and reconstruction (see Figure 3.6). Most of the aid is triggered by appeals in the *emergency relief period*. After the initial search and rescue phase, the priority is for medical support (Beinin, 1985). Some disasters in the LDCs, such as floods, can create disease epidemics out of the ongoing health problems within the country, for example diarrhoeal, respiratory and infectious diseases. Other disasters, such as earthquakes or technological accidents, create quite different medical problems such as bone fractures or psychological trauma. In all these cases, the use of local medical teams is preferable

Plate 4 Effective search and rescue is vital in the 'golden hours' after an earthquake. A specialist team combs the rubble in the city of Erzincan, eastern
Turkey, in March 1992 following an earthquake of magnitude $M = 6.8$
on the active 1,000 km long North Anatolian fault zone. The death toll
was limited to 547 but nearly 5,000 buildings were destroyed or damaged
beyond repair and over 2,000 people were injured. (Photo: International
Federation)

to the deployment of expatriate expertise because they can be mobilised
more quickly and are more likely to be culturally integrated with their
patients. This suggests the importance of emergency preparedness for
medical staff and disaster volunteers.

In order to save lives, it is vital that appropriate technical support and
medical supplies are delivered to the victims as quickly as possible, preferably within the first few 'golden hours' after the disaster. For example,
nearly 90 per cent of trapped earthquake victims brought out alive are
rescued in the first 24 hours after the event. Being trapped under rubble
increases the victim's chance of injury five-fold and, only 2–6 hours after
an event, fewer than half the people trapped are likely to be still alive
(De Bruyker *et al.*, 1985). Even in the MDCs, rapid aid can be difficult to
organise in the aftermath of a major disaster and aid donations – especially
of medical supplies – may well be wasted. Following the Guatemala City
earthquake of 1976, the peak delivery of medical supplies did not occur
until two weeks after the event, by which time most casualties had been
treated and attendance at hospitals had fallen to normal levels (Figure 4.1).
Similarly, the field hospitals supplied to Guatemala by neighbouring
countries were the only fully effective ones and, by the time the main
supply of 'packaged' disaster hospitals arrived from the USA, a surplus of

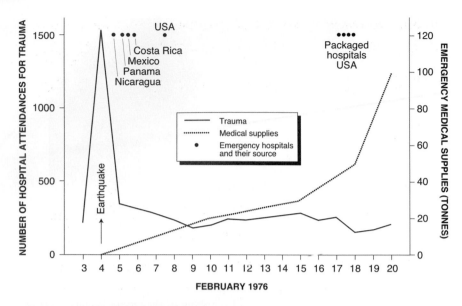

Figure 4.1   The daily number of trauma victims attending hospitals in Guatemala City in relation to the arrival of medical supplies and emergency hospitals from major international donors after the 1976 Guatemala earthquake.
Source: After Seaman *et al.* (1984)

hospital beds was available in the capital. Problems remain even if medical aid is delivered speedily. According to Autier *et al.* (1990), only 30 per cent of the drugs shipped into Armenia after the 1988 earthquake were immediately useful because many were unsuitable, past their expiry date or badly labelled.

Emergency relief is followed by the much longer *rehabilitation period*. Rehabilitation involves the rebuilding of lives as well as of the physical infrastructure and there may well be a need for psychological support. Although many victims will recover with support from family and friends, emergency managers should minimise negative feelings by planning goal-directed behaviour in the rehabilitation period. This involves the facilitation of support networks designed to raise community morale and the empowering of survivors by providing them with roles in decision making and planning. Special attention may be needed for the most vulnerable groups such as women, children and the elderly. The *reconstruction period* is even longer; Haas *et al.* (1977) produced a sequential model of disaster recovery for cities (Figure 4.2), which indicates that the complete process may take up to 10 years. Government support is central to such programmes but, even in developed countries like the United States, a good deal of assistance is provided by voluntary agencies. In the case of disasters in the LDCs,

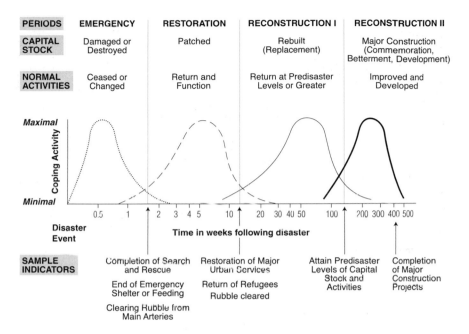

Figure 4.2   A sequential model of disaster recovery for urban areas. The graph shows how characteristic activities and achievements can be related to the emergency, restoration and reconstruction phases. Each recovery period appears almost equal on the logarithmic time-scale.
Source: After Haas *et al.* (1977)

the extent of reliance on foreign aid, rather than the national government, may be almost total.

It is increasingly difficult to draw a distinction between disaster aid and longer-term development aid. Governments see a need to harmonise emergency donations with trade and investment decisions, so there is a growing emphasis on the use of aid for strengthening democratic institutions, building local capacity and encouraging environmental sustainability. Many of the charitable agencies have changed the emphasis of their work from emergency relief to what they term *disaster prevention*. For example, the Save the Children Fund (UK), established in 1919 in response to famine in eastern Europe, now works for the long-term development of health, education and welfare in the LDCs. The aim of such NGOs is to reduce hazard vulnerability, especially amongst the poor, by investing in rural development projects which will rehabilitate basic natural resources, such as soil and water, and lead to more empowerment for women. This policy accords with the structuralist view that environmental hazard is endemic in the LDCs and is created as much by poverty and exploitation as by extreme natural events.

Community aid

After small disasters, voluntary contributions directly from the public, often coordinated by a relevant local authority, may form the bulk of disaster aid. The amount of money collected is often roughly proportional to the degree of devastation, with small events attracting aid from the immediate area and medium events stimulating some regional contributions. If contributions are confined to the local disaster area, there will be no increase in overall resources but financial redistribution provides some compensation for the most affected victims. Such voluntary contributions can rarely provide all the support needed and most estimates suggest that less than one-quarter of individual losses are recouped by this method.

Internal government aid

Following a severe event, it is often left to governments to provide post-disaster relief. Within their own countries, many governments seek to lift the burden of disasters from the shoulders of those who, for whatever reason, have assumed the risk, and spread the load throughout the whole tax-paying population. This is usually achieved by the advance creation of a disaster fund and legislative arrangements for its distribution. For example, following a series of natural disasters in Australia, agreements between the commonwealth government and the states were formalised into the National Disaster Relief Arrangements (NDRA). The individual states receive aid through the National Emergency Operations Centre and stores have been established holding relief items ready for rapid deployment. Not all central government assistance is in the form of grants and a substantial proportion may be given as interest-free repayable loans. Typically, these schemes incorporate a formula whereby the national disaster fund contributes at some agreed ratio to local spending once the disaster impact has exceeded a minimum threshold figure or the local authority has expended some minimum proportion of its annual budget.

Federal disaster aid in the USA costs billions of dollars every year. However, the majority of these costs are created by crop losses arising from weather extremes. As shown in Table 4.1, the total direct cost of agricultural losses over the period 1980–96 was more than double that paid to compensate victims of natural disasters. On the other hand, the cost of disaster assistance rose sharply in the 1990s due to a sequence of unusually costly disasters (hurricane 'Andrew', 1992; the Midwest floods, 1993; the Northridge earthquake, 1994), an increase in the number of presidential disaster declarations and an expansion in eligibility for assistance. There is now considerable debate about how the government can reduce this expenditure.

After a disaster strike, the US President may issue a disaster declaration if so requested by a state governor. Officially, such requests are supported

Table 4.1 Total direct costs to the US federal government of disaster aid and emergency aid for agricultural losses 1980–1996 (values for financial years in US$ billion)

Natural disasters		Agricultural losses	
SBA[a] loans	3.7	FSA[b] loans	7.0
FEMA[c] grants	5.4	Govt relief grants	13.6
Total	9.1		20.6

Source: Data presented in Barnett (1999)

Notes
a SBA – Small Business Administration
b FSA = Farm Service Agency
c FEMA = Federal Emergency Management Agency

by preliminary damage assessments which are then scrutinised by the Federal Emergency Management Agency (FEMA) before an announcement is made. But there is evidence that this procedure can be short-circuited in the interests of political expediency, especially when media pressure exists (Sylves, 1996). A Presidential Disaster Declaration potentially releases aid from some thirteen federal agencies and departments designed to cover 75 per cent of the cost of repairing or replacing damaged public and non-profit facilities, although Presidential authority was used to raise this proportion up to 100 per cent in the case of hurricane 'Andrew' (Czerwinski, 1999).

The rising cost of disaster relief has prompted fresh debate on the extent to which the governments anywhere can continue to provide such funds. According to Barnett (1999), the US federal disaster system – like many national relief schemes – is now seen as:

- *expensive*, because of the rapid rise in payouts and other costs in recent decades;
- *inefficient*, because it allows local governments to avoid a fair share of the costs, for example, through a failure to enforce building codes or to insure public property;
- *inconsistent*, because it does not ensure that equivalent losses are treated in the same way, for example, localised damage may not attain disaster area status and thereby deprive local victims of assistance;
- *inequitable*, because it permits a misallocation of national resources, for example, when wealthy disaster victims are compensated by the general taxpayer.

These concerns are likely to lead to tighter controls on the allocation of central funds plus a search for alternative strategies. The options for reform include: refusing disaster payments to households over higher income thresholds; tying assistance more closely to the local enforcement of building codes; and increasing the extent to which disaster costs are contained within the private sector through insurance or other means. In addition, a two-tier approach to disaster aid has been proposed based on the ability of communities to pay part of the costs themselves (Burby *et al.*, 1991). Thus, the relevant local authority would be wholly responsible for the first $x per capita of infrastructure losses in disaster. Above that threshold, the federal government would cover 100 per cent of the eligible costs. The threshold amount would vary according to the income level in the area. For example, a community with a per capita income double the national average would have a threshold of $2x whilst a community with an income half the national average would have the threshold set at $0.5x. Such a system would ensure that all communities were exposed to the same relative level of financial risk, but determining the value of x may prove difficult.

International aid

Official development assistance is provided to many LDCs by the indus-trialised countries through allocations of *bilateral aid* (donated either directly from government to government or indirectly through NGOs) and *multi-lateral aid* (donated through international bodies such as the EU, World Bank and UN agencies). The agreed UN target for annual government spending on overseas aid is 0.7 per cent of the GDP of the contributing countries. In practice, the global sum donated has rarely been more than half this figure and, during the 1990s, fell to the lowest level ever. In turn, the amount of money spent on disasters is only a fraction of the total over-seas aid budget. For example, the proportion of bilateral aid spent on emergencies is normally less than 10 per cent (IFRCRCS, 1998). In general, disaster aid from governments is declining in real terms both as a percentage of donor GDP and in comparison to direct overseas investment in trade and development policies.

All disasters are first tackled at the local level and the pre-disaster estab-lishment of competent local capacity ensures that international relief can be channelled through an effective partner. Since the 1970s there has been a substantial increase in the aid donated through the NGOs, which are often able to deliver aid to disaster victims faster than government agen-cies can. Before the creation of the International League of Red Cross and Red Crescent Societies in 1922, subsequently re-named the International Federation, the transfer of aid was largely bilateral from one nation to another. As various religious and charitable bodies began to interest them-selves in overseas aid work, further steps towards more permanent and specialised international agencies took place in 1946 with the foundation of the UN Children's Fund (UNICEF) and in 1963 when the FAO World Food Programme (WFP) was set up. Eventually, in 1972, the UN estab-lished the Disaster Relief Organisation (UNDRO), based in Geneva, to coordinate the efforts of the many organisations involved in disaster relief.

UNDRO's original functions were: relief coordination (mobilisation of emergency relief); disaster preparedness (raising the level of pre-disaster planning, especially in developing countries); and disaster prevention (prediction and mitigation of natural disasters, including scientific and technical information). In 1992 the United Nations created a new Department of Humanitarian Affairs (DHA), with offices in New York and Geneva. This was in recognition of the need to deal more effectively with the political constraints which hamper the provision of humanitarian assistance in complex disasters and to ensure a smoother transition from relief to rehabilitation and development. Political leadership is exercised from New York, with a view to improving the coordination with other relevant UN agencies, such as the High Commission for Refugees (UNHCR), World Health Organisation (WHO), UNICEF and the World Food Programme (WFP).

The technical and operational part of the United Nations emergency management is maintained by UNDRO working through its Disaster Reduction and Relief Coordination branches in Geneva. The Relief Coordination Branch operates a 24-hour duty system to alert the world powers before international assistance is formally requested and to issue situation reports when a stricken government does appeal for aid. A central warehouse, maintained at Pisa, Italy, stockpiles relief items for emergency distribution. Standard items include tents, blankets, water containers, generators, tools and cooking utensils. The DHA has an Inter-Agency Standing Committee (IASC) to promote better international collaboration for disaster relief between the UN agencies, the Red Cross and the main NGOs. The International Federation of Red Cross and Red Crescent Societies, also based in Geneva, has national organisations in over 150 countries with some 250 million members. In addition, several western countries have their own coordinating bodies for international disaster reduction, for example, in the USA the Office of Foreign Disaster Assistance (OFDA) operates within the Agency for International Development.

Despite all efforts to organise disaster relief, the results can be disappointing. For example, severe earthquakes and tropical cyclones, which invariably result in a high ratio of deaths to survivors, usually generate a large donor response irrespective of true need. Alternatively, droughts and floods tend to produce comparatively low responses, despite the large numbers of survivors who will be adversely affected. Not only are some disasters apparently more fashionable than others but aid is often highly political and may even be used as a weapon by the powerful donor nations. There are often strong links between the flow of military, development and disaster aid. Development aid, in particular, may be tied to trade agreements rather than targeted at the countries in most need. In Europe a great deal of disaster aid is raised for former colonies in Africa and Asia whilst the USA most actively supports friendly Latin American countries within its sphere of influence. Very abrupt changes in policy occur. For example, the 1988 earthquake in Armenia, formerly USSR, generated the largest ever initial donation up to that time from the British government for a natural disaster, despite the fact that the government had never previously given any disaster aid to the Soviet Union. This act was clearly influenced by recently improved political relationships between London and Moscow. Occasionally a country may refuse aid for political reasons. This happened in 1976 when Guatemala rejected earthquake assistance from Britain because the two countries were then engaged in a territorial dispute over Belize in central America.

Often the nature of relief aid is neither well thought out nor appropriate. In the worst cases aid can simply consist of the unloading of surplus commodities. There have been several well-publicised cases of redundant luxury goods, death-dated drugs, food which is unacceptable for religious or dietary reasons and expensive equipment without technical support or spare parts being despatched to Third World countries. Food aid from the European Community was started largely as a means to dispose of agricultural surpluses. This remains an element of policy, although much greater consideration is now given to the food requirements and development needs of the recipient country. To be fully efficient, food aid has to be sensitively donated. For example, over-generous donations can lower market prices and disrupt the local agricultural economy in some LDCs. Excessive food aid may also deflect the receiving government from giving priority to the development of its own agricultural base.

The local delivery of aid to victims can remain a problem well beyond the first few hours or days following the declaration of a disaster. This is especially so in the LDCs, where aid may be deliberately intercepted on behalf of urban elites before it reaches those in most need. Wells (1984) highlighted many of the logistical difficulties that exist, such as communications, transport and distribution, and made the point that the same mistakes have been repeated from disaster to disaster. For example, the Karamoja famine relief programme mounted in Uganda in May 1980 was marred by a lack of inter-agency coordination and forward planning (D'Souza, 1984). In July 1980, at the height of the starvation, food supplies ran out and the threat of a second famine was only narrowly averted.

There is also concern that disaster aid is counter-productive as a hazard mitigation tool. This concern is partly based on the belief that external relief funds may provide a net economic benefit if damaged community assets are replaced to a higher standard during a long rehabilitation period. Weaver (1968) cited the consequences of hurricane 'Janet' which swept over Grenada in the Windward Islands in September 1955 and created much economic loss in the agricultural sector because the island relied heavily on nutmeg and cacao as export crops. Within a few years, however, agricultural exports were revitalised by the introduction of bananas, together with the development of quicker-growing species of nutmeg and cacao. Foreign aid led to major improvements in communications and public works, although the community was forced into debt by the spending programme. In fact, there is little hard evidence that disaster aid results in net benefit.

Chang (1984) applied an economic model of disaster recovery to coastal Alabama, USA, following a damaging hurricane in 1979, and concluded that outside assistance from federal agencies and insurance companies was not sufficient to replace lost assets. In other words, the inflow of capital into an area immediately after a disaster and the improvement of some facilities does not amount to an overall net gain. A more serious criticism of disaster aid concerns the possibility of dependency arising through the long-term donation of aid. Williams (1986) has drawn attention to the combination of adverse conditions, including drought, in sub-Saharan Africa which produced a prolonged downturn in per capita food production. In the early 1980s, when aid amounted to about US$18 per person per year in sub-Saharan Africa and financed up to 80 per cent of gross domestic investment in these semi-arid countries, there was a real fear that they would become dependent on imported food.

It is difficult to make reliable assessments of disaster aid as an adjustment to environmental hazard. Given the fact that some emergency response will always be necessary, attention should be given to optimising this form of relief. More training of local aid workers would help, especially if continuity could be maintained by retaining core staff from event to event. A better response to emergencies in the first 'golden hours' is also a priority, as is the avoidance of paternalism and dependency in any transfer of aid from the MDCs to the LDCs. Above all, aid needs to be carefully targeted in order to improve the situation of the most vulnerable people. A recurrent dilemma for aid workers is to decide whether to distribute more supplies to fewer victims or fewer supplies to more victims. In order to resolve such problems there is a need for an improved identification of those most at risk plus the development of better early warning systems for severe disasters involving food shortage.

Insurance

Insurance is a key loss-sharing strategy of the MDCs and is now growing as a response of poorer countries (Kunreather, 1995). It takes place when a risk is perceived and an insurance policy is purchased from a financial partner who guarantees that specified losses will be reimbursed if a disaster occurs. Through the payment of an annual premium, the policyholder is able to spread the cost of the disaster over a number of years. *Commercial insurance* exists when the financial partner is a private company and *state insurance* exists when the financial partner is government.

Commercial insurance

The commercial insurance market works by a company underwriting property, such as buildings or crops, against flood, storm or other specified environmental peril. Policy underwriters try to ensure that the property they insure is spread over diverse geographical areas so that only a small fraction of the total value at risk could be destroyed by a single event. By this means, payments to those policyholders suffering loss are spread over all policyholders. Assuming that the premiums are set at an appropriate rate, the money received from policyholders can be used to compensate those suffering loss. The insurance company makes its profits largely by investing the money received from premiums.

Environmental hazards create special problems for the insurance industry. Unlike claims against car insurance, for example, events such as earthquakes or tropical cyclones have a catastrophe-potential by concentrating losses over short time-scales and within relatively small areas. A typical pattern of large claims following years with few losses makes premium setting difficult and the funding of claims unpredictable. For example, in 1994 the insurance industry in California collected about US$500 million in earthquake premiums but paid out US$11.4 billion for property damage caused by the Northridge disaster (Valery, 1995). Unless an individual company has accumulated a large catastrophe fund, it may not survive. In the past, the build-up of such reserves has been prevented in some countries by taxation designed to prevent unscrupulous companies manipulating such reserves in order to hide profits (Lockett, 1980).

An extremely unfavourable temporal and spatial concentration of risk is known as *adverse selection* by the insurance industry. This occurs when the policyholder base is too narrow and is dominated by people who are very likely to suffer loss. Thus, only floodplain dwellers are likely to seek flood insurance whilst earthquake insurance is not likely to be placed immediately after a disastrous event in the (probably correct) belief that another will not occur for some time. Many low-risk individuals will choose not to purchase insurance while high-risk individuals, who believe that disaster costs will exceed premiums, elect to be insured in the expectation that, over time, they will profit. These problems are accentuated by the tendency for some natural hazards to exist in particular landscape settings. Many individual companies have been made bankrupt, such as the one formed to sell flood insurance after the Mississippi valley floods of 1895 and 1896; it was quickly wiped out by the flood of 1899 (Hoyt and Langbein, 1955). After hurricane 'Andrew' in 1992 nine insurance companies became insolvent and others attempted to quit the market in Florida (Barnett, 1999).

Faced with the prospect of large and irregular financial losses, the insurance industry attempts to limit its liability in various ways. For example, the policyholder base in the UK is deliberately widened by incorporating cover against the most common environmental hazards, such as storm and flood, as part of standard comprehensive household structure and contents policies. Individual policies can limit the maximum amount payable under any claim and the company can also control the number of policies which it sells in a particular area. In Japan the possibility of huge earthquake losses in urban areas has led to the introduction of special clauses which limit indemnity to a maximum of US$110,000 on any one earthquake claim. Above certain levels, the government bears a share of the payments (Morimiya, 1985). Insurance companies can also share the risk amongst themselves. This is achieved either by companies joining together to write the primary insurance cover or by passing on part of the risk via *reinsurance*. A typical reinsurance arrangement might be for the primary company to pay the first $5 million in losses from a storm but, for losses in excess of $5 million, the primary company would be entitled to reimbursement of 90–95 per cent of the sum from its commercial partners (AIRAC, 1986). The reinsurance market is international in scope. In this way, risk can be spread from the individual hazard-zone occupant to the world markets, although some companies find it impossible to obtain all the natural hazard reinsurance they wish to buy.

Commercial hazard insurance has a number of advantages:

1   It guarantees the victim some predictable recompense after loss. Such compensation is more reliable than disaster relief and it also appeals to those opposed to excessive government regulation because it depends on the private market.

2   Insurance should also provide an equitable distribution of costs and benefits. This assumes that property owners in hazard areas pay premiums which reflect their actual risk and that insurance payments fully compensate the victims. In practice, this ideal is unlikely to be achieved. One reason is that for many environmental hazards the database is insufficient to devise a realistic premium based on predicted average annual losses at a specific site. A step towards setting premiums more in line with the actual local level of risk has proved possible with the advent of Geographical Information Systems (GIS). This technology has allowed insurers in the UK to accumulate large databanks of past claims. Spurred on by recent increases in claims for storm, flood and subsidence damage, several companies now charge householders a premium based on the post-code district, which comprises a small group of properties. In some cases premiums have fallen but, if the property has been placed in a higher than average risk band, the annual cost of premiums may have risen by 15–20 per cent.

3   Although insurance is designed to redistribute losses, it can also be used to encourage the adoption of measures designed to minimise future damages. If residents in hazardous areas pay the full-cost premium for their risk, insurance provides an economic disincentive to locating in such areas. The difficulty here, especially in the private residential sector, is that a great deal of development is undertaken by speculative builders rather than by the eventual occupants of the property. Only if the insurance premiums became high enough to make the new properties difficult to sell in the first place is it likely that developers would be deterred from building on such sites.

4   Once properties have been built, it is possible for insurers to offer lower premiums to policyholders who take measures to reduce risks to their property. Such measures might include special construction methods (for example, anchoring the structure to the foundation to prevent slippage; using plates to secure roof members to prevent roof lift-off; and using nails rather than staples to secure shingles) and building materials (for example, wind-resistant roofing and walling materials; earthquake-activated gas shut-off valves). In extreme cases insurers could require property owners to retrofit risk-reduction measures before accepting any premium. Kunreather (1978) found that those with either flood or earthquake insurance were more likely to have adopted protective measures than those without insurance, although this combination probably reflected a general concern for hazard rather than any premium incentive or positive attempt by the insurance industry to reduce losses.

The limitations of commercial hazard insurance are as clear-cut as the advantages:

1   Private insurance may simply be unobtainable in very high-risk areas, although this does not necessarily discourage building development. In the USA the insurance industry has been reluctant to offer flood cover without government support and landslide cover is still generally unavailable. Even when available, landslide insurance normally covers the cost of structural repairs to property only and not that of permanent slope stabilisation, because the scale of the latter is uncertain (Olshansky and Rogers, 1987).
2   There is frequently a low voluntary uptake of hazard insurance. For example, only 10 per cent of the buildings damaged in the 1993 Midwest flood were covered by flood insurance (Anonymous, 1995). Such low involvement limits the efficacy of insurance as a hazard management tool, although it may benefit the industry when a major disaster strikes. Japanese insurers tended to survive the Kobe earthquake because only 3 per cent of affected home-owners had earthquake coverage, compared to an estimated 40 per cent of home-owners in Los Angeles. Even when insurance policies are taken out, a significant proportion of policyholders will be under-insured and are unlikely to be fully reimbursed by the company in the event of a claim.
3   Unless premium rates are scaled directly according to the risk, hazard zone occupants are not likely to bear the full cost of their location. For example, UK insurance companies have traditionally charged a flat-rate premium per £1,000 of buildings cover for all houses. This amounts to a subsidy from the low-risk to the high-risk property owners. Even if some link is attempted between premium and risk, the most hazardous locations still benefit from cross-subsidisation, either through the

company charging rates higher than necessary in less-hazardous areas or by switching profits from other classes of business.

4　Although insurance can, in some circumstances, be employed to reduce losses, the existence of *moral hazard* increases damages. Moral hazard arises with the tendency for some insured persons to reduce their level of care and thus change the risk probabilities on which the premiums were based. For example, some people, knowing that they will be compensated, may be reluctant to move furniture away from rising flood water. Similar complacency with regard to building structures, rather than contents, can be lessened by the imposition, and subsequent policing, of local planning regulations designed to strengthen buildings against hazard impact.

At the present time, environmental hazard insurance is undergoing a reappraisal, mainly as a consequence of rapidly rising costs. More than half the insurance claims paid worldwide are now due to natural and technological disasters (El-Sabh and Murty, 1988). On the other hand, according to Mileti *et al.* (1999), only 17 per cent of the US$500 billion losses sustained in the USA between 1975 and 1994 were insured. Before 1988, the insurance industry worldwide had never faced losses from a single disaster that exceeded US$1 billion (Anonymous, 1995). Since then there have been at least fourteen events which have cost more than US$1 billion. These events exclude the Kobe earthquake disaster, for which the total damage cost was estimated at US$150–200 billion, or nearly 4 per cent of Japan's annual GDP. Other estimates place the present-day costs of any repeat of major historic earthquakes, such as San Francisco in 1906 and Tokyo in 1923, in a range from US$100 billion to over US$1 trillion. According to Malmquist and Michaels (2000), insured losses of US$100 billion arising from an intense tropical cyclone or an earthquake in a large city would exceed all the reinsurance capital presently available and would bankrupt many companies.

The current problems within the US insurance industry have several causes. The coastward shift of population over the last 30 years has raised the value of property along the Gulf and East coasts and greatly increased the potential for catastrophic hurricane losses. The rising cost of reinsurance on the world markets has meant that primary insurers now assume direct responsibility for a higher proportion of the potential losses. The consistent failure of local governments to adopt and enforce stringent building codes has created many buildings, especially in coastal areas, that are poorly designed and constructed. The availability of federal disaster assistance for those whose uninsured property has been damaged in a natural disaster has also contributed to this culture of inadequate building construction and maintenance. As a result, the US insurance industry is less willing to provide residential cover in perceived high-risk areas without better preparation for disasters and the adoption of some loss-reducing measures (National Association of Independent Insurers, 1994).

State insurance

To some extent the problems associated with commercial insurance can be overcome by the creation of a national scheme backed by government. Compulsory government insurance for natural disasters not only widens the policyholder base but can also be used to provide the information and technical assistance required for accurate risk estimation and the structural strengthening of buildings. In theory, this would enable premiums to be related more sensitively to the potential loss and thus ensure that those living in very hazardous areas pay a realistic contribution towards the cost of their location. Government would also be in a stronger position than the insurance industry alone to impose restrictions on the sale of insurance in order to reduce adverse selection. For example, a government programme could require that only properties built to specified building standards may be eligible for insurance or could require insurance in hazard-prone areas as a condition for the allocation of federally backed mortgages.

The provision of state insurance for disaster victims is always a political decision influenced by differing political philosophies and social needs. For example, it is possible to detect a long-term trend towards the provision of state insurance in the USA driven by the need to contain the federal costs of disaster relief. The National Flood Insurance Act 1986 was a first attempt to shift federal costs to state and local governments, plus the private sector, on the premise that flood insurance should be sold only in areas where specified floodplain management policies had been adopted. Unfortunately, the scheme has not been wholly successful in halting flood-plain development and flood losses continue to account for the largest slice of disaster funds. Barnett (1999) noted that the US Congress has since introduced legislation with the potential to create a more comprehensive natural disaster insurance programme, but he also cautioned that the potential outlay under such a scheme would be much greater than under the present relief arrangements.

Some countries have had government insurance schemes for many years. New Zealand first established state insurance cover for earthquakes through the Earthquake and War Damage Act 1944, which was subsequently extended to cover damage from storms, floods, volcanic eruptions and land-slips. The scheme was financed by a surcharge on all fire insurance policies of 5 cents per NZ$100 of insured value. Of the extra revenue, 90 per cent was credited to the Earthquake and War Damage Fund and the remainder placed in the Extraordinary Disaster Fund (Falck, 1991). The Commission administering the programme was empowered to set higher premiums in high-risk areas and was also theoretically able to refuse to pay for damages to poorly maintained properties. In practice, the Commission was under political pressure to pay virtually all the claims presented. A similar scheme, embracing floods, landslides, earthquakes and avalanches, was introduced in France in 1982 as part of a comprehensive natural hazard reduction programme.

At the present time, the rising costs of disaster aid and insurance payouts are driving most governments away from comprehensive insurance schemes. The current emphasis is on encouraging individuals to accept more responsibility for their own risk-taking. One of the most radical changes in government attitude has occurred in New Zealand, where the existing state insurance scheme was reformed in 1993 as the government sought to decrease its liability for catastrophic losses (Hay, 1996). Since 1996 the new Earthquake Commission (EQC) stopped providing cover for non-residential property and, although disaster cover for residential property remains automatic for property owners who take out fire insurance, the extent of cover has been limited. The EQC retains a fund of some NZ$2.5 billion as a first call on disaster claims and also has reinsurance arrangements, although the New Zealand government still remains liable for any shortfall in disaster payments.

5 Adjustment to hazard
Reducing the loss

THE OPTIONS FOR REDUCING THE LOSS

According to Anderson (1991), loss reduction is a less costly alternative than disaster recovery and this strategy is widely adopted in the MDCs. Extending disaster reduction to the LDCs will prove more difficult, especially when mitigation is required for disasters – such as drought – that cover large areas and involve severe environmental degradation. Loss reduction can be achieved either by modifying the hazard event itself or by reducing its human impact. These approaches are not mutually exclusive and they often work best in some combination which may also include a loss-sharing measure, such as insurance:

1   *Physical event modification* The aim is to reduce the damage potential associated with a particular hazard by exerting some degree of physical control over the processes involved. Total control, whereby dangerous releases of energy or materials are either suppressed or diffused at lower intensities into the environment, is impossible given the present state of knowledge. However, certain natural threats can be modified by specially engineered structures which provide some protection by buffering individual buildings or limited high-hazard zones, such as parts of a river floodplain, against events of a specified magnitude. This approach is sometimes called 'hazard proofing' which is an unsuitable term because it implies a total level of security that can hardly ever be offered. A better phrase is *hazard-resistance*.

   Hazard-resistance involves more than engineering science. It is operated through building codes and other regulations which imply a high degree of community acceptance and support. To this extent hazard-resistant design lies at the interface between loss reduction measures based on adjusting events to people and those based on adjusting people to events. The critical difference is that event modification relies on some degree of hazard confrontation involving physical protection whereas human vulnerability modification relies on hazard avoidance involving mainly non-structural responses.

2 *Human vulnerability modification* This means creating changes in human attitudes and behaviour towards hazards in order to reduce loss. Such changes may relate either to human responses to a disaster that has already occurred or to the anticipation and warning of disaster. Some of the specific adjustments employ advanced technology, and even structural devices; but, in contrast to event modification, the approach is rooted in social science rather than engineering science. Vulnerability modification covers everything from community preparedness programmes, through forecasting and warning to financial and legal measures designed to promote better land use management. With the possible exception of forecasting and warning systems, fewer resources have been invested in this approach than have been given to event modification.

EVENT MODIFICATION

Environmental control

Event modification is limited as a response by the minor degree of control which humans can exert over the destructive forces of nature. In a single day, the atmosphere receives enough solar energy to generate 10,000 hurricanes, 100 million thunderstorms or 100 billion tornadoes. Expressed relative to this energy receipt (that is, taking the global solar energy received in one day as 1), a very strong earthquake would release 10^{-2} units and an average cyclone 10^{-3} units (Figure 5.1). Compared to this, human energy sources are puny. For example, the detonation of the Nagasaki bomb in August 1945 released only 10^{-8} units. This means that event modification operates most successfully in marginal conditions when human intervention can tip the balance between two finely balanced states. For example, cloud seeding is only likely to be effective in ending a drought if the clouds involved are already naturally disposed towards precipitation. For tectonic hazards, such as volcanoes, earthquakes and tsunamis, there are no known and reliable methods of event control. Other limits exist because of the possible adverse environmental and ecological consequences of interfering with many natural processes. Despite these problems, experiments have been made to modify some atmospheric hazards, such as hurricanes, hail and floods, mainly by cloud seeding methods (see Chapter 9).

The potential scales of human intervention with natural hazards can be illustrated with reference to floods. Theoretically, large-scale environmental control can be attempted through either weather modification or watershed treatment. The aim is either to stop flood-producing rains by cloud seeding, or to reduce flood flows by conservation measures such as afforestation or contour ploughing over large areas of the drainage basin (land-phase management), respectively. On the other hand, hazard-resistant design would involve building structures, such as dams to store the floodwater in the

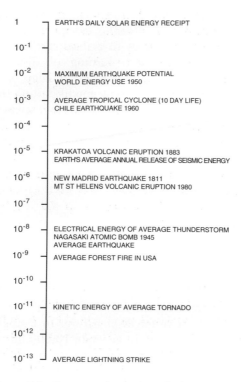

Figure 5.1   Energy release, in ergs, on a logarithmic scale, showing the relationship between certain hazardous geophysical events, together with some
selected human uses of energy, and the Earth's daily solar receipt of
energy.

upper part of the basin or embankments to contain the flood flows further
downstream (channel-phase management). At the smallest scale, hazard-
resistant design can apply to individual buildings through structural
adaptations, such as raising the floor level, that make them and their contents
less susceptible to flooding. However, there is little evidence that floods
can be fully controlled over large drainage basins by any of these adjustments (Smith and Ward, 1998).

Hazard resistance

Hazard resistance occurs when *engineered buildings* are erected in compliance
with local *building codes*. A building code is intended to ensure that a
building is located, designed and constructed so that, if it is subjected to
specified destructive forces of either natural or man-made origin, it will
present no threats either to its occupants or to the general public. For
example, the Uniform Building Code, which is updated annually in the

USA, contains a map of six seismic zones based on ground motions and recorded damage from previous earthquakes. The higher the apparent risk, the more stringent the building regulations. Many buildings can also be retrospectively strengthened to withstand loads more safely.

In most countries, public facilities such as dams, bridges and pipelines are likely to be hazard-resistant because they have been designed by professional engineers. The same is true for large industrial structures such as nuclear power plants and chemical factories. However, design protection can never be total. Building failure will be greatest in data-poor, hazard-prone environments where many structures – older buildings and small rural houses – will have been constructed without any thought to hazard impact. The two most common natural hazards considered in building codes are earthquakes and windstorms, although the adjustment is also applicable to other hazards such as floods (Key, 1995).

Hazard-resistant design

In the case of earthquakes, hazard resistance begins with geotechnical engineers, who apply the principles of rock and soil mechanics to the safe design of earth-supported structures. Other things being equal, buildings on solid rock are less likely to suffer damage than those built on clays or softer foundations. Small-scale maps of the sub-strate can be prepared which suggest local variations in building strength and ensure that major buildings are not located over faults or areas of unconsolidated material. Local building codes tend to undergo continuous improvement and the adoption and enforcement of such improvements is the primary means of increasing the hazard resistance of structures through time.

Today, virtually all building failure under hazard stress can be explained with existing knowledge. This is not to say that engineered structures will never be damaged, because building codes establish minimum standards and rarely specify the maximum possible loads. Instead, they provide against the intensity of hazard event which is considered to have a reasonable chance of occurring during the lifetime of the structure. If an event of greater magnitude than the design standards occurs, then the structure can be expected to fail to some extent. For wind loading codes, common in areas subject to tropical cyclones, design velocities are given for various return periods. Figure 5.2 shows the hypothetical example of a structure designed to cope with a wind stress that occurs on average only once in 100 years (1 per cent probability). As can be seen, windspeeds that occur up to, and even slightly beyond, the design limits are unlikely to cause any real damage but larger stresses, well outside this planned performance envelope, may well result in total structural failure.

Most problems arise because the strict enforcement of building codes and the quality of on-site construction can rarely be guaranteed. This is a special difficulty in the LDCs, where a lack of technical expertise and other

Figure 5.2   A theoretical illustration of the resistance of a well-constructed building to wind stress from storm return intervals up to and beyond the design limit of a 1:100 year event. It is important that such building codes are enforced if the economic losses from natural hazards are to be reduced.

resources commonly produces a failure to meet the design requirements. But the problem also exists in the MDCs and it has been suggested that the cost of the 1994 Northridge earthquake in California, estimated at over US$20 billion, might have been halved if all the damaged buildings had been built to the appropriate code (Valery, 1995). Some buildings may have to withstand events that occur after decades without threat. Rather than construct a new, earthquake-resistant building at high capital cost, owners tend to refurbish the interior many times to keep pace with changing uses. Thus, the structure may have to endure far beyond a theoretical life of, say, 50 years. It follows that inventories should be available in all hazard-prone areas of the engineering status of the built environment. In Los Angeles, California, un-reinforced masonry buildings constructed before the advent of the 1933 building code can be identified as hazardous simply by date. Unfortunately, structural inventories and the associated loss estimation studies are rarely available because of the high technical requirements and large costs of surveying properties.

Hazard-resistant design has had rather limited success in the past. One reason has been the priority given to the functional requirements of buildings during disaster events. Most attention has been paid to public buildings and facilities which are expected to remain operative during emergencies (hospitals, police stations, pipelines) and those which are expected to remain intact in order to house essential items (certain computer centres, warehouses holding emergency supplies). Other public buildings, such as schools, offices and factories, have sometimes been strengthened in the belief that they will shelter large numbers of people seeking refuge. On the other hand, government premises have sometimes been exempted from building codes. In comparison, little attention has been given to privately owned residential buildings.

The consequences of the selective application of building codes may be demonstrated by tropical cyclone 'Tracy', which struck Darwin, northern Australia, in 1974. This event caused structural collapse in only 3 per cent of the engineered buildings that had been constructed to wind-resistant codes over the previous 20 years. In contrast, some 5,000 out of 8–9,000 un-engineered houses (50–60 per cent) were physically destroyed or damaged beyond repair, with the result that three-quarters of the population had to be evacuated (Stark and Walker, 1979). In this rapidly developing area, the construction of new, low-rise housing had been based on building traditions inappropriate for a cyclone-prone settlement. The Darwin disaster prompted a radical review of policy in Australia. It is now accepted that, typically, residential housing accounts for about half the total value of buildings in a community and is, therefore, of equal economic importance to the public buildings previously regarded as more significant. Experience has also shown that, when cyclone warnings are issued, most public buildings close down and the vast majority of the local population seek shelter in their own homes. The general conclusion is that greater attention should be paid to cyclone-resistant design in the residential sector.

However, private property owners are reluctant to pay for strengthening their homes against natural hazards if they believe that any losses will be compensated by insurance or government assistance. In addition, local authorities may resist the introduction of building codes if they feel that the costs of compliance and inspection will hamper economic development. For example, South Carolina, USA, had no state-wide building code by early 1995 despite having suffered appreciable losses from hurricanes (Anonymous, 1995). But, as commercial insurance against natural disasters becomes harder to obtain, and as taxpayers increasingly rebel against property owners who take no anti-hazard actions, hazard-resistant design is likely to become more important. Faced with rising costs, insurers and governments are requiring owners to adopt more responsible attitudes and will become less willing to provide routine compensation for poorly constructed and maintained buildings. Indeed, if taken to its theoretical limit, proper hazard-resistant design could make structural property insurance and related disaster funds largely irrelevant.

Retrofitting

Another limitation exists when hazard-resistant design is restricted to new properties. Because of this, *retrofitting* – the act of modifying an existing building to protect it, or its contents, from a damaging event – is important. Many individual measures can be taken. In the case of earthquakes, for example, brick chimneys can be reinforced and braced onto structural elements to prevent collapse into a living area. Un-reinforced masonry walls can be strengthened and tied to adequate footings, while closets and heavy furniture can be strapped to the walls whenever they constitute a danger or contain valuable items. To protect against floods, measures include making the walls watertight and fitting flood-resistant doors and windows.

Retrofit measures are useful in that they are often quicker to install than some other hazard responses. On the other hand, doubts have been expressed about the economic viability of retrofitting. Hundreds of Californian schools and hospitals have been strengthened for seismic resistance but at a cost up to 50–80 per cent of that for new buildings. Such resources are unlikely to be put into the private sector. For residential property the alternative to retrofitting is to let the occupants of older houses remain at risk until the building deteriorates to a point where the owner or a developer sees some economic incentive to replace the structure at a higher standard.

Retrofit laws passed by local authorities require the identification and strengthening (or demolition) of existing hazardous buildings but the social, economic and political issues involved can make the approach difficult to implement. It has been estimated that a retrofit policy in Los Angeles would substantially reduce the potential hazard to life, with perhaps a five-fold reduction in casualties from earthquakes. However, one disadvantage of the retrofit ordinance adopted in 1981 was that some of the city's lowest-priced housing would be demolished, tenants would have to be relocated while remedial work was carried out and there would be considerable disruption of small business activity. Special provisions also had to be introduced to ensure that some unsafe buildings of historical significance were preserved.

For governments, an attractive feature of retrofitting is that most of the construction and maintenance cost is borne by the property owner, although little official encouragement has been given. One reason for the relative under-employment of retrofitting is that many government agencies are unable to spend any public funds for the direct improvement of private property. Furthermore, it is difficult to disseminate information about measures, which may not be appropriate for all buildings, in terms that are both understandable to the home-owner yet sufficiently technical to ensure that the best method will be chosen and correctly installed. In these circumstances, it may well be that there is a case for more government commitment to retrofitting. With more technical and financial assistance, hazard-conscious property owners could be helped to help themselves to a far greater extent.

VULNERABILITY MODIFICATION

Preparedness

Community preparedness

Disaster preparedness may be defined as 'the pre-arranged emergency measures taken to minimise the loss of life and property damage following the onset of disaster'. It involves the detailed planning – and testing – of prompt and efficient responses to hazard threats. Once a threat has been identified, various groups of people and officials, representing many different interests, become involved in the assembly and transfer of relevant information, as indicated in Figure 5.3. Appropriate loss reducing measures, depending on the nature of the hazard, may include the activation of evacuation plans (often in response to an expected warning message), the

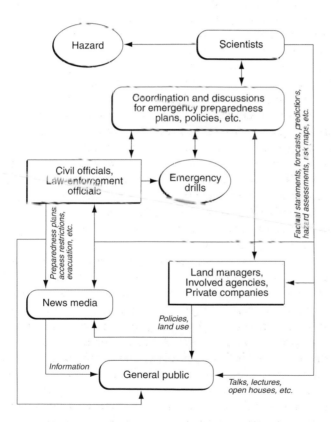

Figure 5.3   The involvement of various interest groups in preparedness for disaster, and the role played by each group in assembling and disseminating information, in advance of a hazard strike.
Source: Modified after Peterson (1996)

provision of medical aid and the preparation of emergency food and shelter, together with alerting the media and the public of the impending threat.

Long-term programmes have proved successful in reducing deaths from hazard. Within individual countries, the degree of preparedness usually depends on the arrangements which have been made for civil defence. For example, the mission of Emergency Preparedness Canada is to ensure that the emergency plans of the Government of Canada are in place and are ready to protect life and property. In the USA, the Federal Emergency Management Agency (FEMA) has the lead responsibility for the coordination and management of all actions to protect the civilian population from a range of hazards. Such organisations are increasingly calling for a change in emergency management from a reactive response to disasters to a more proactive strategy. In Canada the official view is that mitigation should be seen as an investment, rather than a cost, and that a national risk reduction policy should be integrated into the daily lives of all citizens (Emergency Preparedness Canada, 1998).

Plate 5   Preparedness programmes in the LDCs face problems of implementation when illiteracy is widespread and the potential for mass communication via radio or television messages is low. In Bangladesh, cyclone early-warning schemes have relied on the formation of small teams, issued with key equipment such as a radio, torches and a loud hailer, and trained to alert people at the village level. (Photo: International Federation)

Effective preparedness depends on a strong political will and the means to make such measures work. These commodities are less likely to be available in the LDCs but progress is taking place. For example, in India cash-for-work schemes and feeding programmes have rendered all-out famine unheard of for about 20 years. The volcanic emergency plan for Rabaul, Papua New Guinea, which was prepared under the guidance of UNDRO in the 1980s, was largely responsible for the effective evacuation and low loss of life when major eruptions occurred in 1994. Evacuation away from hazard zones is generally effective in saving lives. Figure 5.4 shows part of the evacuation map prepared for the city of Salinas, which is situated on a low, narrow peninsula on the coast of Ecuador (UNDRO, 1990). The western coast of South America is one of the most seismically active zones on Earth and much of the city is only a few metres above sea level and vulnerable to tsunami flooding either from the north or the south. If a tsunami is generated locally, the evacuation time before the arrival of the

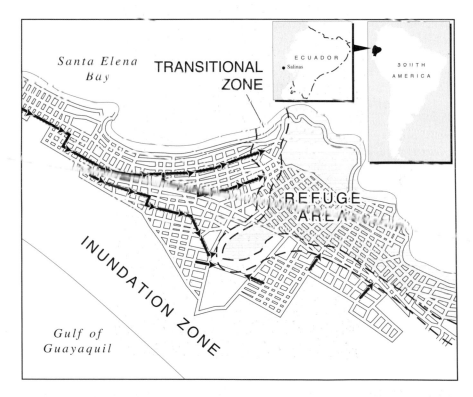

Figure 5.4   Part of the emergency evacuation map for the coastal city of Salinas, Ecuador, showing the expected tsunami inundation zone and the safe refuge area, with the intermediate transitional zone and the main escape routes.
Source: After UNDRO (1990)

first tidal wave is likely to be only 20–25 minutes so routes to the higher refuge area have to be well organised and understood. For those people unable to reach this area – the ill, the elderly and small children – several tall seismically resistant buildings will be used as refuges for vertical evacuation.

A major practical problem arises because comprehensive emergency planning is a long-term, costly exercise. It ties up facilities and people that are apparently doing nothing, other than waiting for an event that no one wants and many believe will never happen. For example, in urban areas threatened by severe earthquake impact, it may not be unreasonable to plan for the emergency sheltering of up to 25 per cent of the community's population. This requires not only the location of usable buildings but also the stockpiling of food, medical supplies and sanitation equipment. Therefore, although there may be great official concern immediately after a disaster, this mood can soon evaporate, and there is usually little political will to fund and maintain disaster plans. In recent years, some developed countries – such as the UK – have disbanded their formal arrangements for national civil defence in favour of reliance on the emergency coordination of more routine public resources housed, for example, in the fire and police departments.

Public understanding and cooperation are vital elements in the successful operation of any disaster plan. In California there is a relatively long history of raising earthquake hazard awareness. Turner *et al.* (1979) found that people were poorly prepared to handle the consequences of a damaging event. Most residents had done little but acquire a working torch, a battery-operated radio and first aid supplies. More recent evidence, based on a survey following publicity about earthquake hazards in the San Francisco Bay Area, suggests that, when preparedness advice is disseminated effectively, residents can be much better prepared (Table 5.1). Most of the people surveyed had stockpiled supplies, whilst the proportion of those who took very specific steps, such as strapping water heaters to walls, bolting the house to the foundations and installing flexible piping to gas stoves, are encouragingly high given that many Bay Area residents live in apartments without individual water heaters to strap or accessible foundations and cook on electric stoves (Mileti and Darlington, 1995).

Good educational programmes are necessary to ensure that widespread public support is available for hazard mitigation and that disaster experience is transferred from one community to another. Ill-conceived awareness exercises may create panic and other counter-productive attitudes, so it is important that such programmes are suitably founded in terms of social feasibility and an understanding of the probable economic consequences of the mitigation measures recommended to the public. Recent experience suggests that carefully prepared advice needs to distributed, both widely and often, to the public through the media and other channels. All preparedness advice should come from an authoritative government agency and

Table 5.1 The proportion of residents in the San Francisco Bay Area of California taking selected loss-reducing actions within the home before and after newspaper publicity about increased earthquake risk

Preparedness action	Pre-publicity (per cent)	Post-publicity (per cent)
Stored emergency equipment	50	81
Stockpiled food and water	44	75
Strapped water heater	37	52
Rearranged breakable items	28	46
Bought earthquake insurance	27	40
Learned first aid	24	32
Installed flexible piping	24	30
Developed earthquake plan	18	28
Bolted house to foundation	19	24

Source: Mileti and Darlington (1995)

Note
The postal sample in this survey consisted of 1,309 households and a total of 806 usable questionnaires were returned. Respondents could report multiple actions.

should be endorsed by well-known local officials and public figures. Each measure should be accompanied with a brief explanation which justifies the recommended action and details how it can best be implemented.

Although community preparedness may appear little more than applied common sense, there are many pitfalls which prevent a less than adequate response. For example, once awareness begins to increase and people actively start to seek information, the responsible agencies may find it difficult to meet the demands for written materials or for speakers to attend public meetings. Effective preparedness involves the home as well as public locations, such as schools, hospitals, offices and theatres. Public bodies and private sector corporations often have an opportunity to build awareness of environmental hazard into existing health and safety programmes but it is difficult to monitor the status of such initiatives within the home. To ensure full effectiveness, there is probably little alternative to regular home inspections in order to determine the level of domestic readiness. This is rarely attempted, not least because of the resource implications.

Over twenty years ago, Stretton (1979) emphasised the need to learn from past disasters. There was little preparedness in Darwin for the 1974 storm in which 65 lives were lost and most houses were rendered uninhabitable. The absence of local emergency shelters and the high risk of disease in this tropical environment led to the evacuation of 35,000 people. Major problems, which have also been found in other disasters, included: the breakdown of the normal communication systems; disputes between various emergency and relief organisations over priorities and responsibilities; lack of coordination in the distribution of relief goods; and an under-availability of trained medical and ambulance personnel. These difficulties can be reduced by pre-designating centralised control of the relief operation. It should also be recognised that basic services, such as roads, water supplies or telephones, are unlikely to be fully available and a wider practical knowledge of appropriate self-help techniques – such as first aid, search and rescue and fire-fighting – should be promoted within communities at risk.

International preparedness

One of the main challenges in disaster prevention is to implement effective preparedness schemes in the developing nations through a sensitive understanding of the different socio-cultural settings which prevail. The lead responsibility for this is taken by DHA/UNDRO in Geneva, supported by a variety of government agencies and NGOs. For example, in Britain, the Overseas Development Administration (ODA) has maintained a Disaster Unit since 1974 which offers government aid for both natural and man-made disasters in the LDCs and for natural disasters in the MDCs.

Several countries maintain specialist rescue and relief groups pledged to UNDRO who can be provided with equipment and transport when disaster strikes. For example, Oxfam maintains an emergency store with cooking equipment and material for constructing temporary shelters. The success of such arrangements depends on UNDRO acting as an effective link between the donors and recipients of aid. UNDRO needs to maintain a register of available expertise and also needs specific information from the victim nation about exactly what aid is required in order to avoid mismatched donations. Regular regional training programmes are held for staff involved at designated Disaster Preparedness Centres worldwide.

Much disaster planning is based on military lines. This stems from a view that disaster relief is similar to running a battle, and great stress is laid on aspects such as communications, logistics and security. These are clearly important requirements but the 'command and control' model, represented by a top-down, rigidly controlled organisation, is not always appropriate, especially for the LDCs. External military forces may be seen as enacting foreign policy on behalf of distant 'colonial' powers, giving the dangers of a confrontation with either the national government or aid agencies striving to be neutral. Poor people are often suspicious of military forces, who may not be completely sensitive in operating refugee camps or dealing with women and children.

On the other hand, specialised forms of military assistance – such as airlifted relief supplies – can be of great value. This happened, for example, when in April 1991 the US Defense Department deployed assets positioned in the Indian Ocean at the time of the Gulf War to transport relief supplies, repair damaged roads and provide purified water for cyclone victims in Bangladesh. One of the traditional deficiencies of military support has been that it was inevitably short-lived because it was diverted from on-going defence duties. It would be an imaginative step if disaster relief work became a more permanent function of western military capability. But, without formal training in relief work, the best role for the armed services is still likely to be in a support capacity for the civil authority or the NGOs.

Beyond the technical aspects of delivering emergency aid, preparedness programmes should seek to encourage more local awareness of disaster prevention and relief. Local governments have the best access to the information needed to determine their own priorities and to manage their own environment. Workshops, pamphlets, brochures, videos and other materials are important tools to create a high level of knowledge among the threatened population. With a greater commitment to self-help and a greater reliance on local initiatives, preparedness plans could ensure a faster and more efficient reaction to disaster events, especially in the LDCs.

Forecasting and warning

Forecasting and warning systems (FWS) have become important due to scientific advances, for example in weather forecasting, and the associated improvements in communications and information technology, such as satellites. Most warnings of a future environmental hazard are based on forecasts. But some threats, notably earthquakes and droughts, are insufficiently understood for forecasts to be routinely issued, so reliance for warning has to be placed on predictions instead. It is important to understand the difference between predictions, forecasts and warnings:

- *Predictions* are largely based on statistical theory and use the historical record of past events to estimate the future probability of similar events. Because the results are often expressed in terms of average probability, there is no precise indication of when any particular event may occur. Some hazard predictions tend to be relatively long-term. For earthquakes they may extend to several decades ahead and it is not usually possible to specify the location or the magnitude of the event with much precision.
- *Forecasts* depend on the detection and evaluation of an individual event as it evolves through a sequence of reasonably well-understood environmental processes. This means that, depending on the ease with which such individual events can be monitored, it is often possible to specify the timing, location and likely magnitude of an impending hazard strike. Forecasts are scientific statements and normally offer no advice as to how people should respond. They tend to be short-term. Indeed, contrary to predictions, the limited lead time for issuing forecasts often restricts the effectiveness of warnings.
- *Warnings* are messages which advise the public at risk about an impending hazard and what steps should be taken to minimise losses. All warnings are based on either predictions or forecasts but for many agencies, such as those involved with national weather services, very few of the routine forecasts are followed by warnings.

Combined forecasting and warning systems (FWSs) are especially useful against the rapid-onset hazards where short-term action, often involving evacuation, can avert disaster. The greatest success has been achieved with atmospheric and hydrologic hazards to the extent that much of the reduced death toll from natural hazards in the MDCs can be attributed to improved hurricane and flood warning procedures. On the other hand, some social changes – such as the growth of international tourism – place increasing numbers of people at risk on hurricane-prone beaches, avalanche-prone ski slopes and flood-prone river banks. Drabek (1995) suggested that the tourist industry is poorly prepared to meet this challenge, especially in terms of emergency evacuation planning. Drought and tectonic hazards remain diffi-cult to forecast, although some success is possible. For example, based on early warning indicators, the Philippine Institute of Volcanology and Seismology advised the government to evacuate residents within a 20-mile radius of Mt Pinatubo before the volcanic eruptions in June 1991. Although hundreds of people were killed, over 10,000 homes destroyed and some US$260 million-worth of damage occurred, a further 80,000 people were saved together with an estimated US$1 billion in US and Filipino assets (OFDA, 1994).

As shown in Figure 5.5, each FWS tends to consist of a number of sequential and interrelated stages:

1   *Threat recognition* covers the preliminary planning phases when a decision is taken to fund, plan and establish a FWS. To be effective, such schemes need to be widely publicised among the community at risk and then tested with mock disaster exercises before a major hazard appears. Ideally, feedback from this experience leads to design improvements in the system. Other revisions should occur as a result of hindsight reviews.

2   *Hazard evaluation* includes several sub-steps, from trained observers first detecting an environmental change that could cause a threat, through to estimating the nature and scale of the hazard, and finally deciding to issue a warning to an endangered community. The task of evalua-tion is a technical operation which is entrusted to a specialised agency such as a national meteorological or geological service. This is because of the need for continuous monitoring by comprehensive networks backed up with heavy investment in scientific equipment and personnel. The priority at this stage is to improve the accuracy of the forecast and to increase the lead time between issue of the warning and the onset of the hazardous event. In order to complete the process, and to retain public confidence, stand-down messages should be issued when the emergency is over.

3   *Warning dissemination* occurs when the warning message is transmitted from the forecasters to the hazard zone occupants. The message is likely to be formulated and conveyed by a third party through an intermediate medium which may involve different communication

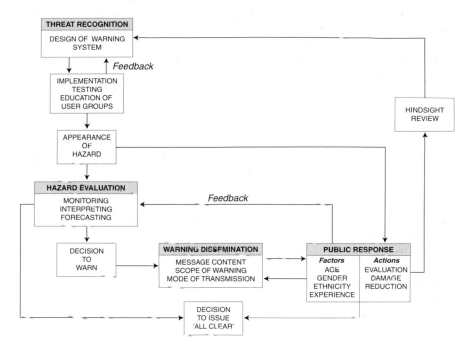

Figure 5.5   A model of a hazard forecasting and warning system showing bypass and feedback loops. The key stages are threat recognition, hazard evaluation, warning dissemination and the public response to the warning.

methods, such as radio or television, and different personnel, such as the police or neighbours. Again, the stage contains several components, such as the content of the message and the way in which it is conveyed, which are known to affect the eventual outcome.

4   *Public response* is the key phase where loss-reducing actions, such as property protection and evacuation, are taken, sometimes on a massive scale. For example, over 2 million residents of the east coast of the USA evacuated inland following warnings of hurricane 'Floyd' in September 1999. From Figure 5.5, it can be seen that the response may be influenced directly through an input based on the public's own knowledge of the evolving hazard and various feedback mechanisms can help to improve later editions of the warning. However, the response is largely determined by the nature of the warning message and a range of factors which bear on the recipient's behaviour.

The value of a FWS depends on both the skill of the forecaster and the effective translation of the forecast into a warning. In other words, an understanding of the social setting is as important as the accuracy of the scientific information if the community responses are to achieve optimum loss reduction. Unfortunately, there has been a tendency for forecasting to become divorced from the remainder of the sequence because it relies so heavily on sophisticated equipment and complex modelling techniques. The decision to warn is crucial and, in marginal situations, forecasters have to make difficult decisions quickly. Problems of interpretation can occur at the interface between the evaluation and dissemination stages where the transition from 'forecast' to 'warning' takes place. Less attention has been given to the warning delivery system. Forecast agencies tend to assume little direct responsibility for their products after they have been issued and it may be left to others to formulate the warning message and to ensure that it reaches the public. This attitude may be aided by a wish on the part of some agencies to avoid legal liability arising from the consequences of either defective forecasts or poor advice about damage-reducing actions.

Public confidence is most likely to be eroded in situations where either no warning is issued or when a false warning is given. Such mistakes can be costly. For example, the erroneous prediction of the eruption of the Soufrière volcano in Guadeloupe in 1976 led to the evacuation of 72,000 people for several months. Glantz (1982) drew attention to the drought forecast issued in February 1977 by the Bureau of Reclamation for the Yakima valley, Washington, which predicted that less than half the normal water supply would be available for the coming irrigation season. By May 1977 it was evident that the forecast, and the attendant water allocations, had been pessimistic. Many farmers, committed to a course of action based on the forecast, took legal action against the Bureau to recover alleged costs and the Bureau's credibility was significantly diminished. In some circumstances, a failure to issue adequate warnings, although criticised after the event, may have had little practical effect. For example, the storm-force winds which devastated south-east England in October 1987 were under-forecast. But, because the storm occurred overnight, few lives were lost and little could have been done to protect the trees and power lines which were the main focus of damage.

The effectiveness of hazard response is influenced by a number of factors, some procedural and some which are due to the characteristics of both the message and its recipients. *Tiered warnings*, incorporating a 'watch' phase before the 'warning' phase, tend to avoid gross errors involving evacuation, but not all hazards, for example, earthquakes, are suitable for tiered warnings. Pre-planning should ensure that basic procedures are understood, such as the advance identification of the people and organisations to be warned. There should also be some alternative means available to distribute messages in adverse environmental conditions, which may include the loss of electrical power and communications systems. Preparedness programmes should also help hazard zone occupants to recognise the threat and to take suitable defensive actions, although there will always be some gap between what people are advised to do, what they say they will do and what they actually do in a stressful situation.

Feedback within the system, involving both an accuracy check on the forecasters and a response check on those being warned, is vital. This is because the onward transmission of the message may be unnecessarily delayed, or even halted, at various points by individual operators seeking confirmation of some aspect. This is most likely to happen with ambiguous messages. Warrick *et al.* (1981) cited the case of the ashfall warning issued by the Washington State Department of Emergency Services before the eruption of Mt St Helens. Although released by an authoritative source, this message was not passed on to the community at risk because it lacked a sense of urgency, was not specific about the areas likely to be affected by ashfall and contained no guidance about precautionary actions which people were expected to take. It is now believed that effective warning messages should contain a moderate sense of urgency, estimate the time before impact and the scale of the event, and provide specific instructions for action, including the need to stay clear of the hazard zone (Gruntfest, 1987). Advice on present environmental conditions and notice of when the next warning update will be issued are also helpful.

According to Quarantelli (1984), planning for effective warnings should start with some knowledge of the perception and likely behaviour of those being warned. In turn, this will depend on the mode of warning. For the general public, the news media act as the primary source of information about hazards and hazard warnings. However, there is evidence that, to be effective, a hazard warning is much more than the transmission of a message from a warning source to a large number of people. Both the mode of warning, as well as the content of the message, need to be considered. For example, it is believed that the best warning messages make the content personally relevant to those who are expected to act on the information (Fisher, 1996). In this context, mass communication may not be the best means of hazard dissemination, and warnings delivered directly by other people are seen as more relevant. Although warnings via the mass media are most likely to be believed if issued by government officials or an emergency organisation, the initial media message is more likely to alert people to the fact that something is wrong rather than mobilise them to an immediate response.

Confirmation of the first warning received by an individual is almost always sought before any action is taken, hence the advantage of tiered warnings. For example, confirmation may be sought from neighbours or the police. This means that the interpretation of the warning message normally takes place as a group response as opposed to that of an isolated individual. Not surprisingly, past experience with the same hazard increases the level of warning belief and there is some evidence that women are more likely than men to interpret a message as valid. Old and infirm people living alone are less likely to make an effective response to hazard warning, either in terms of protecting property or evacuating the premises, and there is a need for special support to be made available to such sections of the community. Often there is a reluctance to evacuate. This may be because the message fails to specify this action, or because people feel that they can cope, or because they fear looting of their empty houses. There is a considerable natural attachment to the home environment and the strength of family ties has been found to be significant in this context (Drabek and Boggs, 1968). For example, family groups are much more likely to evacuate than single-person households, and often go to the homes of relatives rather than to disaster shelters.

Land use planning

Land use planning is a comprehensive approach which seeks to intervene in the process whereby hazard-prone land, initially held in low-intensity uses such as forestry or agriculture, is converted into higher intensity occupation. Such conversion increases unit land values and therefore leads to greater economic losses when disaster strikes. The increasing competition for land makes this strategy important. In the LDCs land competition has been a function of the upsurge in population and the associated unregulated growth of urbanisation. In the MDCs too there has been an accelerating trend towards the invasion of previously avoided hazard-prone areas. In part this has been due to urban growth, but more affluence and leisure time has led to the erection of second homes and recreational facilities in environments such as coasts and mountains which are intrinsically hazardous.

The main purpose of land planning is to guide new residential, commercial and industrial development away from identified hazard zones, but this is not always possible. Therefore, this approach has an additional role to play in reducing losses in areas already occupied whilst it can also help to steer new development away from environmentally sensitive areas, such as wetlands. Because hazard-based land management normally has to function as a planning device within communities that are already there, it frequently seeks to combine the beneficial use of known hazard-prone areas with a minimum of hazard loss and expenditure. It is undertaken mainly at the local government level and, because it is part of the overall political process, it requires broad community support.

Land use planning deploys regulatory tools at a variety of scales, from regional planning through town zoning ordinances down to plot subdivision byelaws. According to Burby and Dalton (1994), land use measures are most likely to be adopted if they are encouraged by the national government. The approach works most obviously by prohibiting new development in very high-hazard areas but it can also deploy measures which are less likely to be opposed by local developers. For example, whilst low-density zoning might be imposed in order to limit the potential property losses in one area, the builder concerned might be compensated by the granting of a permit for a much higher density development in a safer area nearby. More than any other method of loss reduction, land use planning depends for its success on the allied use of other hazard mitigation measures, sometimes of a structural nature. It is also true that the wide-ranging implications of this measure sometimes bring it into conflict with other community objectives. This is because, in the process of converting land to higher intensity uses, several powerful local groups with vested interests are involved. These range from the original land owners, whose motive is often capital gain, to developers and builders, who are also driven by profit. Conflict arises from the fact that, although such land management is undertaken as a public sector policy, it must control private sector 'rent seeking' behaviour before it becomes effective.

The main practical limitations on land use planning are:

1   lack of knowledge about the location, recurrence interval and hazard potential of events which might affect small parts of urban areas;
2   the presence of extensive existing development;
3   the relative infrequency of many hazardous events and the difficulty of maintaining community awareness about the need to avoid hazard-prone land;
4   the high costs of hazard mapping, including detailed inventories of existing land use, structures, occupancy levels, etc.;
5   local resistance to land use controls on political or philosophical grounds (Beatley, 1988);
6   profit-seeking processes which seek to pass the hazard-related costs on to others.

Land use planning is most useful in communities that are still growing and which have undeveloped land. Successful land management techniques also depend on the availability of information with which to identify particularly hazardous locations. Indeed, the accurate delimitation of the hazard zone is crucial because the entire policy is based on the detailed recognition, and the community acceptance, of different degrees of risk which, in turn, justify the implementation of selective development controls. Ideally, variations in risk should be identifiable down to the level of individual properties. For many hazards, such as cyclones and earthquakes, such precision is unattainable. The greatest accuracy is achieved with topographically controlled hazards such as floods, landslides and avalanches.

The first stage is to identify and evaluate potential high-risk areas. Often considerable information already exists in the form of local records, newspaper accounts or other historical sources. Once the general location has been identified, mapping is usually needed at a variety of scales:

- *Macrozonation*, or regional planning, can help to steer broad policy decisions. For example, the regional map of seismic risk in New Zealand (Figure 6.9) could be used to delineate national priority areas for retrofitting existing buildings with anti-seismic measures or for the introduction of anti-seismic building codes for new development.

- *Microzonation* works at the local planning scale when it deals with individual communities and building lots. *Zoning ordinances* are used to implement the regional plan. They can be used to control development through the provision of reports on aspects such as soils, geological conditions, grading specifications, drainage requirements and landscape plans. Relatively large-scale maps (at least 1:10,000) are usually required for zoning in high-risk urban areas, as shown by the seismic survey for Tokyo (Nakano and Matsuda, 1984). Other regulations can then be used for more detailed analysis as individual applications are submitted for permission to develop the land at building plot level. For example, *subdivision regulations* ensure that the conditions under which land may be subdivided are in conformity with the general plan.

Seismic microzonation has been a goal in the contiguous United States for some 50 years, especially in California. It may be used to restrict development behind some minimum set-back distance from an active fault-line. Thus, California's Alquist–Priolo Special Studies Zones Act stipulates that a structure shall not be located across the trace of an active fault and that a uniform 50-foot (15 m) set-back from the fault line is normally required. If development is allowed, zoning can be used to maintain low levels of building density, perhaps by requiring only large lots to be developed or by dedicating areas to open-space uses, such as parks or grazing. Some uses, such as industrial activity, may be prohibited.

Microzoning of land can be highly effective. Thus, before 1952 no studies of landslide potential were required in the Los Angeles area of California before building was permitted, despite the presence of weak surface materials and steep slopes. Since then progressively strict regulations have been introduced, including the need for detailed site surveys since 1963. When heavy rains occurred in 1969, 1,040 of the 10,000 hillside lots developed prior to 1952 were subject to landsliding compared to only 17 out of 11,000 home sites developed in the area since 1963 (Foster, 1975).

Once the high-risk areas have been identified, a number of options can be considered. The public acquisition of hazard-prone land is the most direct measure available to local governments and is one of the most effective long-term strategies. Once acquired, the land can be managed to protect public safety or to meet other community objectives, such as open space or low-density recreational facilities. But land acquisition is expensive and local authorities rarely have the resources for outright purchase.

Another means is for an agency to acquire land through purchase and then to control development in the public interest by selling the land under certain conditions or leasing it for low-intensity use. If public lands are available close to a hazard zone, and if occupants are willing to relocate, it may be possible for privately owned hazardous areas to be exchanged for safer land. Any relocation which involves moving structures and occupants from a hazard area is much more difficult and expensive than the acquisition of vacant land; also, it is often highly controversial within the community. For example, it may well be opposed by advocates of economic development if it is seen to destroy any potential the land might have to promote growth or generate local tax revenues. In some cases public land acquisition could involve the purchase and demolition of buildings which are of historical or architectural importance and thus generate opposition from pressure groups.

Hazard-prone land often appears very desirable. Many landslide areas and floodplain sites have outstandingly scenic views and can command high market prices if there is a low awareness of the hazard threat. Under the ancient legal doctrine of *caveat emptor* ('let the buyer beware'), there is usually no obligation for the owner of such land to disclose the risks to an intending purchaser. However, there is a growing demand that the vendor should have a statutory duty to make a prior disclosure of geological and other environmental hazards in real estate transactions so that the potential buyer can make a more informed choice (Binder, 1998). Such legal impediments to building construction are unpopular with local commercial interests, such as land developers, builders and estate agents. Under pressure from these groups, local authorities may refuse to adopt land use regulations in the belief that they will lose economic initiatives to more lenient communities nearby. For these reasons, any regulations adopted must be seen to be reasonable and capable of defence in a court of law.

Public education and other voluntary methods can sometimes be more useful than legislation in discouraging development in hazardous areas. Some of the oldest methods rely on public information to divert development away from such zones. For example, warning signs that are readily visible help to alert both potential developers and purchasers to the hazard. Since any effective hazard-reduction strategy depends on the understanding and cooperation of the community as a whole, public information programmes are essential aids to awareness. These programmes may operate through a wide variety of dissemination means, including conferences, workshops, press releases and the publication of hazard zone maps.

Financial measures can also be important in discouraging development in hazardous areas, largely because of the great significance of the profit motive in promoting land use conversion. Unlike land acquisition and zoning, which directly control development, the use of financial incentives and disincentives affects development indirectly by altering the relative advantage which people may see in building in a hazard zone. For example, the appropriate local government body may elect to locate any investment in public facilities, such as roads, water mains and sewers, only in those areas deemed hazard-free and zoned for development.

Any national government programme that provides grants, loans, tax credits, insurance or other types of financial assistance has a large potential effect on both public and private development. Tax credits may be used as an incentive to reduce a property owner's tax liability as long as hazard-prone land is either left undeveloped or developed at a very low density. Rather less popular are the financial disincentives which act as a deterrent to land use conversion in hazard areas. For example, the US Congress introduced provisions into the Flood Disaster Protection Act of 1973 to withhold federal benefits from flood-prone communities that did not take part in the National Flood Insurance Program. Other disincentives include the denial of loans by private sources or government lending agencies, and the fact that, under some legal systems, civil liability may be recognised for death, bodily injury, property damage or other losses which might ensue from the erection of buildings on land designated as hazardous.

Part II

The experience and reduction of hazard

6 Tectonic hazards

Earthquakes

EARTHQUAKE HAZARDS

During the twentieth century, well over 1,000 fatal earthquakes were recorded with a cumulative loss of life estimated at 1.5–2.0 million people (Pomonis *et al.*, 1993). About one-third of all these deaths were in China, which suffered badly in the Tangshan event of 1976 when between 250,000 and 750,00 people died. In fact most earthquakes killing more than 100,000 people have occurred along the tectonic collision zone which extends over 12,000 km from Indonesia through the Himalayas, the Middle East and the Alps to the western Mediterranean and north Africa. These recurrent high losses are due to a combination of physical exposure (mountainous topography, frequent earthquake-related ground failure) and human vulnerability (high population densities, poorly constructed buildings).

Generally speaking, the largest earthquakes pose the greatest hazard since they shake the ground more severely, for a longer duration and over more extensive areas than smaller events. But, in terms of disaster potential, event magnitude is often over-ridden by local conditions. For example, the losses at Tangshan were high because the night-time earthquake occurred at a shallow depth directly underneath a city of 1 million people sleeping in un-reinforced houses. Over 90 per cent of residential buildings and 75 per cent of industrial buildings were destroyed. Local geological conditions greatly increase losses when steep slopes help to cause landslides or alluvial soils liquefy and collapse. Most of the estimated 180,000 deaths in the 1920 Kansu, China, earthquake arose from slope failure. In urban areas, fire is an important secondary cause of disaster, due to the rupture of gas and water pipes. For example, more than 80 per cent of the property damage in the San Francisco earthquake of 1906, when up to 3,000 people died, was due to fire. The worst natural disaster in Japan was the Great Kanto earthquake of 1923, which devastated Tokyo and Yokohama, killing nearly 160,000 people. Many were killed because it was almost lunchtime and, in almost a million wooden houses, charcoal braziers were alight to cook the mid-day meal. Clearly, the time of day when an event occurs can be critical. The 1992 earthquake at Erzincan, Turkey, claimed only 547 lives, largely because it happened in the early evening when many people were worshipping in local mosques which proved relatively earthquake-resistant. By contrast, the 1993 earthquake of similar magnitude in the state of Maharashtra, India, killed over 8,000 people sleeping in unsafe houses designed for coolness during the day.

The late twentieth century has provided ample evidence of the continuing threat from earthquakes. The events in 1994 at Northridge (California) and in 1995 at Kobe (Japan) brought disaster to wealthy countries, whilst the 1993 Maharashtra earthquake showed the impacts on less privileged nations. At Northridge and in the port city of Kobe, earth movements along previously underestimated faults brought devastation in the pre-dawn hours to densely populated, urban environments. Although fewer than 60 people died at Northridge, this event was one of the most expensive disasters in US history with total losses over US$20 billion. It has since provided a focus for the study of earthquake impacts and community vulnerability in the MDCs (Bolin and Stanford, 1998). At Kobe more than 5,300 people died, over 300,000 were made homeless and property losses exceeded US$100 billion. Fire, fanned by high winds and inaccessible to fire fighters by rubble-blocked streets, was an important element in these losses. Whilst this scale of material loss is confined to western-style cities, the relative cost of earthquake disaster on rural areas in the MDCs can be even greater. The 1993 Maharashtra disaster occurred in a rural area of central India, thought to be largely free from earthquakes, where the houses were built mainly of stone walls with earth and timber roofs. In almost half of the seventy villages badly affected, over 90 per cent of the houses collapsed and about 8,300 people died. Although no large-scale urban infrastructure was at risk, Table 6.1 shows that the survivors found themselves with much depleted assets. For such people, already very poor, the replacement of these resources and a swift return to their agricultural way of life is a major challenge.

Table 6.1 The percentage of agricultural assets destroyed by the 1993 Maharashtra, India, earthquake

Livestock	Per cent	Implements	Per cent
Cattle	18.3	Buffalo carts	36.6
Buffalo	23.5	Tractors	48.9
Goats/sheep	47.5	Ploughs	50.2
Donkeys	43.5	Pump sets	47.8
Bullocks	12.9	Cattle sheds	67.2
Poultry	65.3	Sprayers	62.3

Source: Compiled from data presented by Parasuraman (1995)

Note
The survey covered 69 affected villages with a population of 170,954 persons.

THE NATURE OF EARTHQUAKES

Earthquakes are caused by sudden movements, comparatively near to the earth's surface, along a zone of pre-existing geological weakness, called a *fault*. These movements are preceded by the slow build-up of tectonic strain which progressively deforms the crustal rocks and produces stored elastic energy. When the imposed stresses exceed the strength of the fault, the rock fractures. This sudden release of energy produces seismic waves that radiate outwards in ever-widening spheres around the fault. It is the fracture of the stressed and brittle crust, followed by elastic rebounding on either side of the fracture, which is the cause of ground shaking. The point of rupture, known as the *hypocentre*, can occur anywhere between the earth's surface and a depth of 700 km. Shallow-focus earthquakes (<40 km below the surface) are the most damaging events, accounting for about three quarters of the global seismic energy release. For example, the San Fernando, California, earthquake of 1971 had only a moderate magnitude (M = 6.4) but, because it occurred only 13 km below the surface and was on the margin of a highly urbanised area, much damage was created. The source point for earthquake measurement is the *epicentre*, which lies on the earth's surface directly above the hypocentre.

The global distribution of earthquakes is far from random. About two thirds of all large earthquakes are located in the so-called 'Ring of Fire' around the Pacific which, in turn, is closely related to the geophysical activity associated with plate tectonics (Bolt 1993). The earth's crust is divided into more than fifteen lithospheric plates (Figure 6.1). These plates move across the globe, at perhaps 20–50 mm yr^{-1}, carried along by convection currents in the mantle. Most of the world's earthquakes and volcanoes occur along the tectonically active margins of these major plates but they can also occur at weak points within plates. The Maharashtra disaster is an example of an intra-plate earthquake, as is the 1989 event at Newcastle, Australia, the first to claim life in that country's history. Most earthquakes are created by movements at the edges of plates. For example, one of the plates may bend down to pass beneath the other in a subduction zone. These subduction zones often give rise to large earthquakes, for example, Chile (1960) and Alaska (1964).

Earthquake magnitude is assessed on one of the scales based on the work of Charles Richter. These scales describe the energy of the seismic waves radiating outwards from the earthquake as recorded by the amplitude of ground motion traces on seismographs, at a normalised distance of 100 km from the source. The original scale was devised by Richter for conditions in southern California in the 1930s. Today, a commonly used Richter scale is known as the *local magnitude* (M_L) scale which incorporates various modifications, including those for modern recording equipment and regional conditions, so that it may be employed worldwide.

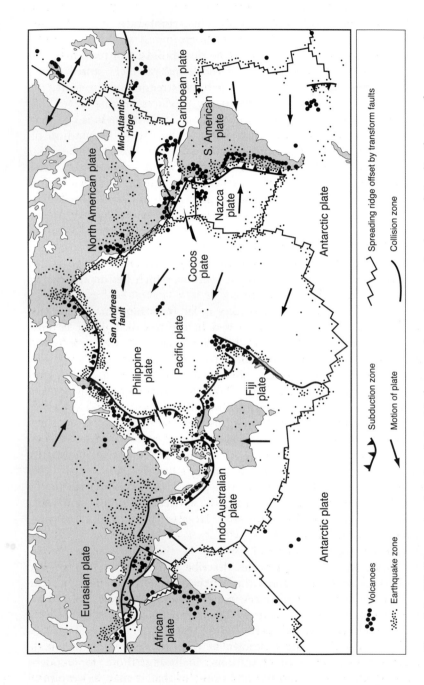

Figure 6.1 World map showing the relationship between the major tectonic plates and the distribution of recent earth-quakes and active volcanoes.
Source: After G.W. Housner and quoted in Bolt (1993)

M_L is the common logarithm of the 'corrected' ground motion in micrometers (10^{-3} mm). For example, an earthquake 100 km away producing a seismograph trace of maximum amplitude 1 mm would indicate a magnitude 3.0 earthquake ($\log_{10}$ of 1,000 = 3). Because the M_L scale is logarithmic, each whole number represents a potential ground shaking at the seismograph site which is ten times greater than the next lower number. An M_L = 7 earthquake produces about ten times more ground shaking than a M_L = 6 event and around 1,000 times more ground shaking than a M_L = 4 event. On the other hand, approximate energy–magnitude relationships show that, as the magnitude increases by one whole unit, the total energy released increases by 31.6 times. Thus, compared to the energy released by a M_L = 5 earthquake, a M_L = 6 event releases over thirty times more energy, a M_L = 7 event about 1,000 times more energy and a M_L = 8 event over 30,000 times more energy. The scale has no theoretical upper limit. Empirical evidence suggests that most shallow earthquakes need to attain a magnitude of at least M_L = 5.5 before a major disaster occurs. Although earthquakes of this size, and greater, account for little over 1 per cent of all events, they are responsible for well over 90 per cent of all the seismic energy which is released.

Large earthquakes with a great rupture length are not well represented by the M_L scale. The reason is that the seismic waves used to measure a large earthquake come from only a small part of the fault rupture and do not accurately represent the total energy release in the event. Therefore, for events over M_L = 6.5 it is more usual to employ the *moment magnitude* (M) scale which is based on the surface area of the fault displaced, its average length of movement and the rigidity of the ground material, rather than on a seismograph trace. One difference is that, whilst no earthquake of M_L = >9 has ever been recorded, several of the largest events have achieved M = >9 status. In this book, the letter M will be used to indicate magnitude. Where M = <6.5 it denotes local magnitude; where M = >6.5 it should be taken as moment magnitude.

Energy release and earthquake magnitude can be a poor guide to hazard impact, partly because the duration of ground shaking is not accounted for in the magnitude concept. But earthquake losses depend on many other factors such as the distance from the epicentre to the damage area, rock and soil conditions, population density and the nature of building construction. For example, the Kobe, Japan, earthquake was a relatively moderate (M = 6.9) event. The scale of impact was due mainly to the fact that the shock ruptured through a densely populated port and industrial city, where some construction was on soft soils and landfill. In addition, most of the housing had been built to withstand tropical cyclones, rather than earthquakes, and the heavy clay-tile roofs offered poor resistance to lateral shaking. The collapse of these roofs accounted for over 90 per cent of the deaths.

Earthquake intensity is a measure of ground shaking that correlates more directly with hazard impact. It can be assessed by allocating a numerical value to human observations of the tremor, plus the extent of physical damage to buildings and undeveloped land. The *Modified Mercalli* scale is most commonly used and ranges from MM = I (not felt at all) to MM = XII (widespread destruction) (see Appendix 1). This intensity scale is probably no less 'scientific' than the Richter scale because seismologists can disagree about the exact rating on the magnitude scale. A prime advantage is that Mercalli intensities can be assigned to earthquakes which occurred prior to the development of seismographs and isoseismal maps can then be produced which show the pattern of areal damage.

PRIMARY EARTHQUAKE HAZARDS

The main environmental hazard created by earthquakes is *ground-shaking* which produces several types of elastic wave (Figure 6.2). The *primary or P wave* is a compressional wave, similar to a shunt through a line of connected rail coaches. It spreads out from the focus, with the ground vibration following the direction in which the wave travels at a fast speed of about 8 km s^{-1}. These P waves, like sound waves, are able to travel through both solid rock and liquids, such as the oceans and the earth's liquid core. The *secondary S waves* move at about half the speed of primary waves and cause vibration at right angles to the direction of wave travel. S waves cannot propagate in the liquid parts of the earth but, when they reach the surface, the resulting vertical ground motion can be highly damaging to structures. However, most structural damage beyond a few kilometres from the epicentre is associated with the surface waves, which are either *Love waves* or *Rayleigh waves*. It is the magnitude of these surface waves which is essentially captured in the M_L scale. They do not possess vertical motion but shake the ground horizontally at right angles to the direction of propagation. Love waves usually travel faster than the Rayleigh waves and pose a special problem for the security of buildings.

The severity of ground-shaking at any point depends on a complex combination of the magnitude of the earthquake, the distance from the rupture and the local geological conditions, which may either amplify or reduce the earthquake waves. It is measured by *ground acceleration*, which refers to the rate at which the earth is moved, both horizontally and vertically, by the force of the earthquake. Acceleration is usually expressed in units of 1.0 g, or the acceleration due to gravity (9.8 m s^{-2}). An acceleration greater than 1.0 g in the vertical plane means that unsecured objects would leave the ground. For some time it was thought that a maximum possible peak acceleration in firm ground might be around 0.5 g but values as large as 0.8 g have been recorded from earthquakes with Richter magnitudes as small as 3.5. The Northridge, California, earthquake in 1994 produced ground motions of nearly 2.0 g and, as suggested by EERI (1986), it now seems possible that peak accelerations may exceed 2.0 g. The peak

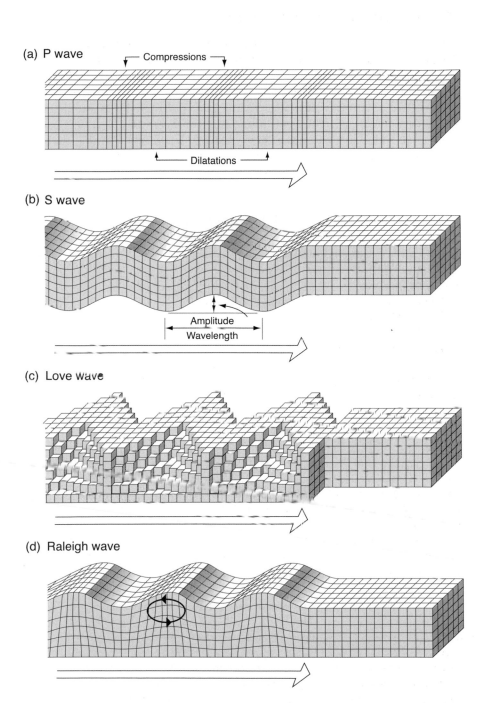

(a) P wave

Compressions

Dilatations

(b) S wave

Amplitude

Wavelength

(c) Love wave

(d) Raleigh wave

Figure 6.2 A schematic illustration of the forms of ground-shaking near the earth's
surface associated with the main types of seismic wave.
Source: Bruce A. Bolt, *Earthquakes*, © 1978, 1988, 1993 by W.H. Freeman
and Company. Reprinted with permission

acceleration decreases quite rapidly within 50 km of the earthquake source, although the detailed pattern will depend on local geology and soil conditions.

The greatest damage to buildings is created by horizontal ground movements. This is partly because all buildings are constructed to resist the pull of gravity and can, therefore, withstand some vertical movement. However, weak structures, such as unreinforced masonry buildings (URMs), may be unable to cope with horizontal accelerations as little as 0.1 g. The significance of horizontal ground-shaking is further increased by the fact that peak horizontal accelerations can be double those in the vertical plane. Figure 6.3 shows that the vertical component of shaking in a Californian earthquake reached a peak slightly above 0.1 g in response to the arrival of the P waves. However, on the east–west axis of shaking, peak horizontal ground-shaking reached just over 0.2 g between 3–4 seconds after the record began following the arrival of the S waves. The north–south shaking shows a similar pattern.

Most strong-motion measurements depict ground-shaking as a function of time. This is because the scale of destruction also depends on the frequency of the vibrations and the fundamental period of the building. The frequency of a wave is the number of vibrations (cycles) per second measured in units called Hertz (Hz). High-frequency waves tend to have

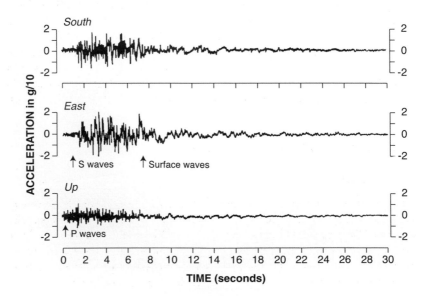

Figure 6.3   Three components of strong ground acceleration recorded about 20 km from the fault rupture during the 1971 San Fernando earthquake, California.
Source: Bruce A. Bolt, *Earthquakes*, © 1978, 1988, 1993 by W.H. Freeman and Company. Reprinted with permission

high accelerations but relatively small amplitudes of displacement. Low-frequency waves have small accelerations but large velocities and displacements. During earthquakes, the ground may vibrate at all frequencies from 0.1 to 30 Hz. If the natural period of a building's vibration is close to that of seismic waves, *resonance* can occur and cause the sway of the structure to increase. Low-rise buildings have short natural wave periods (0.05–0.1 seconds) and high-rise buildings have long natural periods (1–2 seconds). The P and S waves are mainly responsible for the high-frequency vibrations (>1 Hz) which are most effective in shaking low buildings. Rayleigh and Love waves are lower frequency and are usually more effective in causing tall buildings to vibrate. The very lowest frequency waves may have less than 1 cycle per hour and have wavelengths of 1,000 km or more.

Local site conditions have important effects on ground motion. Significant amplifications occur in steep topography, especially on ridge crests. Ground motions in soil are enhanced in both amplitude and duration, compared to those recorded in rock. This agrees with the observation that structural damage is usually more severe in buildings founded on unconsolidated material. In the Michoacan earthquake of 1985 the recorded peak ground accelerations in Mexico City varied by a factor of 5. Strong-motion records obtained on firm soil showed values of around 0.04 g. This compared with observations from the central part of Mexico City, which is founded on a dried lake bed, where the measured peak ground accelerations reached 0.2 g. Similar effects were noted in the San Salvador earthquake of 1986. This had a very modest size (M – 5.4) but produced large-scale impacts, including the destruction of thousands of buildings, 1,500 deaths, 10,000 injuries and a quarter of a million people homeless. The unusual devastation was rooted in layers of volcanic ash, up to 25 m thick, which underlie much of the city. As the 3-second-long earthquake tremor passed upwards through the ash, the amplitude of ground movement was magnified up to five times.

SECONDARY EARTHQUAKE HAZARDS

Soil liquefaction

One of the most serious hazards associated with loose sediments is soil liquefaction. This is the process by which water-saturated sediments can temporarily lose strength, because of strong shaking, and behave as a fluid. Cohesionless, granular sediment situated at depths less than 10 m below the surface is the principal medium. According to Tinsley *et al.* (1985), four types of ground failure commonly result from liquefaction:

1 *Lateral spread* involves the horizontal displacement of surface blocks as a result of liquefaction in a subsurface layer. Such spreads occur most commonly on slopes between 0.3° and 3°. They cause damage to pipelines, bridge piers and other structures with shallow foundations,

especially those located near river channels or canal banks on flood-plains.

2   *Ground oscillation* occurs if liquefaction occurs at depth but the slopes are too gentle to permit lateral displacement. Oscillation is similar to lateral spread but the disrupted blocks come to rest near their original position, while lateral spread blocks can move significant distances. Oscillation is often accompanied by the opening and closing of fissures, and in the 1964 Alaskan earthquake cracks up to 1 m wide and 10 m deep were observed.

3   *Loss of bearing strength* usually occurs when a shallow layer of soil lique-fies under a building. Large deformations can result within the soil mass causing structures to settle and tip. For example, in the Niigata, Japan, earthquake of 1964, four apartment buildings tilted as much as 60° in unconsolidated ground newly reclaimed along the Shinana river.

4   *Flow failure* is associated with the most catastrophic effects of lique-faction. This happens on slopes of >3° when liquefaction occurs at the surface as well as at depth. Flow failure has displaced material by tens of kilometres at velocities of tens of kilometres an hour. Such failures can happen on land or under water, when they are sometimes known as submarine avalanches. The devastation of Seward and Valdez, Alaska, in 1964 was largely caused by a submarine flow failure at the marine end of the delta which carried away the harbour area and raised water waves which swept back into the town, causing further damage.

Landslides, rock and snow avalanches

The severe ground-shaking in an earthquake can cause natural slopes to weaken and fail. The resulting landslides, rock and snow avalanches are major contributors to earthquake disasters, largely because many destruc-tive earthquakes occur within mountainous areas. In a study of large magnitude (M = >6.9) Japanese earthquakes since 1964, Kobayashi (1981) reported that more than half of all earthquake-related deaths were caused by landslides. Correlations between magnitude and landslide distribution show that landslides are unlikely to be triggered by earthquakes less than M = 4.0 but that the maximum area likely to be affected by landslides in a seismic event increases rapidly thereafter to reach 500,000 km^2 at M = 9.2 (Keefer, 1984).

Rockfalls are the most common earthquake-induced form of slope failure but the two leading causes of death are *rock avalanches* and *rapid soil flows*. The greatest hazard exists when high-magnitude events (M = 6.0 or greater) create rock avalanches, which are large (at least 0.5×10^6 m^3) volumes of rock fragments which can travel several kilometres from the source at velocities of hundreds of kilometres per hour. At least 90 per cent of all reported landslide deaths are caused by rock avalanches, rapid soil flows and rock falls.

The greatest landslide disaster ever recorded occurred when an offshore earthquake (M = 7.7) triggered a massive rock and snow avalanche from the overhanging face of the Nevados Huascaran mountain, Peru, in 1970 (Plafker and Ericksen, 1978). At an altitude of 6,654 m, Huascaran is the highest peak in the Peruvian Andes. Its steep slopes have been the source of many catastrophic slides, including a 1962 rock avalanche, involving 13×10^6 m³ of material, which killed some 4,000 people. In 1970 the resulting turbulent flow of mud and boulders, estimated at $50-100 \times 10^6$ m³, passed down the Rio Shacsha and Santa valleys, in a wave 30 m high travelling at an average speed of 70–100 m s⁻¹ in the upper 9 km of its course (Figure 6.4). The wave then buried the towns of Yungay and Ranrahirca plus several villages under debris 10 m deep. At least 18,000 people were killed in the densely populated valley. The huge kinetic energy needed for the long-distance transport of such a large volume of material is produced by the initial fall from steep, high slopes and, in this event, the settlements were overwhelmed less than four minutes after the original slope failure on the mountain.

Tsunamis

The most distinctive secondary earthquake-related hazard is the seismic sea wave or tsunami. The word 'tsunami' comes from two Japanese words, *tsu* (port or harbour) and *nami* (wave or sea). This derivation is appropriate since these waves can inundate low-lying coastal areas. Most tsunamis result from tectonic displacement of the sea bed associated with large, shallow-focus earthquakes under the oceans but they can also be caused by exploding volcanic islands (for example, Krakatoa in 1883) and large rockfalls into confined bays. Tsunamis pose a threat to over twenty countries in the circum-Pacific region. In the past 100 years, over 50,000 coastal residents have lost their lives. According to Iida (1983), a total of 370 tsunamis were observed around the Pacific between 1900 and 1980. The most active source region was along the Japan–Taiwan island arc where over one-quarter of all events were generated. For seventy events, appreciable loss of life and damage was confined near to the source and only seventeen tsunamis caused widespread disasters around the ocean basin.

Japan is especially vulnerable. In eastern Honshu a tsunami wave of 10 m has a return period of about 10 years. The 1933 tsunami which devastated the Sanriku coast was caused by a submarine earthquake (M = 8.5) off the north-east coast of Honshu. This event had an estimated return period of 70 years and produced a wave cresting up to 24 m above mean sea level (Horikawa and Shuto, 1983). The death toll was 3,008, with 1,152 injured, together with 4,917 houses washed away and 2,346 otherwise destroyed. During the present century over 350 people have been killed in the USA and there has been some US$500 million in property damage (Lockridge, 1985). The greatest losses in the USA were sustained from the 1964 Alaska

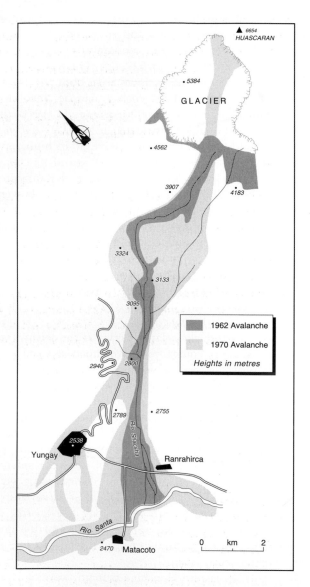

Figure 6.4   Map to illustrate the Mt Huascaran rock avalanche disasters which
occurred in the Peruvian Andes in 1962 and 1970. The map shows
the greater extent of the debris deposited in the 1970 event.
Source: After Whittow (1980)

tsunamigenic earthquake. Crescent City, California, suffered US$7 million
damage associated with a wave 6 m above low tide level which penetrated
over 500 m inland, flooding about thirty city blocks and destroying most
of the waterfront buildings.

The typical Pacific-wide tsunami results from a rapid vertical displacement of the sea floor over many hundreds of square kilometres. In turn, this displaces a column of sea water, which then travels outward from the epicentre at speeds of 140 m s^{-1} and more. In deep water out to sea these seismic waves are very long (100–200 km apart) and very low (only about 0.5 m high). But in shallow, coastal water they slow up and increase in height to produce onshore waves up to 30 m high. Figure 6.5 shows the record of a tsunami in Hawaii. The onset was signalled by an upward swell of about 0.5 m followed by a drawdown of the water surface by nearly 1 m below normal. The tsunami wave action lasted for over four hours. Empirical relationships have been established between the magnitude of relevant earthquakes and the resulting tsunami. One such set of relationships for shallow-focus earthquakes is shown in Table 6.2. The local wave, perhaps confined within a bay, may be the greater threat in some areas because it strikes quickly with little opportunity for warning.

Tsunamis travel at a speed which varies only with water depth:

$$\text{speed} = \sqrt{g \times d}$$

where g = acceleration of gravity and d = depth of the ocean.

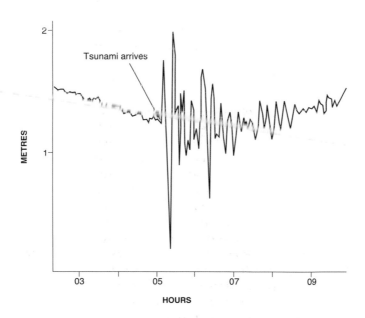

Figure 6.5 Record of a tsunami in Kuhio Bay, Hilo, Hawaii, on 29 November 1979. Note the draw-down of the water level immediately before the arrival of the main wave.
Source: Bruce A. Bolt, *Earthquakes*, © 1978, 1988, 1993 by W.H. Freeman and Company. Reprinted with permission

Table 6.2 Tsunami–earthquake magnitude relationships

Earthquake magnitude	Tsunami magnitude	Tsunami maximum wave run-up (m)
6.00	slight	
6.50	−1	0.5–0.75
7.00	0	1–1.5
7.50	1	2–3
8.00	2	4–6
8.25	3	8–12

Source: After Bolt *et al.* (1975)

It is possible, therefore, to predict with some accuracy the arrival time on any Pacific shoreline once the epicentre of the earthquake has been located. However, when these long-period waves approach a shoreline, it becomes less easy to predict the local wave behaviour. As the wave runs up the sloping sea floor, the increased frictional drag lowers the speed and shortens the wavelength. The energy contained within the wave is crowded into progressively smaller volumes of water and the wave adjusts by increasing its height. During this run-up phase, waves may attain heights in excess of 20 m and have great destructive force. Some tsunamis may be preceded by a draw-down of water along the coast (see Figure 6.5). Knowledgeable residents recognise this temporary withdrawal as a warning and use the short period available for evacuation to higher ground.

LOSS-SHARING ADJUSTMENTS

Disaster aid

Earthquake disasters are usually effective in attracting emergency funds, mainly because of the sudden loss of life. This is not necessarily an efficient response because the number of deaths in an event is a poor guide to the scale of support required for the survivors. The initial 'golden hours' are crucial for the location and rescue of victims trapped in fallen buildings, many of whom will have suffered head and chest injuries. By Japanese standards, Kobe was ill-prepared for the 1995 disaster and the search and rescue activity was hampered by several factors, not least a legal ruling which kept rescue dogs sent from overseas in quarantine until the fourth day after the earthquake (Comfort, 1996). Table 6.3 illustrates the drop in the number of people recovered from buildings over the first five days and the even greater decline in the proportion that survived. In these circumstances, local self-help is vital. After the Michoacan (Mexico City) earthquake of 1985, the official rescue and relief programme appeared to have only a

Table 6.3 The number of people who survived after being rescued from collapsed buildings, by day of rescue, following the Kobe earthquake on 17 January 1995

	Jan 17	Jan 18	Jan 19	Jan 20	Jan 21
Total rescued	604	452	408	238	121
Total who lived	486	129	89	14	7
Per cent of those rescued who survived	80.5	28.5	21.8	5.9	5.8

Source: Comfort (1996)

minor effect (Comfort, 1986). A survey of residents in the badly damaged area showed that few had received assistance and some groups set up their own rescue services to bypass the uneven government effort.

Longer-term assistance is also required. The earthquake which struck the republic of Armenia, former USSR, in 1988 killed at least 25,000 people, made 514,000 homeless and resulted in the evacuation of nearly 200,000 persons. Following the Soviet government's decision to accept international aid, over sixty-seven nations offered cash and services amounting to over US$200 million. A programme was announced to rebuild the cities within a 2-year period on sites in safer areas and with building heights restricted to four storeys. But only two of the 400 buildings projected for construction in Leninakan during the first year were actually completed and many people were still living as evacuees many months after the disaster. Continuing aid is also required for less tangible purposes. Karanci and Rustemli (1995) have shown that, 16 months after the Erzincan, Turkey, earthquake of 1992, many survivors had raised levels of phobic anxiety.

Insurance

Earthquake insurance is available in a number of countries but the capacity of private companies to cope with the potential losses is limited. A catastrophic earthquake is one of the greatest natural hazards faced by the USA, with an estimated 70 million Americans exposed to severe risk and an additional 120 million exposed to moderate risk (Lecomte, 1989). Table 6.4 has a breakdown of losses likely to be suffered by the insurance industry following major (*M* = approx. 8.0) earthquakes along the northern San Andreas and Newport–Inglewood faults in California. There is a real possibility that the USA might experience a major earthquake on the scale of the Kobe disaster with costs exceeding US$100 billion. In an attempt to limit their exposure, private insurers in California have sought government assistance through the creation of the California Earthquake Authority.

Table 6.4 Estimated insurance industry losses for Californian earthquakes (US$ billion)

	Northern San Andreas fault	Newport-Inglewood fault
Residential shaking	3.4	2.3
Commercial shaking	8.1	11.2
Fire following earthquake	11.6	24.3
Workers' compensation	2.4	4.5
General liability	5.5	10.1
Total	31.0	52.4

Source: After Lecomte (1989)

The vast majority of the exposed risk from earthquakes is presently uninsured, even in those countries where government-supported schemes have been introduced. Most policies are on commercial and industrial property rather than residential property, yet private-sector rebuilding is the costliest part of most earthquake recovery programmes. For example, the insured losses after the Northridge earthquake totalled US$12.5 billion, of which approximately two-thirds came from residential claims. At the time of the Loma Prieta earthquake in 1989 only 20 per cent of Californians carried earthquake insurance. Although this figure has since increased to 40 per cent in urban coastal areas and to 30 per cent in inland parts of the state, the majority of residents remain uninsured (Comerio *et al.*, 1996).

Most earthquake insurance policies are intended to cover catastrophes rather than small losses. As a result, they usually have a deductible amount that is either a percentage of the insured value or a fixed cash amount (Gere and Shah, 1984). For homeowners the deduction is often 5 per cent of the insured value. This deductible amount presents a deterrent to the purchase of insurance, especially for the owner of a modern wood-frame house which is unlikely to suffer total destruction. In California, the cost of insurance is also rated according to location, type of construction and soil conditions. Premium rates rise progressively from small wood-frame houses to unreinforced masonry. A building on filled land, for example, might well attract a 25 per cent surcharge. However, not all potentially active faults are known and the insurance industry is suspicious of the degree of compliance with local building codes. Given the limited premiums that many property owners appear willing to pay and the rising costs of earthquake-related damage, commercial insurance will face increasing problems unless a partnership is struck with government which agrees a viable balance of responsibility between private and public responsibility.

EVENT MODIFICATION ADJUSTMENTS

Environmental control

There appears to be little prospect of suppressing earthquakes. Theoretically, the deliberate inducement of small-scale earth movements would help to prevent the build-up of tectonic strain. One approach lies in the manipulation of surface water resources in a fault zone. There is, for example, some evidence of man-made reservoirs inducing small, shallow earthquakes through the extra load of water on the earth's crust along sensitive fault-lines. Another possibility exists in the control of groundwater levels. A large head of groundwater, associated with a high water-table, would tend to increase the pore pressure within saturated rocks thereby reducing the frictional resistance along a fracture. This would encourage regular slippage along the fault rather than the long-term accumulation of tectonic strain. It may be that the injection of water into fault-lines could help to control the build-up of hazardous tectonic strain.

Hazard-resistant design

Most earthquake-related deaths and most of the financial losses are due to the collapse of houses and other buildings. All unreinforced masonry structures are at risk in earthquakes but the most vulnerable buildings are constructed from *adobe* or sunbaked clay bricks. Adobe construction is common in arid and semi-arid regions because it is cheap, easily worked and readily available. For example, in Peru, an estimated two-thirds of rural dwellers, plus over one-third of those in cities, live in adobe houses. In the earthquake of May 1970 over 60,000 of such homes collapsed, killing at least 50,000 people and injuring a further 150,000. Houses built of rubble masonry are also very vulnerable. In the Maharashtra, India, earthquake of 1993 it was the *pucca* houses, built of thick granite walls bonded with mud and with roofs constructed of heavy timbers, that collapsed and caused most of the deaths rather than the thatched huts and the buildings with reinforced concrete frames. Some traditional societies have employed 'weak' structures as a defence against earthquakes. Throughout much of tropical Asia the indigenous house was lightly built with plant matting walls and palm-frond roofs, and houses in Bali have proved resistant to earthquakes (Leimena, 1980).

According to Key (1995), about 60 per cent of all earthquake-related deaths in the twentieth century were due to the collapse of masonry structures in rural areas. This situation will take decades to rectify although some attempts have been made to produce stronger structures with 'weak' materials available locally. For example, Hardy (1988) reported that the insertion of bamboo canes into adobe walls could make this type of house 40 per cent stronger as a result of the greater flexibility

Plate 6 Ruins of an apartment block in Spitak, Armenia, former USSR, following severe ground-shaking in the 1988 earthquake. The pattern of damage shows that the weakest part of the building was the connection between the different structural elements. (Photo: International Federation)

provided by the bamboo. Despite the high losses in rural areas of the LDCs, the earthquake risk is growing most quickly, and poses the largest economic threat, through the expansion of the world's large cities. Here older buildings constructed of unreinforced masonry have been augmented by high-rise, reinforced concrete structures – such as apartment blocks – built to accommodate the recent growth of population. This means that most urban areas present a complex array of risk. Figure 6.6, based on the effects of the Kobe earthquake, shows the varied relationships between earthquake intensity and building damage for different types of structures and also emphasises how the threat of collapse rises with an ageing stock of buildings.

The key to urban earthquake-resistance lies in the appropriate choice of modern building design and construction methods. In this context, strong, flexible and ductile materials are preferred to those which are weak, stiff and brittle. Steel framing is a ductile material which absorbs a lot of energy when it deforms. Indeed, the spread of well-designed, steel reinforced concrete buildings has been the primary factor in increasing earthquake safety in urban areas for many decades. Glass, on the other hand, is a very brittle material that shatters easily. In practice, both types of material have to be incorporated into structures. Some otherwise well-designed structures collapse because of the failure of a single element which lacks sufficient strength or ductility. For example, buildings with flexible frames will often fail if the frames are infilled with stiff masonry brickwork.

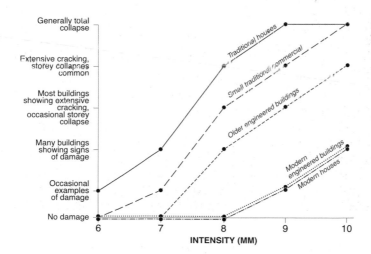

Figure 6.6   The general relationship between earthquake intensity on the Modified Mercalli scale and building damage based on the effects of the 1995 Kobe earthquake.
Source: After Alexander Howden Group Ltd and Institution of Civil Engineers (1995)

The shape of a building will also influence its seismic resistance (Ambrose and Vergun, 1985). A stiff single-storey structure (Figure 6.7a) will have a quick response to lateral forces while tall, slender multi-storey buildings (6.7b) respond slowly, dissipating the energy as the waves move upward to give amplified shaking at the top. If the buildings are too close together, pounding induced by resonance may occur between adjacent structures and add to the destruction. The stepped profile of the vertical mass of the building in 6.7c will offer stability against lateral forces. Most buildings are not symmetrical and form more complex masses (6.7d and e). These asymmetrical structures will experience twisting, as well as the to-and-fro motion. Unless the elements are well joined together, such differential movements may pull them apart. High-rise structures will be vulnerable if they do not have uniform strength and stiffness throughout their height. The presence of a *soft storey*, which is a discontinuity introduced into the design for architectural or functional requirements, may be the weak element which brings down the whole structure. Figure 6.7f shows a soft ground-floor storey, perhaps introduced to ease pedestrian traffic or car parking.

The weakest links in most buildings are the connections between the various structural elements, such as walls and roofs. Connections are important in the case of precast concrete buildings where failure often results from the tearing out of steel reinforcing bars or the breaking of connecting welds. In the 1994 Northridge earthquake a number of multi-storey car parks failed when vertical concrete columns were cracked by lateral ground-shaking to the point where they became unable to support the horizontal concrete beams holding up the different floors. Exterior panels and parapets also need to be anchored firmly to the main structure in order to resist collapse. Architectural style can contribute to disaster if features such as chimneys, parapets, balconies and decorative stonework are inadequately secured.

Difficult construction sites (Figure 6.7g and h) include localities near geological faults and soft soils which amplify ground-shaking. As far as possible these should either be avoided or only built up at low densities so that, for example, buildings cannot collide as a result of downward movement on slopes. Some slopes may have to be reformed by cut and fill to limit the threat from earthquake-related landslides (Figure 6.7i). Methods of building reinforcement include the cross-bracing of weak components, placing the whole structure in a steel frame and the installation of special deep foundations on soft soils (Figure 6.7j–l). Adequate footings are very important. High-rise buildings on soft soils should have foundations supported on piles driven well into the ground. Wood-framed houses should be internally braced with plywood walls tied to anchor bolts linked into foundations 1–2 m deep. Some new buildings can be mounted on isolated shock-absorbing pads made from rubber and steel which prevent most of the horizontal seismic energy being transmitted to the structural components. The technique is expensive but it is claimed to provide maximum protection for the loose contents of buildings, thus making it attractive for hospitals, laboratories and other public facilities. In addition, base-isolated buildings need less structural bracing to withstand lateral forces so that the reduction in construction materials offsets the extra cost of the isolation system.

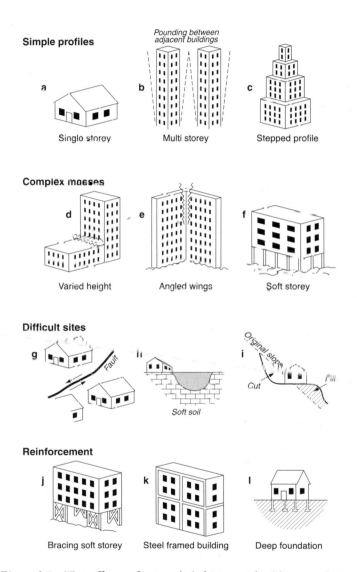

Figure 6.7 The effects of ground-shaking on buildings and some construction methods adopted for seismic resistance: (a–c) simple building profiles; (d–f) complex building masses; (g–i) coping with difficult sites; (j–l) methods of building reinforcement.

Best practice in construction design and materials needs to be applied through *building codes*, and over 100 countries have already adopted such a measure. However, to be effective building codes need to have full legal status, with facilities for up-dating the criteria and provision for the formal inspection of projects during the construction phase. Many countries at risk still do not have a comprehensive earthquake building code in force and many of those that do have one face problems. For example, 40 per cent of all the federal buildings in the USA are located in seismically active areas (General Accounting Office, 1992). In most countries, the vast majority of building stock pre-dates present design standards. California alone has about 60,000 unreinforced masonry buildings which were constructed before 1933 and are now deemed to be unsafe in an earthquake. The Unreinforced Masonry (URM) law passed by the state legislature in 1986 required all cities and counties in Seismic Zone 4, which includes most of the metropolitan areas in California, to identify such buildings by 1 January 1990. This inventory, which includes information on building use and daily occupancy loads, was an important step forward but it will be some time before the legacy of earlier construction can be overcome.

Great importance is attached to the earthquake-resistance of buildings such as hospitals, dams, nuclear power stations and factories with explosive or toxic substances. Urban lifelines pose special problems. The Building Seismic Safety Council (1987) emphasised the importance of transportation, communication, electric power, water and sewerage, and fuel transmission line facilities in the USA. An action plan for seismic hardening has been produced but will take time to implement. Some commercial organisations have taken independent steps to avoid the physical damage and associated loss of business following an earthquake. For example, the IBM manufacturing plant at San José, California, has undergone an extensive retrofit programme (Haskell and Christiansen, 1985). As a result, it was able to get all but one of its Santa Clara buildings back into full operation the day after the Loma Prieta earthquake in 1989. The cost of this event was US$800,000 but, without the contingency planning, it could well have been US$25 million (Valery, 1995).

Specially engineered structures can also offer some protection against tsunamis, as illustrated by the Sanriku coast of Japan. Following the Sanriku tsunami of 1933, the Japanese government offered subsidies for the relocation of some fishing villages to higher ground (Fukuchi and Mitsuhashi, 1983). This policy proved ineffective due to the limited availability of land for redevelopment and the natural desire of fishermen to remain as close to the shoreline as possible. Thus, a policy of on-shore tsunami wall construction was adopted, although it was insufficiently developed to prevent further tsunami losses in 1960. After this event, the government passed a special law to subsidise construction costs up to 80 per cent for walls erected to cope with wave heights equivalent to the 1960 event.

Since then engineers have sought to protect against larger tsunamis and the highest walls stand up to 16 m above tidal datum level. Initially, break-waters were used as well as on-shore walls. For example, the city of Ofunato is protected by a breakwater built across the entrance to the bay and a similar structure was built at Kamaishi to protect the local steel industry and the fishing fleet. Although breakwaters do not take up land and can provide shelter for shipping, they are expensive and were found to inter-fere with tidal circulations to the disadvantage of the local fishing industry. More recently, the trend has been away from the construction of off-shore breakwaters towards the use of more elaborate on-shore tsunami walls designed to protect coastal property and communications, although these lead to some aesthetic degradation of natural seascapes. Figure 6.8 shows an offshore breakwater, an onshore tsunami wall and coastal redevelopment used as a comprehensive tsunami defence.

VULNERABILITY MODIFICATION ADJUSTMENTS

Community preparedness

Community preparedness and recovery planning is a key factor in miti-gating earthquake impact. Formal programmes are necessary because there is usually little spontaneous preparation. For example, a number of earth-quakes along the North Anatolian fault in the late 1930s and 1940s prompted the Turkish government to adopt, in 1944, an early earthquake law specifying measures for emergency assistance, rescue and relief, and temporary housing (Bayulke, 1984). In California, earthquake preparedness planning became more prominent in the 1980s because of long-range predic-tions regarding the potential for large-scale movement along the San Andreas fault. In 1981 the Seismic Safety Commission established the Southern California Earthquake Preparedness Project (SCEPP) to deal with the Los Angeles area whilst the Bay Area Regional Earthquake Preparedness Project (BAREPP) was set up for the San Francisco region.

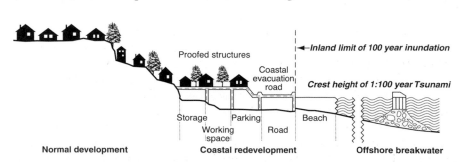

Figure 6.8 Tsunami engineering works, including offshore breakwater and some coastal redevelopment in low-lying areas, as employed in parts of Japan.

Community preparedness is best developed at the local level, within a framework provided by state or national government. In some cases, it may be difficult to prioritise areas successfully. For example, the unexpected devastation suffered in the Kobe earthquake of 1995 was partly attributed to the fact that the Tokyo area was previously seen as more vulnerable and stockpiles of emergency foods and medicines were inadequate in the immediate aftermath of the event. Following the 1988 Armenian earthquake, a lack of preparedness meant that Soviet troops failed to deploy heavy lifting equipment to dig under the rubble for survivors. Emergency services — fire-fighting and medical — were woefully inadequate. Injured survivors subsequently died due to a shortage of hospital beds and specialist treatment.

In 1985 the California Legislature adopted a detailed programme, 'California at Risk', within the context of a wider initiative designed to reduce earthquake hazards significantly by the year 2000. Part of this programme is directed at involving elected local officials, city and county managers, plus other administrators and leaders, in an action plan for earthquake mitigation (Spangle, 1988). The basic checklist provides a useful indication of the wide range of pre-planning which is required (Table 6.5). But it has to be recognised that not all preparedness plans will work smoothly. For example, in the 1995 Kobe earthquake, local emergency responders were overwhelmed by the sheer scale of the task. In addition, the road network failed because key routes either collapsed or were blocked by fallen debris for several days, thus seriously delaying the arrival of specialist teams and medical help in some areas. Better traffic management, through limiting non-essential vehicles and redirecting emergency support, was clearly necessary after this event, together with heavy lifting equipment to clear the streets of rubble. Most disaster experts believe that urban earthquake survivors should be prepared to spend at least three days on their own and be given training in basic first aid, search and rescue and fire-fighting techniques. In these circumstances, increased preparedness at the family level is important (Russell *et al.*, 1995). In the aftermath of disaster, neighbours and strangers often form spontaneous groups which are organised to undertake search and rescue.

Forecasting and warning

Earthquake prediction and warning, based on either probabilistic or deterministic methods, is a measure that has attracted much research, especially in California and Japan. Yet, despite considerable investment, credible predictions are not yet available. For example, neither the Northridge earthquake ($M = 6.8$) of January 1994, which affected a suburb of Los Angeles, nor the Kobe, Japan, earthquake of January 1995 were predicted. Indeed, both earthquakes occurred on fault systems where the seismic potential was incompletely understood. On the other hand, warnings offer life-saving

Table 6.5 Earthquake safety self-evaluation checklist

1 *Existing development*
  Inventory hazardous buildings
  Strengthen critical facilities
  Reinforce hazardous buildings
  Reduce non-structural hazards
  Regulate hazardous materials

2 *Emergency planning and response*
  Determine earthquake hazards and risks
  Plan for earthquake response
  Identify resources for response
  Establish survivable communications system
  Develop search and rescue capability
  Plan for multijurisdictional response
  Establish and train a response organisation

3 *Future development*
  Require soil and geologic information
  Update and improve safety element
  Implement Special Studies Zones Act
  Restrict building in hazardous areas
  Strengthen design review and inspection

4 *Recovery*
  Plan to restore services
  Establish procedures to assess damage
  Plan to inspect and post unsafe buildings
  Plan for debris removal
  Establish programme for short-term recovery
  Prepare plans for long-term recovery

5 *Public information, education and research*
  Work with local media
  Encourage school preparation
  Encourage business preparation
  Help prepare families and neighbourhoods
  Help prepare elderly and disabled
  Encourage volunteer efforts
  Keep staff and programmes up-to-date

Source: After Spangle (1988)

potential if people can be evacuated from unreinforced buildings or areas prone to secondary hazards, such as landsliding.

Where a good historical database exists, statistical methods have been applied to earthquake prediction. In a country like New Zealand, where the pattern of earthquake activity does not correlate well with the surface geology, the historical record is useful in assessing the short-term risk (Smith and Berryman, 1986). The method assumes statistical stationarity for estimates of the return periods which are of the order of the observation period or less. Figure 6.9a uses shallow focus earthquakes of M = >6.5 from 1840 to 1975 to map the return periods for intensity MM VI and greater, which is the level at which significant damage begins. Figure 6.9b shows the intensities with a 5 per cent probability of occurrence within 50 years. Such regional macrozonations are always uncertain because they are based on comparatively short human records and are subject to modification according to ground conditions (Fahmi and Alabbasi, 1989).

One problem in making probability assessments arises from the fact that fault lines move in many different ways. For example, the San Andreas fault, California, is known to consist of both *locked* segments and *creeping* segments. Locked segments allow sufficient strain to build up to cause major earthquakes. Creeping segments are characterised by continuous sliding at a rate of about 30 mm yr^{-1} which results in thousands of small vibrations only. Such creep appears to result from the presence of finely crushed rock and clay within the fault zone. This has a low frictional resistance, especially if saturated, and limits the build-up of tectonic strain. However, it is possible that the more competent rocks at greater depth may be accumulating some potentially damaging strain. Dolan *et al.* (1995) claimed that, in the geologically complex Los Angeles region, far too few moderate earthquakes have occurred during the last 200 years to account

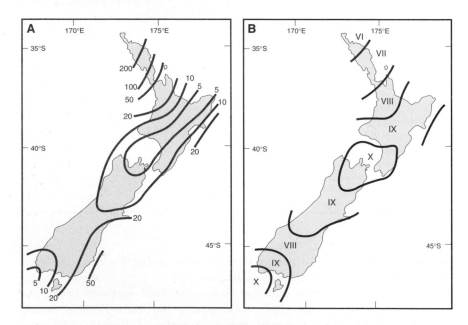

Figure 6.9   Examples of earthquake prediction in New Zealand: (a) return periods (years) for earthquake intensities of MM VI and greater; (b) MM intensities with a 5 per cent probability of occurrence within 50 years.
Source: After Smith and Berryman (1986)

for the observed accumulation of tectonic strain. One possible explanation is that the historic record may simply reflect a period of unusual quiescence and that much larger events – the so-called 'Big One' – will occur in the future.

Some sections of the San Andreas fault have fairly regular movements which are sufficiently frequent to give some confidence to statistical predictions several years ahead. Near the town of Parkfield, California, a 25-km stretch of the San Andreas fault has been intensively studied, partly because it slips with a recurrence interval of about 22 years to give moderate (M = 6.0) earthquakes and partly because over 120,000 households are at risk from seismic activity in the area. The ongoing Parkfield prediction experiment involves close monitoring of the seismological behaviour along the fault line. This is done through a dense network of sensitive seismographs, the use of tiltmeters to detect ground surface changes and geodetic lasers which accurately measure any changes in distance across the fault. Based on this information, prediction experiments have been made, such as that for an earthquake equivalent to M = 5.0–6.0 occurring on the Parkfield segment of the San Andreas fault between 1985 and 1993 with 95 per cent probability (Bakun and Lindh, 1985). This led to the US Geological Survey issuing a public announcement which subsequently proved false.

More deterministic methods of prediction rely on the detection of physical precursors near the active fault. One of the most promising avenues involves the application of the dilatancy theory (Bolt *et al.*, 1975). As the crustal rocks deform under the influence of tectonic strain, local cracking starts to occur, causing the volume of rock to increase or dilate. If the cracking occurs quickly, ground water may not immediately flow into the fissures. These spaces then become filled with water vapour with a reduction in the pore pressures within the rock. In turn, this could lead to a temporary reduction in the velocity of the P waves, which would subsequently increase again following the eventual diffusion of groundwater into the voids. Before the San Fernando, California, event of 1971 the ratio of the velocity of the P to the S waves decreased by about 10 per cent before re-establishing itself to the more normal level. One difficulty with this approach is that the potential lead-time for predicting the major events extends to years (Table 6.6). Many active faults are now surrounded by intensive monitoring schemes (Bakun, 1988). These range from low-technology methods, such as noting any unusual behaviour in animals, to the use of complex instruments able to transmit information directly to seismic laboratories by satellite. Despite close surveillance, it is still impossible to predict earthquakes accurately. In Japan there are about 100 monitoring stations but no public warning has yet been issued, mainly because many earthquakes happen without reliable and observable precursor signals.

If credible earthquake predictions eventually become a reality, problems will continue to exist if long lead-times are employed. Many people find

Table 6.6 Duration of the precursory changes in velocities of P and S waves
determined by the dilatation cycle

Earthquake magnitude	Duration of precursory phase
2	1 day
3	1 week
4	1 month
5	3 months
6	1 year
7	6 years

Source: After Bolt *et al.* (1975)

long-lead warnings difficult to handle psychologically, especially because
earthquakes – unlike major storms – offer no environmental signs by which
a lay person can tell that a damaging event is imminent. In some cases,
long-term predictions could provoke an economic recession in areas subject
to earthquakes. Insurance companies might stop selling cover, investment
might cease and people might move away only to return after the event.
All this may be acceptable if the prediction is reliable and lives are saved,
but the potential for false warnings is likely to continue for many years.
For example, in 1980 it was erroneously predicted that, nine months later,
Peru and northern Chile would suffer their worst earthquake this century
(Echevarria *et al.*, 1986). The prediction was given media coverage in Lima
to the extent that nearly all the city's population were aware of the threat
and over half took some precautions. It has been estimated that, including
lost tourist revenue, the total cost of this false prediction was US$50 million.
Vague threats of earthquake activity, and an ongoing awareness of hazard,
may produce no better response, as illustrated by the 'Palmdale bulge'
(Turner *et al.*, 1986). In February 1976 a major ground uplift centred on
this Californian town was widely viewed as a possible precursor of an earthquake. The scale of ground deformation suggested that the subsequent event
could attain a magnitude of $M = 8.0$ which, in turn, could claim around
12,000 lives. Surveys revealed that, although nearly half the local population expected a damaging earthquake within one year, few saw it as a major
community problem or took precautionary household measures.

In contrast to the hazard of earth-shaking, tsunami forecasting and
warning systems are well established. A tsunami warning system was
established in 1948 for the Pacific whereby seismograph stations around
the Pacific relay information to a Warning Centre near Honolulu, Hawaii
(Lockridge, 1985). This international monitoring network, operated by
the US National Oceanic and Atmospheric Administration, now relies
on about thirty seismic stations and seventy tide stations throughout the
Pacific basin. Following the disastrous tsunamigenic Alaska earthquake
of 1964, the Alaska Tsunami Warning Centre was established in 1967 to

provide more localised warning and in 1982 this regional responsibility was extended to British Columbia, Washington, Oregon and California. Such piecemeal development has been criticised by those who argue for a fully comprehensive approach to tsunami warning in the Pacific basin (Dohler, 1988).

At present, tsunami warning is operated on two levels. The first level of cover provides warnings to all Pacific nations of large, destructive tsunamis which are Pacific-wide. Following a high magnitude earthquake (M = >7.0), tide stations near the epicentre are alerted to watch for unusual wave activity. If this is detected, a tsunami warning is issued. The aim is to alert all coastal populations at risk within a time period of 1 hour or beyond 750 km from the source about the arrival time of the first wave with an accuracy of ± 10 minutes. In comparison, it would take 10 hours for a tsunami generated off northern Japan to reach the US west coast and a similar 10–12 hours for the first waves to travel north from Chile or Peru. This allows ample time to warn shipping and activate evacuation plans on low-lying coasts.

The second level of cover is regional only, based on warning systems serving specific tsunami-prone areas. Local tsunamis may well pose a greater threat than Pacific-wide events because they can strike very quickly. Regional systems rely on local data obtained in real time via telephone lines or dedicated communication links. Typically, they aim to issue a regional warning within minutes for areas between 100 and 750 km from the source. For example, the Japanese Meteorological Agency has maintained its own warning service since 1952 designed to issue a warning within 20 minutes of a tsunamigenic earthquake occurring within 600 km of the Japanese coast. These systems have proved to be effective in saving lives. Before the establishment of the Japanese regional system, more than 6,000 people were killed by fourteen tsunamis. Since 1952, twenty local tsunamis have claimed 215 people only.

However, many coastal areas remain unprotected and, in the LDCs, there is a clear need for reliable, low-cost local warning systems. Project THRUST (Tsunami Hazards Reduction Utilising Systems Technology) is a comprehensive approach to risk mitigation which is based on rapid warning via satellite communication links (Bernard *et al.*, 1988). This system has been trialed for the city of Valparaiso, Chile, where the worst-case scenario would imply a wave arrival within a few minutes. The system is activated instantly by the triggering of an accelerometer and water level sensors which send messages via NOAA's GOES West satellite to the Valparaiso Tsunami Warning Centre and to the Pacific Centre near Honolulu. Less than 3 minutes after an earthquake has triggered the system, information is available locally for the implementation of emergency action. Since the basic hardware costs for the THRUST system are about US$15,000 and the GOES satellite covers the whole Pacific basin, the scheme appears to offer great potential for rapid tsunami warning in future.

Land use planning

Avoidance of high-risk earthquake areas is the most direct land use adjustment. The microzonation of land is expensive but necessary in urban areas where the aim is to convert already developed areas to parkland or other open spaces and to prevent the further development of hazardous sites. The highest priority is to map those urban areas susceptible to enhanced ground-shaking, as a result of the presence of soft soils or land fill, because this process is often the major factor in property damage. Analysis of damage following the 1989 Loma Prieta earthquake in Table 6.7 shows that, whilst 98 per cent of the total property loss from the earthquake was attributed to ground-shaking, enhanced ground-shaking, caused by the local amplification of seismic waves in soft-soil deposits, was directly responsible for about two-thirds of the total (Holzer, 1994). In a separate study, French (1995) examined the claims submitted to FEMA for reimbursement under the Disaster Relief and Emergency Assistance Act. These claims, totalling almost US$600 million, related essentially to public facilities and were found to be clustered. For example, whilst infrastructure systems such as roads, sewers and electrical lines performed well in the earthquake, other structures, such as hospitals and universities, accounted for a large proportion of the damage. This suggests that land use planning should be sensitive to building function as well as geological conditions.

In many counties and cities of California, *setback ordinances* are a major means of enforcing seismic safety. Thus, building setbacks can be recommended where proposed development crosses known, or inferred, faults and slope stability setbacks can be established where unrepaired active landslides, or old landslide deposits, have been identified. Setbacks can also be used to impose the appropriate separation of buildings from each other in order to reduce pounding effects. This phenomenon is most common in urban areas where structures of different heights, resulting from different construction methods, are combined in close proximity. Another type of

Table 6.7 Losses in the 1989 Loma Prieta earthquake according to hazard process

Earthquake hazard	Total damages (millions US$)	Loss (% of total)
Ground-shaking		
Normally attenuated	1,635	28
Enhanced	4,170	70
Liquefaction	97	1.5
Landslides	30	0.5
Ground rupture	4	0
Tsunami	0	0
Total	5,936	100

Source: After Holzer (1994, p. 301, table 3) © American Geophysical Union

setback regulates the distance from buildings to sidewalks or other areas which are heavily trafficked by pedestrians. The main purpose of such setbacks is to reduce the loss of life and injury arising from collapsing buildings during an earthquake.

Successful land use planning depends on the availability of hazard information, making the information available and ensuring a satisfactory public response. In California, state law requires that estate agents inform all home purchasers if properties are located near mapped fault lines. It is likely that newcomers to such hazard areas will be most in need of such information and will also be most likely to heed the advice, assuming that they have few preconceived ideas about the desirability of different residential areas. However, Palm (1981) showed that estate agents were not effective communicators of hazard information, mainly because the hazard potential of property was not disclosed until sale negotiations were well advanced. Consequently, the price and sales of hazard-prone property have not declined as anticipated. In the most desirable residential areas, other attributes – such as schools, shops, investment potential, view – appear more important to buyers than uncertain risks, especially if the purchaser intends to relocate in a few years' time.

The rezoning of low-lying coastal land at risk from tsunamis, in association with the structural strengthening of buildings, can be an effective defence. For example, Crescent City, California, was badly damaged by a tsunami after the 1964 Alaska earthquake. Since then the waterfront has been redeveloped into a public park and the beach area has been rezoned with business premises located back from the shore on higher ground. Preuss (1983) has emphasised the need for tsunami mitigation to be explicitly integrated into the planning of hazard-prone coastlines so that evacuation routes, for example, can be prepared and protected. Figure 6.10 illustrates a variety of measures which could be incorporated into a comprehensive anti-tsunami scheme, including the protection of structures and the provision of a coastal evacuation route.

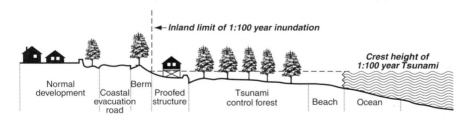

Figure 6.10   Coastal land use planning designed to mitigate tsunamis. The beach and the forest are used to dissipate the energy of the onshore wave whilst development and the coastal evacuation route are located above the predicted height of the 1:100 year event.
Source: After Preuss (1983)

7 Tectonic hazards

Volcanoes

VOLCANIC HAZARDS

There are about 500 active volcanoes in the world. In an average year, around fifty erupt. Since only about 5 per cent of eruptions result in human fatalities, the relative infrequency of hazardous volcanic events is one of their most dangerous features. Traditionally, volcanoes have been classified as active, dormant or extinct but in 1951 Mt Lamington erupted in Papua New Guinea killing 5,000 people although considered extinct (Chester, 1993). To be prudent, all volcanoes which have erupted within the last 25,000 years should be regarded as at least potentially active.

Like earthquakes, the distribution and behaviour of volcanoes is controlled by the global geometry of plate tectonics, and active volcanoes exist in every continent, except Australia. They are found in three tectonic settings:

1 *Subduction volcanoes* are located in the subduction zones of the earth's crust where one tectonic plate is thrust and consumed beneath another. They comprise about 80 per cent of the world's active volcanoes and are the most explosive type with the typical form of a *stratovolcano*, or composite cone, composed of alternating layers of ash and lava (Figure 7.1). Many of the world's most famous volcanoes – Fujiyama in Japan, Mayon in the Philippines, Mt Hood in Oregon and Mt Vesuvius in Italy – are of this type.
2 *Rift volcanoes* occur where tectonic plates are diverging. They are generally less explosive and more effusive, especially when they occur on the deep ocean floor.
3 *Hot spot volcanoes* exist in the middle of tectonic plates where a crustal weakness allows molten material to penetrate from the earth's interior. The Hawaiian islands in the middle of the Pacific plate are a good example.

All volcanoes are formed from the molten material within the earth's crust, called *magma*. This is a complex mixture of silicates which contains

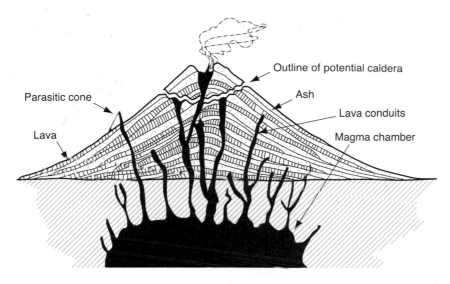

Parasitic cone

Lava

Outline of potential caldera

Ash

Lava conduits

Magma chamber

Figure 7.1   Section through a stratovolcano showing a parasitic cone growing on the steep flank. The outline of the potential caldera indicates the capacity for reshaping the main cone after a major eruption.

dissolved gases and, often, crystallised minerals in suspension. As the magma works upwards towards the surface, the pressure decreases and the dissolved gases come out of solution to form bubbles. As the bubbles expand, they drive the magma further into the volcanic vent until it ultimately breaks through weaknesses in the earth's crust. For a moderately large eruption, the total thermal energy released lies in the range 10^{15}–10^{18} joules, which compares with the 4×10^{12} joules liberated by a 1 kilotonne atomic explosion. There is no agreed international scale against which to measure the size of individual volcanic eruptions but Newhall and Self (1982) drew up a semi-quantitative *volcanic explosivity index* (VEI) which combined the total volume of ejected products, the height of the eruption cloud, the duration of the main eruptive phase and several other items into a basic 0–8 scale of increasing hazard (Table 7.1). On average, an eruption with VEI = 5 occurs every 10 years and with VEI = 7 every 100 years.

The violence of the eruption is determined by the magnitude and the intensity of the event which, in turn, depends largely on the effervescence of the gases and the viscosity of the magma. High effervescence and low viscosity lead to the most explosive eruptions. Thus, subduction zone volcanoes draw on magmas that are a mixture of upper-mantle material and melted continental rocks rich in feldspar and silica. These are called *felsic* (acid) magmas which produce thick, viscous lavas containing up to 70 per cent silicon dioxide (SiO_2) and lead to the most violent eruptions. On the other hand, rift and hot spot volcanoes draw on magmas that are high in

Table 7.1 Selected criteria for the Volcanic Explosivity Index (VEI)

VEI number	Volume of ejecta (m^3)	Column height (km)[a]	Qualitative description	Tropospheric injection	Stratospheric injection
0	$<10^4$	<0.1	non-explosive	negligible	none
1	$10^4–10^6$	0.1–1.0	small	minor	none
2	$10^6–10^7$	1–5	moderate	moderate	none
3	$10^7–10^8$	3–15	mod-large	substantial	possible
4	$10^8–10^9$	10–25	large	substantial	definite
5	$10^9–10^{10}$	>25	very large	substantial	significant
6	$10^{10}–10^{11}$	>25	very large	substantial	significant
7	$10^{11}–10^{12}$	>25	very large	substantial	significant
8	$>10^{12}$	>25	very large	substantial	significant

Source: Modified from Newhall and Self (1982)

Note

a Column height: for VEI 0–2 uses km above crater; for VEI 3–8 uses km above sea level.

magnesium and iron but low (<50 per cent) in silica content. Such *mafic* (or basic) lavas are fluid, retain little gas and erupt less violently. These characteristics allow a broad recognition of volcanic eruptions by type. For example, the *Plinian type* produces the most violent upward expulsion of gas, as the pressure on the underlying felsic magma is relieved. The volcanic plume may extend well into the atmosphere, as when the eruption of Mt Pinatubo, Philippines, in 1991, sent a plume of tephra more than 30 km into the atmosphere. The *Peléan type* is potentially the most disastrous because the upward escape of highly explosive material is prevented by an obstructive dome of solid lava above the main conduit. Compressed felsic magma then forces a new opening at a weakness in the volcano flank. The result is a lateral blast of great force which sweeps downslope to devastate most objects in its path, for example, the Mt St Helens, USA, eruption of 1980.

Volcanic eruptions are the source of multiple hazards. Worldwide they were responsible for killing, on average, about 650 people per year during most of the twentieth century (Blong, 1984). At present values, volcanoes created an estimated total of US$10 billion in property damage over the same period. However, mean values have to be treated with caution because catastrophic eruptions occur very irregularly in space and time. More than half the deaths recorded in the twentieth century occurred in a single event when the 1902 eruption of Mont Pelée, on the island of Martinique in the West Indies (VEI = 4), killed the 29,000 inhabitants of the port of St Pierre. There were only two known survivors of this event. However, countries such as Indonesia, which lies at the junction of three major plates with a population of 150 million, face the greatest overall threat (Suryo and Clarke, 1985).

In the past, most volcano-related deaths have been due to indirect causes, especially famine following the destruction of crops by ashfall. Today, deaths are more directly associated with highly explosive eruptions. Lahars (volcanic mudflows) account for at least 10 per cent of all fatalities. On the other hand, gentler, more effusive eruptions of molten lava and ashfalls (tephra) are likely to be a greater threat to property. For example, only one person was killed in the twentieth century by a volcanic eruption in Hawaii, despite the fact that, over the same period, some 5 per cent of the island was covered by new lava flows (Decker, 1986). Ashfalls rarely claim lives directly but an estimated 2,000 people died when many domestic roofs collapsed under a 200 mm deposit of ash following the 1902 eruption of the Santa Maria volcano in Guatemala.

The flanks of volcanoes often attract human settlement, and the hazard impact of eruptions depends heavily on the local population density and building type. The Etna region contains about 20 per cent of Sicily's population, with rural population densities between 500–800 per km^2 on rich agricultural soils. Such population concentrations are even more typical of the wet tropics where the potentially fertile soils are well-watered and explain why some two-thirds of all volcano-related deaths have been recorded in Indonesia. In 1815 a massive eruption (VEI = 7) shook the Tambora volcano. The height of the mountain was reduced by 1,400 m, leaving a caldera 12 km in diameter. About 12,000 people died immediately after the event, although a further 80,000 persons eventually perished through the disease and famine which followed.

More than most other hazardous phenomena, volcanoes also provide natural resources. Apart from the fertility of some ash, they also supply energy, building materials and enhanced opportunities for tourism. As Ollier (1988) pointed out, about one-third of the energy output in Italy and over 10 per cent of New Zealand's requirements are met from geothermal sources. The hot water needs of the population of Reykjavik, Iceland, are practically all supplied by geothermal heat. Basaltic lava flows provide masonry material, although it can be rather difficult to work, and volcanoes provide important scenic resources. Several national parks are centred on volcanoes and tourists flock to sites such as Etna and Fujiyama.

PRIMARY VOLCANIC HAZARDS

These are associated with the products ejected by the volcanic eruption. The most explosive volcanic eruptions are accompanied by *pyroclastic flows*, sometimes called *nuées ardentes* or 'glowing clouds'. These flows result from the frothing of molten magma in the vent of the volcano. The gas bubbles then expand and burst explosively to fragment the lava. Eventually, a dense cloud of lava fragments is ejected to form a turbulent mixture of hot gases and pyroclastic material (volcanic fragments, crystals, ash, pumice and glass shards) which then flows down the flank of the volcano. Pyroclastic bursts flow downhill because, with a heavy load of lava fragments and dust, they are appreciably denser than the surrounding air. Such clouds may be literally red hot (up to 1,000°C) and may be ejected many tens of kilometres into the atmosphere. However, they pose the biggest hazard when they are directed laterally by explosive blasts (Peléan type) and remain close to the ground. These directed blasts are capable of advancing in surges at speeds well beyond 30 ms^{-1}. In historical eruptions, pyroclastic flows have travelled some 30–40 km from the source.

Very little can survive in the path of a pyroclastic surge and such flows have been responsible for more than 70 per cent of all deaths in volcanic eruptions during the twentieth century. During the Mont Pelée disaster on the island of Martinique the town of St Pierre, some 6 km from the centre of the explosion, suffered a surge temperature around 700°C borne by a blast travelling at around 33 m s^{-1}. Such surges can melt plastic, metal and glass. People exposed to the surge are immediately killed by a combination of severe external and internal burns together with asphyxiation through inhalation of the searing hot *nuée*. The surge itself is usually preceded by an air blast with sufficient force to topple some buildings.

Air-fall tephra comprises all the fragmented material which is ejected by the volcano and subsequently falls to the ground. For subduction zone volcanoes, it represents a high proportion of the total material emitted. Most ashfalls produce less than 1 km^3 volume of material but the largest explosions produce several times this amount. The material may range in size from so-called 'bombs' (>32 mm in diameter) down to fine ash and dust (<4 mm in diameter). The coarser, heavier particles fall out first close to the volcano vent while, depending on wind conditions, the finer dust may be deposited hundreds of kilometres away.

The degree of hazard created by air-fall tephra varies greatly. For example, if injected high into the atmosphere, it can circle the globe and affect weather patterns worldwide. Although ashfalls account for less than 5 per cent of the direct deaths associated with volcanic eruptions, the choking dust greatly reduces visibility and can disrupt air traffic. Within six hours of the comparatively modest eruption of Mt St Helens in 1980, ash clouds had drifted 400 km downwind. The ash cloud was sufficiently dense to ensure that the street lights remained on all day in the darkened towns of Yakima and Spokane. People experienced breathing difficulties. Road traffic was halted partly because of the poor visibility and partly because the air filters on car engines became choked with fine dust. Radio communication was lost for a time, with serious implications for emergency response, because of atmospheric interference. Flat-roofed, un-reinforced buildings found in rural areas tend to collapse when total ash accumulation approaches 1 m and an ashfall imposing a surface load of 200 g m^2 is deemed sufficient to produce the collapse of many un-reinforced roofs.

Ashfall has a variety of effects on agriculture. Even very light falls of ash and other debris may cause injury to grazing animals if the tephra contains fluorine or other toxic chemicals which can contaminate pasture and water supplies. Heavier falls of *scoria* or cinder also blanket and destroy vegetation and crops. Following the Mt St Helens eruption (VEI = 5), up to 30 kg m² of ash fell on cropland in eastern Washington state and reduced subsequent yields (Cook *et al.*, 1981). On the other hand, it has been asserted that light falls of tephra in the tropics renew soil fertility which otherwise would be lost from the leaching of nutrients in a high rainfall environment. But ash is unlikely to contribute many elements which are either available to plants or are present in sufficient concentrations to be useful. After the eruption of Mt Pinatubo in June 1991, the livelihood of 500,000 farmers was directly disrupted as agricultural land up to 30 km distant was covered in ash and well over 1 million people were affected. During 1963–5, ashfalls from the Irázu volcano, Costa Rica, almost completely destroyed the coffee crop and ruined agricultural land causing losses estimated at US$150 million. In some cases the tephra will be sufficiently hot when it reaches the ground to start fires in both vegetation and man-made structures.

Lava flows pose the greatest threat to human life when they emerge rapidly from fissure eruptions rather than from central-vent volcanic eruptions. The fluidity of lava is determined by its chemical composition, especially the proportion of silicon dioxide (SiO_2) which is present. If silicon dioxide forms less than about half the total, the lavas are termed mafic. They are very fluid and release their gases much more easily than viscous, acid lava flows. These characteristics have led to the recognition of two broad types of lava flow: *Pahoehoe lava* flows are the most liquid, with a relatively smooth but wrinkled surface; *aa lava* is blocky, spiny and slow moving with a rough, irregular surface. The more fluid streams are potentially highly mobile. On steep slopes low-viscosity lava streams can flow downhill at speeds approaching 15 m s⁻¹. In the 1977 eruption of Nyiragongo volcano, Zaire, five fissures opened up on the flanks of the volcano and released a lake of molten lava contained within the summit crater. The lake drained in less than one hour creating a wave of highly fluid lava which killed 72 people and destroyed over 400 houses.

Thick lava blankets sterilise agricultural land for many years. The rate of land recovery in wet, tropical countries such as Hawaii is relatively fast compared to that in Iceland, where large areas have remained virtual deserts for centuries. Around Mt Etna, Sicily, aa-type lava flows have done much damage in historic times (Figure 7.2). Some flows have been guided with little damage by the large depression of the Valle del Bove but others have destroyed valuable vineyards and orange groves. The city of Catania was partially destroyed in 1669 when lava breached the old defensive walls, whilst the 1928 eruption overran the outskirts of the town of Mascali. Vast areas of Hawaii and Iceland are made up of lava flows. The greatest lava-related disaster in historic times occurred in 1783 when lava flowed out of a 24 km long fissure in Iceland for a period of more than five months (Thorarinsson, 1979). The Lakagigar lava flow covered some 560 km² and its volume was estimated at 12.3 km³. It overran two churches and

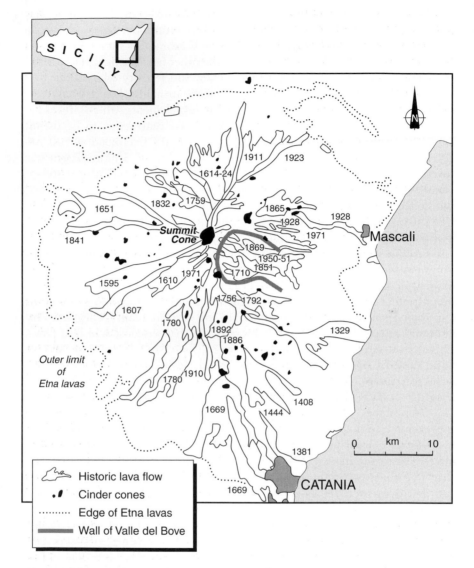

Figure 7.2   Mt Etna, Sicily, showing the date and the path taken by lava flows recorded since AD 1300 in relation to the city of Catania and the town of Mascali. In 1669 and 1928 lava streams partly destroyed Catania and Mascali respectively.
Source: After Bolt *et al.* (1975)

fourteen farms, whilst a further thirty farms were badly damaged. Few casualties were caused directly by the lavas but it has been estimated that 10,521 people – nearly 22 per cent of Iceland's population at the time – subsequently died in the resulting famine.

Volcanic gases are released by explosive eruptions and lava flows. The gaseous mixture commonly includes water vapour, hydrogen, carbon monoxide, carbon dioxide, hydrogen sulphide, sulphur dioxide, sulphur trioxide, chlorine and hydrogen chloride in variable proportions. Measurement of the exact gas composition is made difficult by the high temperatures near an active vent and by the fact that the juvenile gases interact with the atmosphere and each other, thus constantly altering their composition and proportions. Despite the toxic nature of some of these gases and the unpleasant sulphurous smell which accompanies most volcanic activity, they have only rarely been the direct cause of a major disaster.

Carbon monoxide has caused deaths because of its toxic effects at very low concentrations but most fatalities have been associated with carbon dioxide releases. Carbon dioxide is dangerous because it is a colourless, odourless gas which has a density about 1.5 times greater than air. Thus, it accumulates in low-lying places, such as topographic hollows and the basements of houses, without being detected. Once inhaled, it can cause death in 10–15 minutes at atmospheric concentrations as low as 10 per cent by volume. Isolated disasters have occurred. For example, in 1979, 142 people evacuating a village in Java, Indonesia, at night to escape a threatened eruption simply walked into a dense pool of volcanically released carbon dioxide and were immediately asphyxiated (Le Guern *et al.*, 1982).

It is now known that the release of carbon dioxide from previous volcanic activity also poses a threat. In 1984 a cloud of gas, rich in carbon dioxide, burst out of the volcanic crater of Lake Monoun, Cameroon, and killed 37 people by asphyxiation (Sigurdsson, 1988). Almost exactly two years later, in 1986, a similar disaster occurred at the Lake Nyos crater, also in Cameroon. This time 1,746 lives were lost, together with over 8,300 live-stock. Some 3,460 people were displaced from their homes to temporary camps as a result of the disaster. Field evidence at the site indicated that the burst of gas created a fountain that reached over 100 m above the lake surface. The dense cloud then flowed down two valleys and affected a total area over 60 km^2.

Such hazards are very rare and are a function of high levels of carbon dioxide in the waters of these lakes. It is possible that the gas derived from an active vent on the lake floor, but it is more likely that it was supplied to the lake by warm springs and built up over a long period of time from CO_2-rich groundwater flowing at depth into the submerged crater. The small seasonal variation in temperature in this low latitude area may well prevent the annual turnover of waters that is characteristic of temperate lakes and thus allow the long-term accumulation of gas. Under normal circumstances the dissolved CO_2 would remain trapped in the highly stratified, stagnant water. In the case of Lake Monoun, the sudden gas release could have been triggered by a disturbance of the lake waters by a landslide originating on the crater's rim, but no evidence exists for such a trigger at Lake Nyos. Such gas bursts may well reoccur. The only remedy appears to be a pumping programme designed to remove the mineralised bottom water from the flooded crater.

SECONDARY VOLCANIC HAZARDS

Ground deformation occurs widely as volcanoes grow from within by magma intrusion and as new layers of lava and pyroclastic material accumulate on the surrounding slopes. Real-time measurements of this process can now be made with GPS technology. The deformation is not in itself a hazard but it provides a destabilising process by over-steepening hillslopes. This may lead eventually to a catastrophic failure of the volcanic edifice and various mass movement hazards. For example, the collapse of the north flank of Mt St Helens in 1980 produced a massive debris avalanche which advanced more than 20 km down the North Fork of the Toutle River and filled the valley to an average depth of about 40 m. According to Siebert (1992), such structural failures of a volcano have occurred worldwide, on average, four times per century over the last 500 years, although few deaths have resulted compared with those caused by lahars and smaller landslides. Major structural instability is most likely on large polygenetic volcanoes, such as Mauna Loa and Kilauea, Hawaii. Volcanoes such as Etna are also prone to instability because of their complex construction of inter-bedded lavas and pyroclastic deposits and their steep flanks.

Lahars are volcanic mudflows in which at least 50 per cent of the sediment is of sand grain size or smaller. They occur widely on the flanks of volcanoes in the wet tropics and the term is of Indonesian origin. Apart from pyroclastic surges, lahars present the greatest threat to human life. For example, about 5,500 people were killed in a mudflow following the eruption of the Kelut volcano, Java, in 1919. Lahars may occur in association with any volcanic event, whether explosive eruption or effusive lava flow, whenever large quantities of water are present on the steep sides of a volcano. Sometimes this water comes from violent electrical rainstorms, which may be triggered by the eruptions, or from the collapse of a crater lake. Some of the most destructive events have been linked to the rapid melting of snow and ice. This happens when pyroclastic flows cause incandescent lava fragments to fall over a wide area of snow and ice at the summit of the highest volcanoes. The water mixes with soft ash and volcanic boulders to produce a debris-rich fluid, sometimes at high temperatures, which then pours down the mountainside at speeds which commonly attain 15 m s^{-1} and may reach in excess of 22 m s^{-1}.

The lahar threat is prominent along the volcanic chain of the northern Andes. Andean volcanoes result from the Pacific ocean floor (Nazca plate) descending beneath the continent of South America (South American plate). There are at least twenty active volcanoes in the resulting mountain arc which extends for 1,000 km and straddles the equator from central Colombia in the north to southern Ecuador in the south (Clapperton, 1986). The highest peaks exceed 5,000 m in altitude and are permanently snow capped. Many of the mountain tops are structurally very weak due to the action of hot gases over time. Lahars have caused several historic disasters. For example, Cotopaxi in Ecuador has erupted at least fifty times since 1738. During an eruption in 1877, so much ice and snow was melted that enormous lahars, 160 km long, discharged simultaneously to the Pacific and Atlantic drainage basins.

The worst volcanic disaster in the world since the eruption of Mont Pelée occurred as a result of lahars following the 1985 eruption of the Nevado del Ruiz volcano in Colombia. Nevado del Ruiz (5,200 m) is the most northerly active volcano in the Andes. It has generated large lahars in the past, notably in 1595 and 1845, but additional settlement has taken place in the surrounding valleys over the last century. Fresh volcanic activity started in November 1984, accompanied by some glacier melting, but the main eruption did not take place until one year later. This caused large-scale, rapid melting and a huge lahar rushed down the Lagunillas valley sweeping up trees, buildings and everything else in its path (Sigurdsson and Carey, 1986). Some 50 km downstream it overwhelmed the town of Armero. Over 5,000 buildings were destroyed by a deposit of mud 3–8 m deep and almost 22,000 people lost their lives within a few minutes. Some of the survivors were trapped up to shoulder height in the ash slurry for two days before being rescued.

The widespread accumulation of volcanic ash in lowland valleys commonly results in an increased threat of river flooding and sediment redeposition, especially in countries subject to tropical cyclones or monsoon rains. At any break in the hillslope, pre-existing river courses are quickly filled by spreading ash fans deposited by lahars. Tephra offers decreased infiltration opportunities for rainfall, thus producing more sediment-rich surface flows. This, in turn, can lead to channel aggradation so that the capacity of rivers to carry flood flows is reduced and there is a greater risk of the lateral migration of rivers across floodplains (Pierson, 1989). Tayag and Punongbayan (1994) reported that the 1991 eruption of Mt Pinatubo produced 1.53×10^6 m^3 of new lahar source material capable of being washed out and transported downstream. If this ash is mobilised with older valley deposits, an estimated $1.2–3.6 \times 10^9$ m^3 of sediment could wash down on to low-lying areas around the volcano within a decade. Such sediment can create major problems if it is deposited in rice fields, irrigation systems or fish farms, all of which rely on regular supplies of fresh water.

Landslides and debris avalanches are a common feature of volcano-related ground failure. They are particularly associated with eruptions of dacitic magma, which is siliceous with a relatively high viscosity and has a large content of dissolved gas. This material can intrude into the volcano, creating so much stress that the mountain fractures along massive cracks.

This happened in the Mt St Helens eruption of May 1980. Mt St Helens is one of at least seven active subduction volcanoes in the Cascade Range of the Pacific Northwest, USA. Swarms of small earthquakes ($M = 3.0$) and minor ash eruptions gave the first signs of a major event and were followed by ground uplift on the north flank of the volcanic cone. Over a two-month period the uplift continued at a rate of about 1.5 m per day. More than one month before the main eruption the bulge was nearly 2 km in diameter and had swelled by as much as 100 m (Foxworthy and Hill, 1982). Large cracks were evident in the cover of snow and ice. All this evidence was consistent with the injection of viscous magma under the volcano at shallow depth. On 18 May, when the bulge became 150 m high, an earthquake ($M = 5.1$) shook the walls of Mt St Helens' summit crater and started many small avalanches. Then a huge slab of rock and ice on the over-steepened northern slope of the volcano broke away from the main cone along a crack across the upper part of the bulge. The earthquake triggered a debris avalanche containing 2.7 km³ of material and pressure in the shallow intrusion was further relieved by an explosive eruption and massive ash cloud. Volcanic ash blanketed much of eastern Washington State, 57 people were killed and property damage was estimated at US$1 billion.

Tsunamis are much rarer volcanic hazards than landslides. They are mainly triggered by earthquakes but can occur after catastrophic eruptions. The most quoted example is that of the island volcano of Krakatoa, between Java and Sumatra, in 1883 (VEI = 6). A series of enormous explosions, audible at a distance of almost 5,000 km, produced an ash cloud which penetrated to a height of 80 km into the atmosphere. The dust was carried round the world several times by upper level winds. Such was the force of the eruption that the volcanic cone collapsed into the caldera. The resulting tsunamis swept through the narrow Sunda Straits creating onshore waves over 30 m high in places. It has been estimated that over 36,000 people were drowned.

LOSS-SHARING ADJUSTMENTS

Disaster aid

Volcanic emergencies have attracted international aid for many years. Even so, mistakes are made through a failure to appreciate the true nature of the hazard. The Galunggung, Indonesia, eruption in 1982 eventually led to the evacuation of over 70,000 people and UNDRO was called in to help with the coordination of the international relief effort (Tomblin, 1988). One problem was that the tents initially provided for shelter proved inadequate. Some were perforated by falling cinders and others simply collapsed under the weight of ashfall. The Cameroon volcanic gas disaster of 1986 illustrated the fact that uncoordinated aid is often inappropriate. Othman-Chande (1987) noted that, within a matter of days, Cameroon received more than 22,000 blankets (five for each displaced person), 1,430 tents and more than 5,000 gas masks (supplied without some necessary components and cylinders). Large quantities of unsolicited food aid included canned sardines, instant and packet soup, cakes, war rations, mineral water and protein biscuits. One donor government sent 5,000 kg of jam; another nearly 11,000 frozen chickens. Most of these foods were unusable because of their unfamiliarity in local diets and the impossibility of storing them adequately.

One special feature of volcanic emergencies is that eruptions may not come singly, and repeated activity can continue over many months. This means that refugees need housing and care for much longer periods than the victims of other disasters. The period of eventual rehabilitation and reconstruction will inevitably be delayed. In the case of Galunggung no less than twenty-nine explosive phases occurred over a six-month period. During the first three months the evacuees were provided for on a rather makeshift basis. In the following four months the Indonesian Red Cross organised a regular feeding programme.

EVENT MODIFICATION ADJUSTMENTS

Environmental control

There is no known method of preventing volcanic eruptions. Similarly, there is no known defence against the primary threat from pyroclastic flows and comparatively little can be done to protect standing crops and exposed water supplies from air-fall tephra. Therefore, lava flows moving at comparatively slow speeds are the volcanic hazard over which most physical control can be exerted. The first known attempt to divert a lava flow took place in Sicily in 1669, when men used iron bars to try to stop an advancing flow from Etna reaching the town of Catania. A breach was opened in the flank of the flow which then began to take another direction. Unfortunately, the new course threatened a neighbouring village and the attempt was abandoned. Control of lava flows has also been attempted in Hawaii, mainly to protect the city of Hilo, which was reached by a flow in 1881 and is at risk from future events.

Bolt *et al.* (1975) identified three possible methods of diverting and controlling lava flows:

1   *Bombing*, or the use of ground explosives, can be used in three situations. First, bombing of lava high on the volcano may cause the flow to spread there and halt the advancing lava front by depriving it of supply. This method was first tried with limited success in 1935 on a fluid lava stream advancing on Hilo, although some scientists believe that more modern techniques of aerial bombing could achieve better results (Lockwood and Torgerson, 1980). Second, control of *aa* flows has been attempted by breaching the levee-like walls which form along the edges of the flow so that the lava will flood out locally and starve the advancing front of material. This method was tried on the

aa flow of Mauna Loa's 1942 eruption and in the 1983 eruption of Etna, when it proved possible to divert some 20–30 per cent of the blocky flow from its natural channel (Abersten, 1984). The third possibility, not yet tried, is to bomb the walls of the cone at the vent so that the very fluid lava there spills out over a relatively wide area and is unable to contribute to a definite stream. These methods involve an element of risk, even when the topography is suitable, and are best attempted with good atmospheric visibility, which is unusual during volcanic eruptions.

2   *Artificial barriers* can be used to divert lava streams away from valuable property if the topography is favourable. Barriers must be constructed from resistant, large-calibre material, such as massive rocks, with a broad base and gentle slopes. The method is most appropriate for thin and fluid lava flows which exert a relatively small amount of thrust. It is doubtful if diversion would work with more powerful blocky flows which may attain heights of 30 m or more.

Several diversion barriers have been proposed to protect Hilo, Hawaii (Figure 7.3). The topographic setting is favourable because flows can only approach the city through a narrow corridor, allowing intercepting barriers to be located in advance of an eruption with a high degree of confidence. The walls suggested would be around 10 m high and the channels created by the barriers would accommodate a flow about 1 km wide, which would hold the volume of lava resulting from any nearby eruptions in historic times. This method has applications elsewhere. For example, in the Krafla area of northern Iceland, the land has been bulldozed to create two barriers to protect a village and a factory respectively from the free-flowing lava.

3   *Water sprays* were first employed to control lava flows during the 1960 eruption of Kilauea, Hawaii, in a spontaneous experiment by a local fire chief. The method was used on a larger scale during the 1973 eruption of Eldfell to protect the town of Vestmannaeyjar on the Icelandic island of Heimaey. It has been estimated that 1 m^3 of water will cool about 0.7 m^3 of lava from 1100°C to 100°C when totally converted to steam. On Heimaey, special pumps were shipped to the island so that large quantities of sea water could be pumped from the harbour. At the height of the operation, the pumping rate was almost 1 m^3 s^{-1}, effectively chilling about 60,000 m^3 of advancing lava per day. The exercise was expensive, lasting for about 150 days, but it appeared to be successful. Soon after spraying started, the lava front slowed up into a solid wall some 20 m in height. Measurements of lava temperature, made in specially drilled boreholes after the eruption was over, confirmed that where water had not been applied, the lava temperature was 500–700°C at a depth of 5–8 m below the surface. In the sprayed areas an equivalent temperature was not attained until a depth of 12–16 m below the lava surface (UNDRO, 1985).

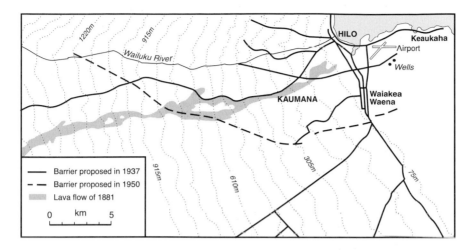

Figure 7.3 A map to show the location of barriers proposed to divert lava flows from the town of Hilo, Hawaii, after the settlement was threatened by a large volcanic eruption in 1881.
Source: After Bolt *et al.* (1975)

Some protection against lahars can be achieved by the construction of artificial diversion barriers similar to those for lava flows. But this is only possible where previous paths are well defined and the method would not work for the major, destructive flows which are confined to deep valleys. Some Indonesian villages are now provided with artificial mounds so that people can quickly climb to a higher, safer level while a lahar passes. The effectiveness of such a scheme relies on adequate warning of an approaching mudflow.

Siltation problems caused by the deposition of river-borne volcanic sediments can be alleviated by trapping the material behind check dams, although such storages are expensive and have a limited life. For example, it has been estimated that around US\$350 million was spent after 1980 to counter channel aggradation and downstream flooding near Mt St Helens by a combination of channel dredging, dyking and the construction of sediment-retention basins. Proposals for the large-scale diversion of lahars into wetland areas used for seasonal floodwater storage and fishing in the Phillipines have proved highly controversial (Tayag and Punongbayan, 1994).

A remarkable attempt to stop lahars at source has been undertaken on Kelut volcano on Java. In 1919 an explosive eruption threw some 38.5×10^6 m^3 of water out of the crater lake and created a lahar which claimed about 5,000 lives. To avoid a repetition of this disaster, engineers made a series of siphon tunnels which enabled the volume of stored water to be reduced from about 65×10^6 to 3×10^6 m^3 (Figure 7.4). This was so successful that a similar event in 1951 resulted in no large lahars and a minimal loss of life. Unfortunately, this eruption destroyed the tunnel

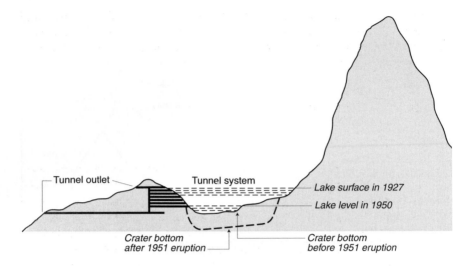

Figure 7.4   Diagrammatic section showing the tunnel system constructed at Kelut volcano, Java, to lower the water level in the crater lake and reduce the threat of lahars.
Source: From Bolt *et al.* (1975)

entrances and added to the water storage capacity by deepening the crater by some 10 m. Even with repair of the original lowest tunnel, the lake soon accumulated a volume of 40×10^6 m^3, which was considered a serious lahar threat. The Indonesian government then financed an even lower tunnel but stopped it short of the crater wall in the hope that seepage would help to drain the lake. This did not happen because of the low permeability of the volcanic cone. Further lahars in the 1966 eruption, associated with the expulsion of 20×10^6 m^3 of water killed hundreds of people and damaged much agricultural land. Since then, the lake volume has been reduced to about 4.2×10^6 m^3 of water (Suryo and Clarke, 1985).

Hazard-resistant design

One problem associated with heavy ashfalls is the potential collapse of unstrengthened buildings, especially those with a flat roof. This is much more likely if the ash is wet. Thus, the bulk density of dry ash ranges from 0.5 to 0.7 t m^{-3}, whereas that of wet ash may reach 1.0 t m^{-3}. Under these loads, many types of roof can suffer structural failure. In this context, the wide variations in roof design between those high-latitude areas of the world, subject to heavy snowfalls as well as volcanic activity (such as Iceland or Japan) and tropical countries is highly relevant. After the 1991 eruption of Mt Pinatubo, ashfalls accumulated to a depth of 8–10 cm in Angeles City, with a population of 280,000 about 25 km from the volcano, resulting in the collapse of 5–10 per cent of the roofs. The only way to strengthen buildings against ashfall is to make an inventory of building design and type with a view to retrofitting existing structures and building new ones to higher standards. Priority should always be given to the construction of strong, steeply pitched roofs.

VULNERABILITY MODIFICATION ADJUSTMENTS

Community preparedness

As with earthquakes, the cost of monitoring volcanic activity and pre-disaster planning is small compared to the potential losses. Given the existence of a monitoring programme and effective preparation, some warning can usually be given to permit evacuation of the most dangerous areas before the eruption occurs. Until recently, emergency planning in volcanic hazard zones was not well developed in the LDCs. Before the Nevado del Ruiz disaster in Colombia in 1985, there was no national policy in place for the systematic monitoring of volcanic hazards or for the management of such hazards. According to Voight (1996), policy failures were compounded by delays in hazard mapping and the unwillingness of the authorities to accept the economic and political costs of early evacuation.

The ideal sequential elements for volcanic preparedness are shown in Figure 7.5. The length of time available for the alert phase differs widely. In some cases volcanic activity may start to increase months before a violent eruption; in other events only a few hours may be available. For effective evacuation, it is essential that the population at risk is clearly informed well in advance about the evacuation routes and the refuge points to which they should go. To some extent these directions will have to be flexible depending on factors such as the expected scale of the eruption, which might influence the pattern of lava flow, and the wind direction at the time, which will influence the pattern of ashfall. Some local roads may be destroyed by earthquake-induced ground failures. Steep sections of highway can become impassable with even small deposits of fine ash, which make asphalt very slippery.

The evacuation of densely populated areas creates special problems of transportation, including the peak capacity of road networks and the balance of public and private vehicles available. During the 1991 eruption of Mt Pinatubo, the total number of evacuees extended to well over 200,000, about three times more people than previously evacuated in any volcanic emergency. In some cases, off-shore evacuation may be necessary for small volcanic islands and coastal communities at risk. Rabaul township lies within a volcanic caldera on the coast of New Britain island, Papua New Guinea, and about 70,000 people live within 15 km of the centre of the caldera. McKee *et al.* (1985) concluded that the existing road network, extending no more than 50 km from the caldera, could not guarantee safe landward

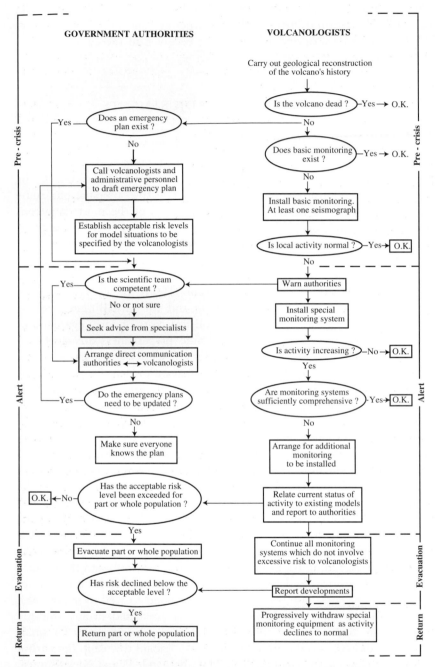

Figure 7.5 A flow chart to illustrate the requirements of a volcanic emergency plan. Close liaison between volcanologists and government authorities is necessary at all stages to ensure an effective response to disaster. Source: After UNDRO (1985)

evacuation during all eruptions whilst seaborne evacuation was limited by the absence of suitable wharf facilities for large ships. In small-to-moderate eruptions, it was suggested that the best option might be to shelter the population in an extensive system of tunnels which were excavated in the tephra deposits around Rabaul during the Japanese occupation of the area during the Second World War.

After evacuees have reached the refuge points, they require support services. These include medical treatment (especially for dust-aggravated respiratory problems and burns), shelter, food and hygiene. Volcanic emergencies may last for many months as eruptions are repeated. This implies that the 'temporary' arrangements planned for refugees may have to function for some time, perhaps from one crop season to another. Another special feature of volcanic eruptions is that, depending on the prevailing wind conditions, ashfall has the potential to disrupt communities several hundreds of kilometres away. This was illustrated for the Mt St Helens event by Warrick *et al.* (1981) from a study of four communities up to 700 km distant. Few residents at these sites claimed any prior knowledge of ashfall hazard and warning messages were largely ineffectual in promoting a response. Hazard mitigation specialists clearly have a difficult task in persuading such distant communities that they face a volcanic risk.

Increasing efforts are now made to encourage the local population in seismically active areas to become more involved with disaster preparedness. Evidence from the western USA suggests that, whilst residents do make responses to such long-term threats, there is little prioritisation of the adjustments (Perry and Lindell, 1990). In the Philippines the Institute of Volcanology and Seismology has adopted a programme whereby residents are given a training course and then encouraged to look out for possible precursory signs of volcanic activity, such as crater glow, steam releases, sulphurous odour and dying vegetation (Reyes, 1992). In Ecuador about 3 million people live within the two main volcanic mountain ranges and are at some degree of risk from lahars. The principal threat is in the Chillos and Latacunga valleys where an ever-growing population of some 30,000 has settled on lahar deposits from the 1877 eruption (Mothes, 1992). Again, public education programmes, including field trips and evacuation exercises involving 5,000 people in a simulated eruption scenario, have been used to raise awareness and encourage better precautionary attitudes.

Forecasting and warning

Major volcanic eruptions do not occur spontaneously. They are preceded by a variety of environmental changes which accompany the rise of magma towards the surface. The monitoring of these changes provides the best hope of developing reliable forecasting and warning systems (Decker, 1986). However, according to Scarpa and Gasparini (1996a), only twenty volcanoes worldwide are monitored by well-equipped local observatories whilst a further 150 have limited instrumentation, mainly seismometers. UNDRO (1985) classified the various unusual physical and chemical phenomena that have been observed to occur before eruptions (Table 7.2). Unfortunately, such phenomena do not always appear and the highly explosive volcanic eruptions are generally the most difficult to forecast. The most reliable monitoring techniques are seismic and ground deformation measurements, although lahar monitoring by automatic raingauges and flow sensors on the upper slopes of volcanoes can provide some warning of this hazard. The most important precursors of eruption are:

1   *earthquake activity*, which is commonplace near volcanoes, although it is not fully understood whether earthquakes trigger eruptions or vice versa. For predictive purposes, it is important to gauge any increase in activity in relation to local background levels. This means that it is essential to have good seismograph records, preferably over many years, for the volcano in question. Immediately prior to an expected eruption these records will be supplemented by data from portable seismometers. There is some evidence for a precursive seismic signature which has been incorporated into a tentative earthquake swarm model for the prediction of volcanic eruptions (McNutt, 1996). As shown in Figure 7.6, the onset and subsequent peak of a 'swarm' of high-frequency earthquakes reflects the fracture of local rocks as the magmatic pressure increases. This phase is followed by a relatively quiet period, when some of this pressure is relieved by cracking in the earth's crust, before a final tremor results in an explosive eruption.

Table 7.2 Precursory phenomena that may be observed before a volcanic eruption

---

Seismic activity
Increase in local earthquake activity
Audible rumblings

Ground deformation
Swelling or uplifting of the volcanic edifice
Changes in ground slope near the volcano

Hydrothermal phenomena
Increased discharge from hot springs
Increased discharge of steam from fumaroles
Rise in temperature of hot springs or fumarole steam emissions
Rise in temperature of crater lakes
Melting of snow or ice on the volcano
Withering of vegetation on the slopes of the volcano

Chemical changes
Changes in the chemical composition of gas discharges from
    surface vents (e.g. increase in SO_2 or H_2S content)

Source: After UNDRO (1985)

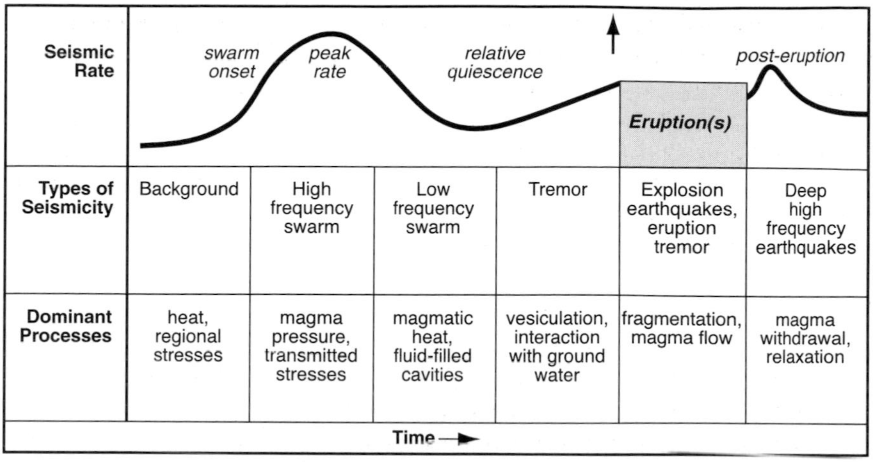

Seismic Rate	swarm onset	peak rate		relative quiescence		↑		post-eruption

Types of Seismicity	Background	High frequency swarm	Low frequency swarm	Tremor	Explosion earthquakes, eruption tremor	Deep high frequency earthquakes
Dominant Processes	heat, regional stresses	magma pressure, transmitted stresses	magmatic heat, fluid-filled cavities	vesiculation, interaction with ground water	fragmentation, magma flow	magma withdrawal, relaxation

Time ➤

Figure 7.6   A schematic diagram of the stages of a volcanic earthquake-swarm model. The observed earthquake swarms have an average duration of five days and may provide a future basis for the forecasting of eruptions. Source: After McNutt (1996)

2   *ground deformation*, which is sometimes a reliable precursive sign of an explosive eruption as magma moves towards the surface, but the relationships are complex and not easy to fit into a forecasting model. The method is also difficult to employ for the explosive subduction volcanoes because they erupt so infrequently that it is difficult to obtain sufficient comparative information. In rare cases, such as the 1980 event at Mt St Helens, the deformation is sufficiently large to be easily visible but it is usually necessary to detect movements with standard survey equipment or the use of tiltmeters. These instruments are very sensitive but can only record changes in slopes over short distances. The use of electronic distance measurement (EDM) techniques provides a more accurate picture of relative ground displacement, although it is less usually available and requires a series of visible targets on the volcano. Global positioning system (GPS) measurements, obtained from satellites, are now also available to reveal the surface displacement of volcanoes.

3   *thermal monitoring*: as magma rises to the surface, it might be expected to produce an increase in temperature. But many volcanoes have erupted without any detectable thermal change. The temperature of hot springs and steam emissions can be fairly easily monitored but it provides only an indirect picture of what is happening beneath the surface. Also, any small rise in surface temperature associated with a greater geothermal heat flux can be obscured by rainfall. There is also a problem of thermal inertia when heat conduction may be too slow for forecasting purposes.

Where a crater lake exists, thermal changes have been meaningful. UNDRO (1985) cited the example of the crater lake on Taal volcano, in the Philippines, which increased in temperature from a constant 33°C in June 1965 to reach 45°C by the end of July. The water level also rose during this period and, in September 1965, a violent eruption occurred. Such observations can increasingly be supplemented by thermal imaging from satellites (Rothery *et al.*, 1988). Heat emission was one of the first volcanic features to be sensed remotely and it has proved a valuable means of hazard assessment.

4   *geochemical monitoring*: any predictive interpretation of the chemical composition of the juvenile gases issuing from volcanic vents is a difficult task. Gas samples taken only a short time, or distance, apart often show considerable variation. It is, therefore, not usually possible to know how representative any changes in composition might be of more general conditions in the volcano. Visual observations of steam emissions or ash clouds depend on meteorological conditions, as well as volcanic activity, but volcanic plumes can be monitored by AVHRRs carried on weather satellites (Malingreau and Kasawanda, 1986).

At present there is no fully reliable forecasting scheme available for volcanic eruptions, although some success has been achieved. For example, a high-confidence forecast of the 1991 Mt Pinatubo eruption allowed the evacuation of people from an area that, at maximum, covered a 40 km radius. By the time the 1996–7 eruptions on the island of Montserrat had destroyed the main town of Plymouth, all the residents had been evacuated to the safer, northern part of the island. But uncertainty often leads to practical problems; this was well illustrated by the events at La Soufrière, on the island of Guadaloupe, Lesser Antilles, in 1976. Abnormal seismic activity over a twelve-month period eventually led to the evacuation of 72,000 people from around the volcano. This evacuation of around one-fifth of the population was one of the largest, and most costly, ever undertaken for a volcanic emergency and lasted for over three months. When the expected eruption failed to occur, there was considerable criticism of the scientific community, who had been divided in their original assessments of the risk.

The managerial problems of responding to an uncertain volcanic hazard prediction have also been apparent in the Cascade Range in the Pacific Northwest of the USA. Following steam discharges at least ten times above the normal level from Mt Baker, Washington, during March 1975, the US Geological Survey foresaw the possibility of a destructive mudflow or avalanche. The US Forest Service closed public access to the Baker Lake Recreation Area in June 1975 (Hodge *et al.*, 1979). The restrictions remained in force for nearly one year, during which time no hazardous event occurred. Subsequent surveys of both residents and recreationalists showed a widespread belief that the authorities had over-reacted and over 70 per cent of residents claimed that they would ignore any future hazard warnings and respond to mandatory controls only. In comparison, the 1980 eruption of Mt St Helens was more accurately forecast but still took the authorities partly by surprise. This was because the main explosion was not immediately preceded by any specially abnormal phenomena and the explosive blast was directed laterally rather than vertically. As a result 57 people who had been allowed to enter the danger area were killed. Although the surrounding area was sparsely populated, it has been estimated that perhaps as many as 1,000 lives might have been lost if free access had been allowed to residents and tourists.

Land use planning

Land planning has a role to play in reducing volcanic disasters, both in terms of restricting development in hazardous areas and in the preparation of emergency evacuation plans. Its absence in the past has created some alarming cases of increased threat. For example, the volcanic complex west of Naples, Italy, has become one of the most densely populated areas of active volcanism in the world following the last major eruption in 1538. All previous eruptions of this volcano have been explosive and it has been estimated that 200,000 people are currently at risk and would require evacuation in the event of an emergency (Barberi and Carapezza, 1996).

Land use zoning and the selection of safe sites depends on long-range predictions of the probability of volcanic activity and the identification of areas at risk. Infrequent eruptions in the past require accurate geological-scale dating techniques, such as radiocarbon dating, tree-ring analysis, lichenometry and thermo-luminescence. Volcanic-hazard maps can then be prepared which show the possible areal extent of volcanic phenomena in the future inferred from the evidence of past events. The major limitations of such mapping are created by lack of knowledge of the size of future eruptions which, for example, make it difficult to assess the extent of a pyroclastic surge or the length of travel of a valley lahar. The problem is most severe in those LDCs which do not have high-quality topographic maps, well-constructed geological records or a pool of trained earth scientists. Environmental conditions at the time of eruption will also be important. For example, the degree of seasonal snow cover will affect the lahar and avalanche hazard whilst the speed and direction of the wind will determine the airborne spread of tephra.

Despite these limitations, the effectiveness of volcanic hazard mapping can be demonstrated with reference to Mt St Helens. This is the youngest volcano in the Cascade Range and, because of its long history of spasmodic explosive activity, was widely regarded as the most dangerous in the western USA before the 1980 eruption. Consequently, Crandell *et al.* (1979) had mapped the potential hazard in the surrounding area (Figure 7.7a). It can be seen that pyroclastic flows were expected to extend over broad areas for

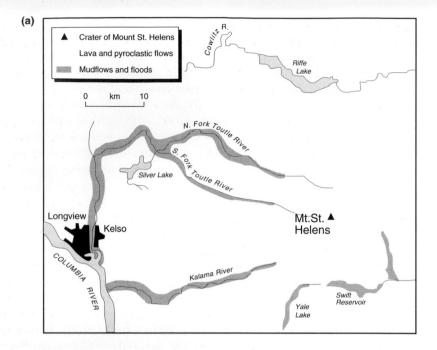

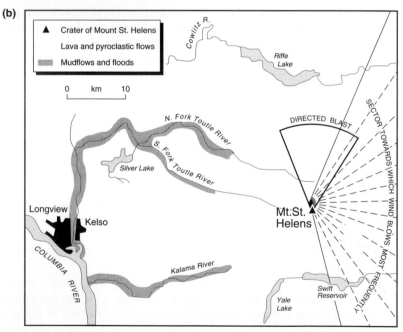

Figure 7.7 (a) Potential volcanic hazards in the area around Mt St Helens, USA, as mapped before the 1980 eruption; (b) The modified hazard zones as drawn after the 18 May eruptions showing, in particular, the area potentially at risk from future directed (lateral) blasts.
Sources: (a) after Crandell *et al* (1979); (b) after Miller *et al.* (1981)

at least 5 km and to flow down the upper valleys for up to 15 km. Mudflows and floods, which follow topographically better-defined paths, were expected to continue down the valleys for many tens of kilometres. Indeed, it was predicted that lahars up to 110×10^6 m^3 in volume could reach the local reservoirs and create additional flooding if storage was not reduced in advance of the wave. Tephra deposits were expected to occur over a broad 155° sector extending away from the volcano from about north-northeast to south-southeast based on the wind direction which prevails for about 80 per cent of the time. Some ashfall was anticipated at a distance of 200 km, well within the range of the town of Yakima. Probability estimates suggested that Yakima had a 1 in 10,000 risk in any one year of being affected by ashfall from Mt St Helens and that the thickness of tephra expected in the town from such an eruption could be slightly more than 100 mm.

Figure 7.7b summarises the actual pattern of volcanic hazards in the local area from the eruption of May 1980. Major mudflows, laden with logs and forest debris, were channelled down the valleys. A flood surge entered the upper Swift reservoir and caused an initial water level rise of almost 2 m. Since the water level had been lowered previously, the added volume did not overtop the dam and flooding anticipated along downstream parts of the Lewis river was avoided. The North Fork Toutle river mudflow rose to high levels and the Cowlitz river flooded low-lying areas near Castle Rock but stayed within its banks further downstream. On the other hand, noticeable ashfalls occurred as far east as Nebraska and the Dakotas, while at Yakima the depth of tephra was around 250 mm. The scenario in the hazard map was generally accurate apart from the actual magnitude of the landslides and the severity of the lateral blast.

8 Mass movement hazards

LANDSLIDE AND AVALANCHE HAZARDS

The downslope movement of large volumes of surface materials under gravitational influences is an important environmental hazard, especially in mountainous terrain. Rapid movements cause most loss of life and damage; slow movements, including human-induced land subsidence, have less potential to kill but can be costly. Depending on the dominant material, these movements tend to be grouped into *landslides* (rock and soil) or *avalanches* (snow and ice). Mass movements may be triggered by either seismic activity or atmospheric events. To that extent, this hazard lies at the interface between endogenous and exogenous earth processes. Some of the largest recorded events have been earthquake-related; these are treated in Chapter 6.

There can be few countries where mass movement processes do not exist, and the landslide risk is increasing worldwide as land hunger forces new development on to unstable slopes. According to Jones (1992), it is an under-recognised threat because the impacts tend to be frequent and small-scale, whilst the process itself is often attributed to other hazards, such as earthquakes and rainstorms. This view was echoed by M.F. Thomas (1994) who drew attention to the neglect of landsliding in tropical and sub-tropical environments. Mass movements also add considerably to the wide range of hazards found in mountainous areas throughout the world (Hewitt, 1992).

During the early 1970s, an average of nearly 600 people per year were killed by slope failures worldwide but, twenty years later, the figure was several thousand (Brabb, 1991). Perhaps as many as 90 per cent of these deaths occur on the Pacific ocean rim which is particularly susceptible to mass movements because of the varying combinations of rock type, steep terrain, heavy typhoon rainfall, rapid land use change and high population density. The main cause of increased deaths has been the expansion of unregulated *barrio* settlements on to unstable slopes in many Third World cities. For example, in Caracas, Venezuela, the number of urban landslides increased from less than one per year up to about 1950 to reach 35–40 per year in the 1980s (Jiminez, 1992). The death toll from mass movements is still comparatively low in most MDCs. In the USA annual mortality runs at

25–50 people and it has been estimated that, for landslides alone, some 22 per cent of the population are exposed to high hazard conditions while another 20 per cent are exposed to moderate hazard conditions (Petak, 1984). As with many other environmental hazards, it is urban areas which are most vulnerable because of the large populations at risk (Alexander, 1989).

Economic losses due to landslides total more than US$1 billion per year in several countries, including Japan. Ground failure associated with landslides, subsidence, swelling clays and construction-induced rock deformation causes property losses in every US state at an annual cost of US$1–3 billion. The losses are concentrated in the Appalachian, Rocky Mountain and Pacific Coast regions. In each year of heavy storms, it is estimated that landslide damage averages US$500 million in the Los Angeles area alone. Other countries with large but unquantified losses include Indonesia, China and the former Soviet Union (Schuster and Fleming, 1986). In Italy it has been estimated that over 1,000 urban centres are threatened by landslide activity. In addition to direct damage, mass movement hazards cause a variety of indirect losses such as road blockages, flooding due to landslide dams across rivers, reduced agricultural and industrial production, and often lead to lower property values.

Snow avalanches are a special type of mass movement. They are common features of mountainous terrain throughout arctic and temperate regions whenever snow is deposited on slopes steeper than about 20°. The USA alone suffers 7–10,000 potentially damaging avalanches per year, although only about 1 per cent harm humans or property. In the past, most casualties were suffered either by travellers passing through the mountains or by miners located in permanent, but badly sited, mining settlements. The Andean countries are notable for avalanche-related mining disasters. The worst avalanche disaster in the USA occurred in 1910 in the Cascade Range, Washington, when three snowbound trains were swept into a canyon with the loss of 118 lives. Historically, the avalanche problem has always been more severe in Europe than North America because the population density is higher in the Alps than in the Rockies. Switzerland has a relatively large number of avalanche deaths amounting to some 20–30 fatalities per year.

Snow avalanche problems have risen in recent decades. This is mainly due to the greater use of alpine areas for winter recreation and the associated development of ski centres and other holiday resorts. For example, the town of Vail, Colorado, located at an elevation of 2,500 m, was founded as a resort community only in 1962. The construction of alpine facilities often requires the removal of timber from the surrounding slopes. If left intact, the trees would help to stabilise the snow cover and protect the new roads, railways and power lines which are invading these areas. Avalanche problems in the Rockies beset the Canadian Pacific Railway and the Trans-Canada Highway together with sections of US Highway 2. The Trans-Canada

Highway alone crosses nearly 100 avalanche tracks in the 145 km section between Golden and Revelstoke in the vicinity of Rogers Pass and it has been estimated that at least one motor vehicle is under a major avalanche path at any given time.

Armstrong (1984) showed that annual avalanche deaths had been on the increase in the United States since the early 1950s (Figure 8.1a) but there is some evidence that, in the winter sports areas of North America and Europe, the annual loss of life may have reached a peak in the mid-1980s. Data for Canada indicate that, although the number of reported avalanche incidents may be still increasing, the number of fatalities shows little upward trend (Figure 8.1b). This pattern has been attributed to the improved

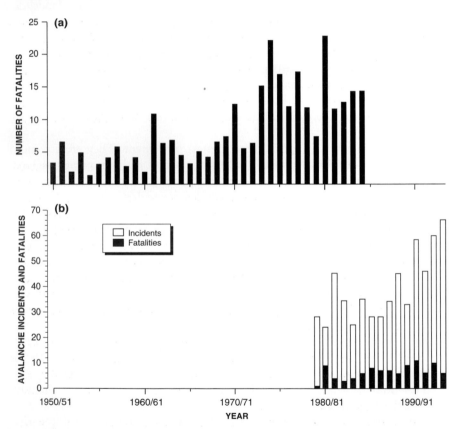

Figure 8.1   (a) Avalanche-related deaths from the 1950s to the 1980s in the USA. Fatalities increased with mountain development, mainly in tourist towns and transport corridors; (b) The number of reported avalanche incidents and fatalities for the winters 1979–80 to 1993–4 in Canada. This diagram suggests that the rise in reported incidents has not been matched by the number of deaths, possibly due to greater hazard awareness.
Sources: (a) after Armstrong (1984); (b) data supplied by the Canadian Avalanche Centre

reporting of incidents combined with better education and risk prevention. About 70 per cent of all avalanche fatalities in the MDCs are associated with the voluntary risk exposure activities of ski touring and mountain climbing. Much less is known about avalanche hazards in the mountainous areas of the LDCs, such as the Himalayas. In the Kaghan valley, Pakistan, large avalanches pose a real threat to the local residents forced to overwinter on the valley floor, and twenty-nine people were killed in a single event in the winter of 1991–2 (de Scally and Gardner, 1994).

LANDSLIDES

The term *landslide* covers most downslope movements of rock and soil debris that have become separated from the underlying part of the slope by a shear zone or slip surface. The type of movement, which may include falling, sliding and flowing, depends largely on the nature of the geologic environment, including material strength, slope configuration and pore water pressure. Jones (1995) asserted that slope failure will become an increasingly important hazard, especially in the LDCs, and identified several types of landslide terrain:

1   *Areas subject to seismic shaking* Earthquakes can promote widespread landsliding, which often occurs in thousands of individual slides, as in the 1950 Assam, India, earthquake when over 50 billion m³ of material was displaced over an area of 15,000 km². Major landslides also occurred after the 1988 Armenian and 1990 Iranian earthquakes.
2   *Mountainous environments with high relative-relief* High energy terrain, such as the Himalayan or Andean mountain chains, produces perhaps one catastrophic rock fall per decade worldwide. These spectacular slope failures comprise huge masses of material (up to 100×10^6 m³) which, at least in the initial stages, travel near-vertically at high velocities over long run-out distances.
3   *Areas of moderate relief suffering severe land degradation* Readily erodible soils on slopes subject to land degradation caused by deforestation or overgrazing have the potential for gully expansion and land slipping. Over the centuries, about 100 villages in southern Italy have been abandoned because of this process.
4   *Areas covered with thick sheets of loess* Any mantling of an existing ground surface with finely grained deposits, such as wind-blown loess or tephra, is likely to lead to a shear zone at the junction of the two materials and the formation of flow slides in the loose deposits. The loess plateau of central China is a classic location, and Sassa (1992) described the Saleshan landslide of 1983 which killed at least 200 people.
5   *Areas with high rainfall inputs* In areas which regularly experience rainfall from monsoons or tropical cyclones, rock weathering can penetrate

tens of metres below the ground surface. For example, in parts of Hong Kong weathered material has moved downslope to cover the bedrock to a depth of more than 20 m. Throughout the humid tropics, these deep and porous mantles are prone to landslides.

The classification of landslides by Varnes (1978), summarised in Table 8.1, is one of the most widely accepted. Some of the major movements are considered below.

Rock falls

These are movements of debris (mainly rock) transported through the air. They are the simplest type of rock movements and occur on steep faces where bedrock weaknesses, such as joints, bedding and exfoliation surfaces, are present. Rock falls are presumed to fall directly off cliff faces, rather than to slip along a joint or bedding plane, although both types of movement may occur. The presence of water in clefts and fissures is highly influential, especially in the mid-latitudes where regular freeze–thaw cycles progressively weaken the rock mass by increasing such openings.

Many of the largest rock falls are induced by earthquakes but more spontaneous slope instability also occurs, especially in closely jointed or weakly cemented materials on slopes steeper than about 40°. The greatest rock fall hazard exists when joints and bedding planes are inclined at a steep angle, as in the

Table 8.1 Classification of landslides

Type of movement	Type of material		
	Bedrock	Engineering soils	
		Predom. coarse	Predom. fine
Falls	Rock fall	Debris fall	Earth fall
Topples	Rock topple	Debris topple	Earth topple
Slides			
(a) rotational	Rock slump	Debris slump	Earth slump
(few units)	Rock block slide	Debris block slide	Earth block slide
(b) translational	Rock slide	Debris slide	Earth slide
Lateral spreads	Rock spread	Debris spread	Earth spread
Flows	Rock flow (deep creep)	Debris flow (soil creep)	Earth flow
Complex	Combination of two or more principal types of movement		

Source: After Varnes (1978)

highly folded rocks common in major mountain chains like the Himalayas, Andes and Rockies. The Frank rockslide, which occurred in Alberta, Canada, in 1903, was a classic example (Cruden and Krahn, 1978). In this case the slide took place across bedding planes in a steep anticline formed in the well-jointed limestone of Turtle Mountain, which was also subject to mining activity. Groundwater seeping into the joints dissolved the limestone and enlarged the fractures. During the winter this water froze and wedged the rock apart, further weakening the structure. The resulting debris destroyed the southern end of the small town of Frank, killing about seventy people. The present town has since been relocated about 2 km north of the original site.

Landslides

Landslides are downslope movements of rock and soil along slip surfaces. They are associated with a disturbance of the equilibrium which normally exists between stress and strength in material resting on slopes. The relationship between stress and strength is determined by factors such as the height and steepness of the slope and the density, strength cohesion and friction of the materials on the slope. Hillslope instability occurs when the strength of the material comprising the slope is exceeded by a downslope stress. The *shear strength* of the material is the maximum resistance to *shear stress* and depends on:

1   *internal cohesion*: this is produced by the interlocking, or sticking together, of granular particles, particularly in clayey soils and rocks, that enables the material to rest at an angle. Some materials, such as dry sand, are cohesionless. Cohesion is independent of the weight of material above the surface;
2   *internal friction*: this is the resistance of particles of granular soil to sliding across each other. The friction component of shear strength depends on the weight of material above the surface.

In turn, these factors will depend on the weight, or loading, on the slope and the moisture conditions. The basic mechanics are illustrated in Figure 8.2 which shows a sliding block on an inclined plane. The weight of material in the block (w), resolved at an angle (a) parallel to the slope, creates a driving force (d) or shear stress. Sliding is resisted by the shear strength (s), which is derived from the cohesion of clay-rich materials and the static friction between the block and the slide plane, which increases as the normal force (n) increases. The block will remain in place as long as the driving force does not exceed this combined shear strength. If the slope becomes steeper, the shear stress exerted on potential slip surfaces increases because the downslope component of gravity increases. If these stresses eventually exceed the shear strength along a 'critical' slip surface, the mass above the surface will move downslope.

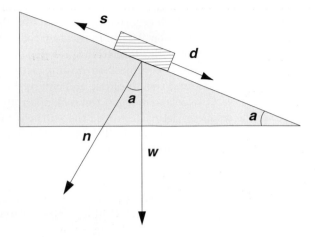

Figure 8.2   The simplified relationship between the driving forces and the resisting forces on a natural slope demonstrated by means of a sliding block on an inclined friction plane.

There are two main types of landslide:

1   *Rotational slides* have curved slip surfaces. The result is a pattern of scars and depositional features, of which the most common is the spoon-shaped scar associated with shear failures along arcuate planes. Figure 8.3 shows the main features of such a rotational slip. The slipped material will be deposited on the slope in either a hummocky or a lobate form depending on the water content. This type of slope movement can cause a great deal of property damage if the slope has been built upon but warnings can often be given for evacuation. A good example is the landslide disaster which affected the city of Ancona, central Italy, in 1982 (Alexander, 1987b). The major short-term cause was an increase in pore-water pressure associated with heavy rainfall in the previous month. Apart from losses to road and rail communications, 280 dwellings were damaged. Although the slide occurred in an area of known slope instability, no formal warning was given on this occasion. Nevertheless, about 4,000 people spontaneously evacuated their houses.

2   *Translational slides* have relatively uniform, planar surfaces of movement and are sometimes known as *block glides* and *debris slides*. Intersecting planar slip surfaces form wedges of rock, which are a relatively common type of translational slide. The destructive landslide that hit the Vaiont dam in the Piave valley, northern Italy, in October 1963 was a combined natural and technological disaster. The Vaiont dam, over 260 m high, the key construction in a chain of hydroelectric dams, was completed

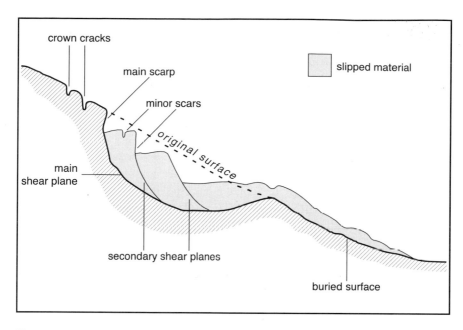

Figure 8.3   The characteristic rotational scar caused by a landslip. Surface changes, such as the prior opening of crown cracks, can often be used as warning signs and permit evacuation to take place.

in 1960 across a narrow canyon set within a broader valley. The slopes of the upper valley were underlain by layered sedimentary rocks dipping towards the valley floor and traversed by fractures parallel to the slope. The sedimentary sequence included limestones with inter-bedded clays which became significantly weaker when wet. As the water stored in the reservoir backed up behind the dam, it entered pre-existing fractures and wetted the vulnerable layers. Following heavy rain and a minor earth tremor, the southern slope of the valley failed. A large volume of material collapsed into the reservoir and displaced some 200×10^6 m^3 of water, mud, rock and timber, which fell almost vertically over the dam into the valley below. Although the dam remained intact, several villages were destroyed with the loss of 1,189 lives.

Debris flows

These are downslope movements of fluidised soil and other material acting as a viscous mass. This occurs when loose slope materials become saturated, resulting in a loss of cohesion and internal friction between the granular particles, so that an unstable slurry mixture is produced. Debris flows tend to be less deep-seated slope failures than landslides. They are a major feature in the tropics, where they are triggered by either the prolonged rainfall associated with slow-moving low pressure troughs or the more intense rainfalls, sometimes exceeding 100 mm h^{-1}, created by tropical cyclones (Thomas, 1994). The high water content means that the slope material moves faster and further from the original source than with landslides. Although the course of debris flows is often guided by stream channels, and to that extent is predictable, the speed and range of movement of these events mean that they tend to claim more lives than landslides. Because of their high density, up to 1.5 to 2.0 times the density of water, debris flows have great destructive force and can remove large boulders and houses from their path.

Several tropical cities, such as Rio de Janeiro and Hong Kong, are at risk from both landslides and debris flows. In Hong Kong, for example, almost all the fatal landslips have been associated with property development involving earth cuts, fills and retaining walls during major rainstorms in the summer months. Jones (1973) documented the effects of exceptionally heavy rainfall, often linked to stationary cold fronts, around Rio de Janeiro, Brazil. In 1966, landslides produced in excess of 300,000 m^3 of debris in the streets of Rio and more than 1,000 people died when many slopes, over-steepened for building construction, failed. One year later, further storms hit Brazil and mudflows caused a further 1,700 deaths and the disruption of the power supply to Rio. In February 1988 more debris flows in Rio de Janeiro claimed at least 200 lives and made 20,000 people homeless (Smith and de Sanchez, 1992). Most of the victims were living in unplanned squatter settlements erected on deforested hillsides. Such hazards are not confined to the tropics and sub-tropics. During a three-day period in January 1982, some 500–600 mm of rain fell in the San Francisco Bay area and debris flows killed twenty-six people, ten of whom died in a block glide incident in Santa Cruz County (Cotton and Cochrane, 1982). In most cases the flows followed stream courses and began as flow-slides before turning into fluidised masses that were able to attain velocities of 10 m s^{-1}. The flows occurred at night and most victims died in their homes, indicating the need for effective evacuation as well as better planning controls.

Causes of landslides

Landslides result from a variety of events that combine either to increase the driving force or to reduce the shear resistance on a slope. Factors that increase the *driving forces* on a slope may be either physical or human and include:

- *an increase in slope angle* which may occur if a stream erodes the bottom of a slope or if the slope is steepened by building work. Jones *et al.* (1989) have described how the cutting of a road into the base of a slope during 1984, which left exposed faces 25 m high and colluvium standing

at an angle of 55° unsupported by anything other than a 3 m masonry
wall, led to the Catak landslide disaster in Turkey in which sixty-six
people died in 1988;

- *removal of lateral support* at the foot of a slope, again caused either by
  natural mass wasting processes or by building activity;
- *additional weight* placed on the slope by the dumping of waste or house
  construction. Residential development not only adds weight to the slope
  through the buildings themselves but also through excess water supplied
  from landscape irrigation and seepage from swimming pools and sewage
  effluent systems;
- *removal of vegetation* either by wildfires or through human activities such
  as logging, overgrazing or construction. Surface materials become looser
  because of the loss of soil binding by roots and the slope is also more
  exposed to the erosive action of surface water through the loss of plant
  cover;
- *local shocks and vibrations* which can occur naturally from seismic activity
  or from the operation of heavy construction machinery.

Factors that lead to a reduction in the *shear resistance* on a slope are:

- *an increase in pore water pressure* in the slope materials, especially along
  a slip surface. This is the most important single factor and explains the
  close relationship which exists between shallow-seated landslides, debris
  flows, and rainstorms. Unfortunately, the detailed interaction of rainwater and soil behaviour is not fully understood and it remains difficult
  to predict landslides on a site-specific basis. In unsaturated material
  that is not totally dry, the internal voids or pores will be filled with
  gas (air and water vapour) and some liquid water. If the slope then
  experiences additional loading, perhaps as a result of building construction, the mineral grains will be able to slide into a more compact
  arrangement. Such compression increases the soil density and additional
  strength will result. However, if there is resistance to a denser configuration due to water in the void space, and rapid surface loading occurs
  relative to the permeability of the soil, then the additional load is transferred into the pore water causing an increase in the pore-water pressure.
  In turn, this reduces the friction component of strength and downslope movement may occur;
- *an increase in slope angle*, which often occurs when developed slopes are
  over-steepened by cutting into the base, a process which increases the
  driving force. Figure 8.4a depicts the initial development of a slope
  with a safe angle of 3:1. In Figure 8.4b the same slope has been locally
  over-steepened to a 1:1 angle by a commercial developer seeking
  to raise the density of building lots, which have been increased in
  number from two to five (Figure 8.4c). After landslipping on the slope
  (Figure 8.4d) the area has to be re-engineered to a safer 2:1 overall

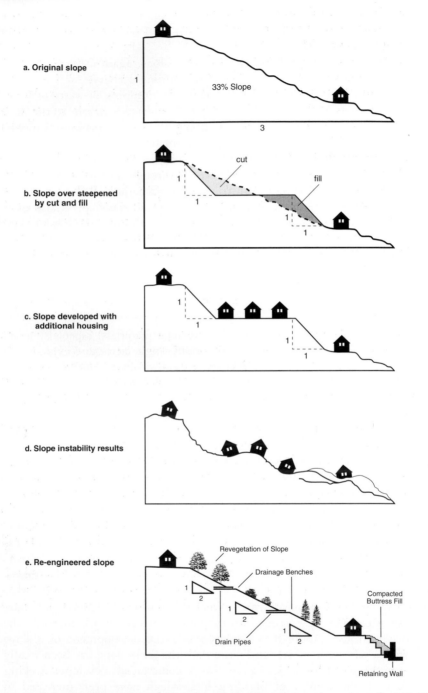

Figure 8.4 An illustration of how the overdevelopment and cutting of a hillside for house construction can create slope failure and the subsequent need for measures designed to restore slope stability.

angle combined with restoration work which includes improved drainage, revegetation and strengthening of the toe of the slope (Figure 8.4e). Cooke (1984) showed that, several decades ago, 25–30 per cent of landslides in southern California were related to construction activities. Improved grading codes have reduced the amount of slope movement damage but there is still a general trend, in the Los Angeles area and elsewhere, towards building on hillslopes which is driven by the decreasing availability of flat land sites and the fashion for houses with extensive views;

- *weathering processes*, which promote the physical and chemical breakdown of slope materials. Certain clay minerals, such as montmorillonite, expand when water is present and the behaviour of these expansive clays has been implicated in the failure of many southern Californian hillsides (Griggs and Gilchrist, 1977). In addition, other natural processes may be involved. The burrowing action of soil animals or soil piping developments on slopes will lead to weakness and the possibility of landsliding.

In most urban areas, landslides may be attributed to a combination of the above factors. One example is the 1979 Abbotsford landslide which created a slope failure covering 18 ha in a suburb of Dunedin, New Zealand, and severely damaged sixty-nine houses. In this case the failure was attributed to the removal of some slope support at the toe by excavation for building, the introduction of additional water to the site and the extensive removal of natural vegetation all superimposed on an unstable geological setting.

The progressive human invasion of landslide hazard zones is not confined to the developed world. The need for improved transportation is leading to new road construction in terrain with a high probability of slope movement throughout the LDCs. In these areas limited resources may lead to inadequate hazard protection. For example, the 52 km long Dharan–Dhankuta road, completed in 1981, provides a key north–south link within Nepal between the Ganges lowlands to the south and the hill villages to the north. The road crosses the unstable Himalayan foothills of East Nepal and is surrounded by long, steep valley-side slopes angled at 30–45°. Engineering is difficult and expensive in such terrain and the road was built to a relatively low-cost specification. The road has since proved difficult to maintain because of cut-slope failures and the blocking of culverts by debris (Hearn and Jones, 1987).

SNOW AVALANCHES

As with slope failures in rock and soil, a snow avalanche results from an unequal contest between stress and strength on an incline (Schaerer, 1981).

The strength of the snowpack is related to its density and temperature. Compared to other solids, snow layers have the unique ability to sustain large density changes. Thus, a layer deposited with an original density of 100 kg m^{-3} may densify to 400 kg m^{-3} during the course of a winter, largely due to the weight of overlying snow, pressure melting and the recrystallisation of the ice. On the other hand, the shear strength decreases as the temperature approaches 0°C. As the temperature rises further and liquid meltwater exists in the pack, the risk of movement of the snow blanket grows.

Most snow loading on slopes occurs slowly. This gives the pack some opportunity to adjust by internal deformation, because of its plastic nature, without any damaging failure taking place. The most important triggers of pack failure tend to be heavy snowfall, rain, thaw or some artificial increase in dynamic loading, such as skiers traversing the surface. For a hazardous snowpack failure to occur, the slope must be sufficiently steep to allow the snow to slide. Avalanche frequency is thus related to slope angle, with most events occurring on intermediate slope gradients of between 30–45°. Angles below 20° are generally too low for failure to occur and most slopes above 60° rarely accumulate sufficient snow to pose a major hazard. Most avalanches start at fracture points in the snow blanket where there is high tensile stress, such as a break of ground slope, at an overhanging cornice or where the snow fails to bond to another surface, such as a rock outcrop.

Plate 7   The small settlement of Vent, in the Austrian Alps, showing steep, unforested slopes directly above the tourist resort. Physical protection against snow avalanches is offered by an array of avalanche rakes but the smooth terrain still poses a threat and the presence of a tower crane (lower right) indicates that new building is taking place. (Photo: K. Smith)

Whatever their individual characteristics, all avalanches follow an avalanche path which comprises three elements: the *starting zone* where the snow initially breaks away, the *track* or path followed and the *runout zone* where the snow decelerates and stops. Because avalanches tend to recur at the same sites, the threat from future events can often be detected from the recognition of previous avalanche paths in the landscape. Clues in the terrain include breaks of slope and eroded channels on the hillsides and evidence from damaged vegetation. In heavily forested mountains, avalanche paths can be identified by the age and species of trees and by sharp 'trim-lines' separating the mature, undisturbed forest from the cleared slope. Once the hazard is recognised, a wide range of potential adjustments is available, some of which are shared with landslide hazard mitigation.

In practice, avalanches result from two quite distinct types of snow pack failure:

1 *Loose snow avalanches* occur in cohesionless snow where inter-granular bonding is very weak, thus producing behaviour rather like dry sand (Figure 8.5a). Failure begins near the snow surface when a small amount of snow, usually less than 1 m³, slips out of place and starts to move down the slope. The sliding snow spreads to produce an elongated, inverted V-shaped scar.

2 *Slab avalanches* occur where a strongly cohesive layer of snow breaks away from a weaker underlying layer, to leave a sharp fracture line or *crown* (Figure 8.5b). Rain or high temperatures, followed by refreezing, can create ice-crusts which may well provide a source of instability when buried by subsequent snowfalls. The fracture often takes place where the underlying topography produces some upward deformation of the snow surface which leads to high tensile stress and the creation of associated surface cracking of the slab layer. The initial slab which breaks away may be up to 10,000 m² in area and up to 10 m in thickness. Such large slabs release considerable amounts of energy and represent the most dangerous type of avalanche (Perla and Martinelli, 1976). When a slab breaks loose, it may bring down as much as 100 times the initially released amount of snow which is then deposited in a rather chaotic heap.

The character of avalanche motion also depends on the type of snow and the terrain. Most avalanches start with a gliding motion but then rapidly accelerate on slopes greater than 30°. It is common to recognise three types of avalanche motion:

1 *Powder avalanches* are the most hazardous and are formed of an aerosol of fine, diffused snow behaving like a body of dense gas. They flow in deep channels and are not influenced by obstacles in their path. The speed of a powder avalanche is approximately equal to the prevailing

(a)

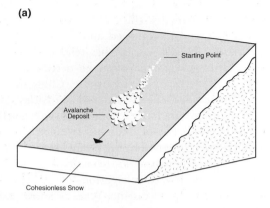

(b)

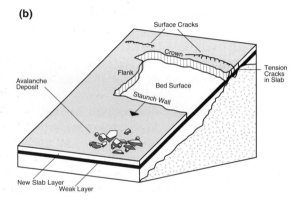

Figure 8.5 The two main types of snow-slope failure: (a) shows the loose-snow avalanche and (b) shows the slab avalanche. Slab avalanches normally create the greater hazard because of the larger volume of snow released.

wind speed but, being of much greater density than air, the avalanche is much more destructive than wind storms. At the leading edge its typical speed is 20–70 m s^{-1} and victims often die by inhaling snow particles.

2 *Dry flowing avalanches* are formed of dry snow travelling over steep or irregular terrain with particles ranging in size from powder grains to blocks of up to 0.2 m diameter. These avalanches follow well-defined surface channels, such as gullies, but are not greatly influenced by terrain irregularities. Typical speeds at the leading edge range from 15–60 m s^{-1} but can reach speeds up to 120 m s^{-1} whilst descending through free air.

3 *Wet-flowing avalanches* occur mainly in the spring season and are composed of wet snow formed either of rounded particles (from 0.1 m

to several metres in diameter) or a mass of sludge. Wet snow tends to flow in stream channels and is easily deflected by small terrain irregularities. Flowing wet snow has a high mean density (300–400 kg m^{-3} compared to 50–150 kg m^{-3} for dry flows) and can achieve considerable erosion of its track, despite reaching speeds of only 5–30 m s^{-1}.

Snow avalanche movements translate into extremely high external loadings on structures. Using reasonable estimates for speed and density, it can be shown that maximum direct impact pressures should be in the range of 5–50 t m^2, although some pressures have been known to exceed 100 t m^2 (Perla and Martinelli, 1976). Table 8.2 provides a guide to the relationships which exist between avalanche impact pressures and the damage to man-made structures. In addition to the direct impact, avalanches may exert upward and downward forces some of which have been known to lift large locomotives, road graders and buildings.

The Galtür disaster in Austria, which occurred in February 1999, was the worst in the European Alps for thirty years and illustrated many of the features of massive powder avalanches. In this event, thirty-one people were killed and seven modern buildings were demolished in a winter sports village previously thought to be located safely at least 200 m from the largest avalanche runout tracks (BBC, 1999). However, a series of major storms earlier in the winter deposited nearly 4 m of snow in the starting zone. This previously unrecorded depth was further increased in places by snow redistributed over the upper slopes by strong winds. By the time the highest level of avalanche warnings were issued, the snow mass in the starting zone had grown to approximately 170,000 tonnes. During its track down the mountain, at an estimated speed in excess of 80 m s^{-1}, the avalanche picked up sufficient additional snow to double the original mass. By the time it reached the village the leading powder wave was over 100 m high and had sufficient energy to cross the valley floor and reach the village with destructive force.

Table 8.2 Relationships between impact pressure and the potential damage from snow avalanches

Impact pressure (tonnes m^2)	Potential damage
0.1	Break windows
0.5	Push in doors
3.0	Destroy wood-frame houses
10.0	Uproot mature trees
100.0	Move reinforced concrete structures

Source: After Perla and Martinelli (1976)

LOSS-SHARING ADJUSTMENTS

Insurance

Private insurance against landslides and other mass movement hazards is not generally available in the USA, largely because of the risk of high losses. The unavailability of insurance can discourage development in hazardous areas but, because information about landslide hazards is not widely disseminated, many people remain unaware that they are at risk (Kockelman, 1986). Some limited insurance cover in the USA is provided through the National Flood Insurance Program, which requires areas subject to 'mudslide' hazards associated with river flooding to have insurance cover before being eligible for federal financial assistance. Unfortunately, technical difficulties in mapping 'mudslide' hazard areas have led to comparatively little use of this provision.

In some countries, legal liability forms a growing basis for financial recompense after landslide losses. American jurisprudence recognises civil liability for death, bodily injury and a wide range of economic losses which may be associated with landslides. The classic defence of 'Act of God' carries decreasing credibility, and recent court judgements have tended to identify the developer, or the consultants, as mainly responsible for damage due to land failure. In some areas, such as Los Angeles County, local government agencies have shared the liability because it has been argued that the issue of a permit for residential development implied a warranty of safe habitation. On the other hand, it is difficult to envisage litigation as an adequate substitute for proper hazard-reduction strategies.

EVENT MODIFICATION ADJUSTMENTS

Landslides

The ability to assess the probability of landslide risk at specific sites is of considerable assistance in implementing mitigation strategies. General indicators include the structure and lithology of slopes, including the presence of weak rock types, clay-rich soils and slopes generally in excess of 25°. It is a challenge to translate these factors directly into terms suitable for risk assessment because of the high spatial variability in soil shear strength, which may be greatly affected by plant root systems stabilising part of a slope, and the usual absence of any piezometric information which would warn of increasing pore-water pressures. Bernknopf *et al.* (1988) divided the Cincinnati metropolitan area into 100 m² cells and devised a probability model from a combination of variables that represent the existing physical state of a hillside, the dominant landslide mechanism in the area and the types of construction activities that can trigger landslides. The results showed that an uncritical application of the Uniform Building Code to the whole area would not be cost-effective.

Property damage from landslides usually leads to demands for engineering works to stabilise the slope. However, the human response to slope failure is often complicated by the statutory and funding distinctions which are made between emergency and permanent works. Emergency responses designed to protect public safety and prevent further immediate damage are usually undertaken satisfactorily but government funds are made available only very reluctantly for permanent slope stabilisation. This may be because the specialised geotechnical information required is not available or because of the high potential cost to the public purse. Alexander (1987b) drew attention to inadequate geological advice and political muddle as contributory factors to the Ancona landslide disaster in Italy. It is a recurrent feature of all hazard mitigation that few publicly funded authorities are willing to pay for expensive defence work for private undertakings when large profits are to be made from property and land speculation.

If these problems can be overcome, the stability of the slope may be improved by a variety of engineering techniques:

1 *Excavation and filling* methods can be used to produce a more stable average slope. This type of reshaping is usually successful but becomes more difficult and expensive as the slide area increases. Specific techniques include unloading the head of a slide and loading the toe, with the replacement of failed material with lighter loads.

2 *Drainage* – especially sub-surface drainage – can be equally effective where changes in pore water pressure have been caused by a rise in the water table. Drainage methods range from the removal of surface water and the drainage of tension cracks to the insertion of trenches filled with gravel or horizontal drains. Properly designed and constructed drainage systems work well but others soon become clogged by fine particles.

3 *Revegetation of slopes* performs several functions: plant roots help to bind soil particles together, the vegetation canopy protects the soil surface from rain splash impact and transpiration processes aid in drying out the slope. Whilst evergreen species are better at providing an all-year canopy, deciduous trees are generally better than conifers at removing excess soil moisture because they have higher rates of transpiration during the summer. However, it may be unwise to rely on vegetation for slope stability because of the possibility of fire or disease during the lifetime of a project.

4 *Restraining structures* such as piles, buttresses and retaining walls can be helpful for slides covering limited areas. But they are generally too expensive for large, unstable slopes and the location of property boundaries may also restrict this approach. Guiding structures near the base of the slope, such as diversion walls, can deflect small debris flows effectively.

5   *Other methods* include the chemical stabilisation of slopes and the use of grouting to reduce soil permeability and increase its strength. In some high-risk urban areas, like Hong Kong, slopes may be covered with materials such as chunam or gunite to reduce infiltration and keep pore-water pressures low. On some construction sites the freezing of a mass of moving soil has been successfully accomplished, the freezing plant being left in operation until the soil-retaining structures were completed.

Slope stabilisation, along with hazard-resistant construction techniques, appears to be the most effective preventative strategy for controlling new development. In this context grading ordinances, such as the *Uniform Building Code* adopted in the USA, are important tools. Along with soil compaction and surface drainage requirements, this Code generally specifies a maximum slope angle of 2:1 for safe development. The basis for such a specification, which means a 27° slope, is that the natural angle of repose for dry sand is 34°, and therefore a 2:1 slope allows for an element of safety over this. Building codes normally require developers to obtain permits before they embark on earthmoving projects on hazard-prone slopes. Ideally, they also require reports from geotechnical engineers and engineering geologists on proposed building sites before plans are approved by a local authority. To work properly, this sort of system needs technically trained inspectors to enforce the regulations and a levy on development fees to become financially self-supporting.

Olshansky and Rogers (1987) cited the success of the city of Los Angeles, which introduced a grading ordinance as early as 1952. Before this date more than 10 per cent of all building lots were damaged by slope failure. Initially, the ordinance required only soil testing but it has subsequently been strengthened. In 1965 the requirement for geological reports was added and in 1973 further inspections were made obligatory, along with final certification of completed earthwork by the city engineer. The benefits have been impressive. For example, during the severe storms of 1978, for a comparable number of sites, the losses at sites developed before 1963 were more than ten times greater than those at sites developed after 1963. Losses at new construction sites are now estimated at less than 2 per cent.

Landslide control is most successful when combined with urban risk assessment and land use planning. In an early programme, begun in 1958, the Japanese government started to enact strong legislation to prevent landslides and debris flows triggered by typhoon rainfall. Mitigation has been pursued through the construction of check dams, drainage systems and other physical controls in combination with development restrictions. As indicated in Figure 8.6, these measures have met with success. In 1938 nearly 130,000 homes were destroyed and more than 500 lives were lost in landslides. In 1976 – the worst year for landslides in that country for about two decades – only 2,000 homes were lost and fewer than 125 people died.

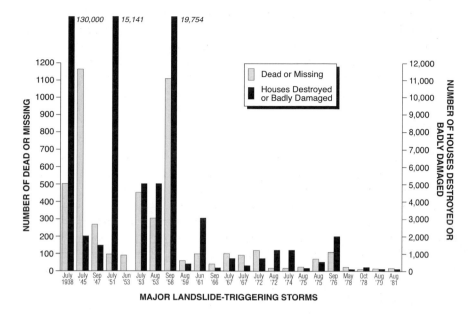

Figure 8.6 The progressive success of the Japanese landslide mitigation programme between 1938 and 1981 in reducing deaths and damages.
Source: Reprinted from *Confronting Natural Disasters*, 1987, with permission from the National Academy Press, Washington, DC

Until the 1970s, hillside development was not effectively regulated in Hong Kong. Since then various improvements have been made culminating in the introduction, in the 1990s, of a comprehensive slope safety system which is managed by a specialist agency (Morton, 1998). Figure 8.7 shows that the average annual fatality rate, which peaked at about twenty during the 1970s, has since declined to around three despite the continuation of land development and the same incidence of tropical rainfall.

Avalanches

Like landslides, snowpacks can be stabilised by a variety of techniques but they also offer opportunities for artificial release. The advantages of artificial avalanche creation are twofold: first, they can be released at pre-determined times, when ski runs and highways are closed, so that snow can then be cleared away with minimum inconvenience; second, and more importantly, the snowpack can be released safely as several small avalanches rather than allowing the snowcover to build up to become a major threat.

Small explosions are used to test the stability of the snowcover but are also employed to trigger controlled avalanches, including the destruction of cornices overhanging ski slopes. Explosives are most effective when placed

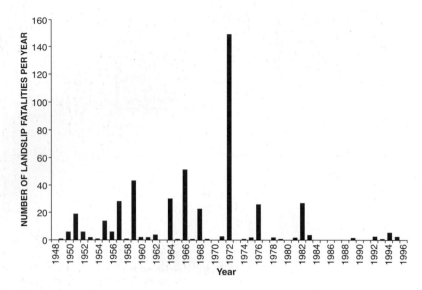

Figure 8.7   The annual number of landslip fatalities in Hong Kong, 1948–96. The
graph shows a reduced incidence from the late 1980s onwards.
Source: After Morton (1998)

in the starting zone, that is, near the centre of a potential slab avalanche,
when the relationship between stress and strength within the snowpack is
delicately balanced but before the snow is deep enough to produce a major
avalanche. Clearly, these requirements can only be met through liaison with
a snow stability monitoring and forecast service. Several methods are used
to place explosives in snow; for example, hand-placed charges of up to 1 kg
of TNT equivalent can be used successfully in accessible ski areas.

Where more remote avalanche paths endanger highways and other valley-
bottom facilities, it is possible to use military weapons. However, this is
expensive and has obvious safety implications, and it is best undertaken
when artillery is deployed by military units on winter training assignments.
Thus, near Rogers Pass, which funnels both the Canadian Pacific rail route
and the Trans-Canada highway through the Selkirk Mountains of British
Columbia, over 1,000 mm of snow falls each year. Avalanches are frequent
and circular gun positions along the road are used to station 105 mm
howitzers. Parks Canada and the Canadian armed forces work together to
trigger avalanches in one of the most comprehensive avalanche-control
programmes in the world. To generate smaller avalanches in remote areas,
the 75 mm recoilless rifle is a popular weapon because it is light enough
to be moved around on the snow slopes.

Despite the advantages of the controlled release of snow, it is often preferable for the snowpack to remain intact. To this end, experiments have been
made with chemicals to retard avalanches. In small starting zones, even

Plate 1 Shanty housing on the edge of Guatemala City, Guatemala. The housing has spread from the settlement centre on the ridge down an increasingly steep slope to the edge of a ravine where landslips are likely after major storms. (Photo: Paul Smith/Panos Pictures)

Plate 2 A homeless family living with few basic amenities on the site of a house destroyed by a lahar following a volcanic eruption near Angeles, Luzon, Philippines. (Photo: Marc Schlossman/Panos Pictures)

Plate 3 A damaging landslide in the humid sub-tropical environment of eastern Brazil in January 1991. Fifteen people in a bus were killed when deeply weathered surface material swept away part of the road from Rio de Janeiro to Tenesopolis. (Photo: M.F. Thomas)

Plate 4 A house on the California coast perched precariously on a disintegrating cliff above the Pacific ocean, 2 March 1998. A total of nine houses were red-tagged and evacuated as this stretch of cliff was severely eroded by high rainfall and strong wave action during winter storms associated with a major El Niño event. (Photo: Lou Dematteis/Popperfoto)

Plate 5 Some local residents survey the damage caused to their homes by strong winds associated with the tornadoes which devastated parts of Oklahoma City, Oklahoma, on 4 May 1999. A category F-5 storm killed 45 people and created economic losses estimated at US$ 1 billion. (Photo: Paul Buck/Popperfoto)

Plate 6 Urban devastation in Tegucigalpa, the capital city of Honduras, after major floods and landslides caused by hurricane 'Mitch' in 1998. The lack of heavy lifting equipment prevented the rapid recovery of urban centres whilst longer-term reconstruction will take many years in a country with a foreign debt burden estimated at US$4 billion. (Photo: Betty Press/Panos Pictures)

Plate 7   A boatman ferrying women and a child around the flooded streets of Dhaka, Bangladesh, during the 1998 floods. Disruption of urban life due to flooding from monsoon rains is a regular feature of large cities in several south-east Asian countries. (Photo: K. McCulloch – Christian Aid/Still Pictures)

Plate 8   Rush-hour traffic congestion and the associated air pollution on the Ratcha-Prarop road, Bangkok, Thailand. People living in stressed urban environments in the Third World are at increasing risk from natural and technological disasters and highlight the growing concern about the future sustainability of many mega-cities. (Photo: Hartmut Schwarzbach/Still Pictures)

trampling the snow by ski or boot can be an effective way of densifying and strengthening the snow. But the use of defence structures has been the most commonly adopted adjustment to avalanches throughout the world (Ramsli, 1974). These structures seek either to retain the snow on the mountainsides or, if an avalanche should occur, to deflect the snow moving across the lower slopes away from humans and their works. The wide variety of defence structures can be classified into (a) snow fences above the starting zone, (b) supporting structures in the starting zone, (c) deflecting and retarding structures in the track and runout zone, and (d) direct protection structures in the track and runout zone.

Above the starting zone, *snow fences* are used to trap and hold snow that would otherwise fall on the avalanche slope (Figure 8.8). On relatively flat ridges and gentle slopes considerable volumes of snow can be intercepted and retained in this way. In the starting zone *snow rakes* or *arresters* are used to provide external support for the snowpack, thus reducing internal stresses. They may also stop small avalanches before they gain momentum. This method has been employed in Europe since at least the early nineteenth century. The earliest structures were massive walls and terraces made of rocks and earth whilst today they can be made of a combination of wood, steel, aluminium and prestressed concrete.

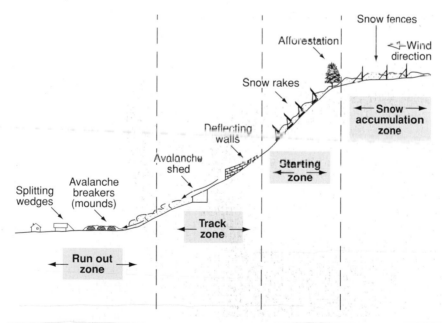

Figure 8.8 Idealised slope showing the various methods used for the physical modification of avalanches. Snow retention is the purpose of the structures in the accumulation and starting zones whilst avalanche deflection away from people and facilities is the purpose of the structures in the track and runout zones. Not all devices will be in place in any one locality.

In the avalanche track and runout zone, various *deflectors and retarding devices* may be located. Large walls, built of earth, rock or concrete, can be used to divert moving snow from its chosen path. The scope for diversion is limited, however, and deflections steered no greater than 15–20° from the original avalanche path have proved the most successful. In addition, wedges pointing upslope can be used to split an avalanche and then divert the sections around vulnerable facilities, for example, electrical transmission towers or isolated buildings. Towards the runout zone other retaining structures, represented by earth mounds or small dams, can be useful as the slope angle declines and the avalanche loses energy. Mounds are generally ineffective on slopes steeper than 20°. It should be stressed that all these responses operate best with wet-snow avalanches and may be of little use in major powder snow events.

The most complete defence against all types of avalanche is obtained by *direct-protection structures*, such as snow sheds and galleries, designed to pass the flow over key facilities such as transport lines. Avalanche sheds typically act as protective roofs over roads or railways and allow avalanches to pass without threat to safety or interruption to traffic. But these structures are expensive and need careful design to ensure that they are properly located and can bear the maximum snow loading on the roof.

In the longer term, there is some incentive to control the avalanche hazard by reafforesting the slopes at risk. This avoids the placing of unsightly structures in highly scenic alpine areas. The main difficulty is that most existing avalanche paths offer poor prospects for successful tree planting and growth. Erosion by previous events means that only thin soils, with limited water retention, are available. Thus, expensive site preparation, coupled with soil fertilisation, is frequently required. It may still be necessary to stabilise the snow in the starting zone while the tree cover establishes itself. When the trees have been established, it could well take 75–100 years before slower-growing species have developed to the point where they are strong enough to resist avalanche forces.

VULNERABILITY MODIFICATION ADJUSTMENTS

Community preparedness

The most formal arrangements for mass movement hazards exist in avalanche-prone areas where a variety of organisations often exists to reduce the risk. For example, in Canada, local offices of Parks Canada maintain observations of the snow stability and avalanche risk in their area, along with the British Columbia Ministry of Highways. The provincial government coordinates most local search and rescue groups in the Canadian Rockies and the Royal Canadian Mounted Police have men and dogs trained for avalanche work. Specialised weather forecasts are available from the

Atmospheric Environment Service and avalanche awareness is promoted through bodies like the Canadian Avalanche Centre with technical courses, films and videos. Although such methods may increase general knowledge of avalanche threats, Butler (1997) found that local residents were often unaware of the immediate danger at a given time.

The need for a locally based, rapid-response search and rescue team is crucial because avalanche victims die quickly if buried beneath the snow. Thus, the chances of survival decline rapidly after 1–2 hours, even when the victim is trapped close to the surface. The overall survival rate after complete burial is less than 1 in 5. An avalanche search is a complex operation, with most victims being found by probing the snow with metal rods. The chances of a victim being found are increased if search dogs are available. Detection is also improved if the victim has used an avalanche beacon or beeper or has been attached to a brightly coloured avalanche cord which extends up to the snow surface.

Forecasting and warning

Various types of forecasting and warning system exist for mass movement hazards. Remote sensing applied to mass movement hazards is limited to the production of preliminary large-scale maps of previous debris tracks, for example, for avalanches from aerial photographs whilst band 5 imagery shows vegetational changes sometimes associated with landslides (Sauchyn and Trench, 1978). This reconnaissance information can then be followed up with low-level air photography (Penn, 1984). Vertical aerial photographs at scales of 1:20,000 to 1:30,000 are often suitable, especially if taken at times of the year when tree foliage and other vegetation cover is at a minimum. Site-specific information is more difficult to obtain. Many landslides are preceded by a period of soil or surface creep before slope failure occurs, often giving rise to surface cracking. This process can be monitored with a view to providing a warning. The most common forms of monitoring include the use of inclinometers and tiltmeters to record evidence of increased hillslope activity but this information is rarely formalised into official warning messages.

For landslides it is possible to issue generalised regional warnings of debris and mudflows following storms and heavy rainfall based on some locally relevant threshold criteria, such as storm rainfall intensity per hour or the cumulative total of rain over a few days. For example, a real-time regional landslide warning system was developed for the San Francisco Bay region using known relations between rainfall and landslide generation and telemetered rainfall data in association with weather forecasts (Keefer *et al.*, 1987). It was used to issue the first regional, public landslide warnings in the USA in February 1986. But the site-specific prediction of landslides remains elusive. Work in Hong Kong by Lumb (1975) emphasised the importance of antecedent precipitation and demonstrated that the total

24-hour rainfall on the day of the landslide, together with the total rainfall over the previous fifteen days, was the best predictor of major mass movements. In contrast, more recent work has claimed that the key lies more in the meteorological characteristics of the triggering short-burst rainstorms, which in the Hong Kong area act in the context of steep terrain and the presence of soils that are mainly non-cohesive when saturated (Au, 1998).

For avalanches, useful warnings are more common and, according to Buser *et al.* (1985), 20–30 countries have systems which employ both forecasts and predictions. Forecasts tend to be used increasingly for day-to-day management of winter sports areas whilst predictions are normally used as an aid in hazard land zoning. Avalanche forecasting involves conducting regular snow stability tests and evaluating the results in conjunction with weather forecast information. As a consequence, forecasting schemes vary greatly in method and sophistication. Sometimes simple visual observations may be made of snow-cover distribution and recent avalanche activity. But, for the most reliable forecasts, a detailed diagnosis has to be made of the snowpack structure, with special emphasis on the presence of weak layers. Local meteorological data form another important input into the evaluation phase of the avalanche forecast.

These monitoring operations are conducted on a daily basis for major ski resorts and other vulnerable areas. Despite this, there are few reliable precursors of slab failure, although Perla (1978) suggested that many releases may be preceded by a gradual increase in tensile stress and that it should be possible to use acoustic signals from within the snow as a means of warning. In conditions of severe risk it is normal practice to clear ski slopes and to stop traffic from entering dangerous sections of highway or railway track. As with all forecasting and warning schemes, there is a basic need for public education and awareness. For example, in southern Glacier National Park, Montana, where major avalanches disrupt traffic about once every five years, it was found that almost 60 per cent of the residents were unaware of the local avalanche warning system (Butler, 1987). It is not surprising, therefore, that few people took extra precautions when driving in the area during times of avalanche hazard.

Land use planning

The recurrence of many landslides and avalanches at the same topographic site means that land use zoning offers a practical method of hazard mitigation. The qualitative recognition of sites susceptible to multiple mass movements is often possible. For example, many avalanche tracks also function as landslide gullies during the spring and summer. Stream channels are the most common paths for debris flows which occur after periods of heavy rain. Although different from floods, debris flows can aggravate flood conditions by blocking the channel and causing waters to overspill the banks.

For landslides, geologists or geotechnical engineers can make a stability assessment for individual sites. Keaton (1994) described a probabilistic approach to site selection. This depends on geological investigations to determine the magnitude–frequency relationships of potentially hazardous processes combined with estimates of the probability that damaging events will occur during a specified exposure period, such as the anticipated life of a structure. The development of information technology with the use of GIS databases and spreadsheet calculations makes such approaches increasingly feasible. Gupta and Joshi (1990) outlined a method employed in the lower Himalayas where landslide activity is related to rock lithology, land use, distance from major tectonic shear zones and slope aspect. Geologic hazard zoning maps at scales around 1:20,000 are still the most common form of hazard identification. Seeley and West (1990) provided an example for a forest park in the western USA where slope instability, including seismically induced rockfalls and avalanches, is the most important hazard.

Once the hazard has been identified, planning law should explicitly encourage local communities to consider mass movement processes when undertaking land use changes. Avalanche zoning employs historical data of avalanche occurrence for the identification of hazardous locations and supplements this information with terrain models and models of avalanche dynamics to determine more detailed degrees of risk. Where sites are near established settlements, avalanche frequency will be a matter of local knowledge. At more remote locations, with insufficient records and maps, other methods are necessary. Gruber and Haefner (1995) have reported the developing use of satellite imagery and a digital elevation model to map large areas of avalanche risk in the Swiss Alps. Sometimes the long-term pattern of avalanche activity can be compiled from trees which remain standing in the track or runout zone but which have been physically damaged by previous events. Sometimes the resulting scarring of the tree rings can provide an accurate means of dating avalanches and producing reliable frequency estimates over the past 200 years or so (Hupp *et al.*, 1987). Where trees have been destroyed by large events, close inspection of the residual damaged vegetation, including height and species, can be a useful guide. Table 8.3 shows how this evidence can be used.

Once potential sites have been identified and frequency estimates made, initial mapping is usually undertaken at a scale of about 1:50,000 with the aid of air photographs. In British Columbia snow avalanche atlases are published primarily as operational guides for highway maintenance personnel (Ministry of Transportation and Highways, 1981). The maps are accompanied by a detailed description of the terrain and vegetation for each avalanche site, together with an assessment of the hazard impact.

Where avalanches threaten settlements, it is necessary to zone the area at a larger map scale (perhaps 1:5,000) and adopt related planning regulations. The length of the runout zone is a critical factor here since it determines whether or not a particular site will be reached by moving snow.

The zoning methodology is well established in many countries (Frutiger 1980; Hestnes and Lied, 1980). For example, in Switzerland and other alpine countries, there is a three-zone, colour-coded system, as detailed in Table 8.4. Similar planning regulations have been adopted for communities in the Rocky Mountains of the USA, as described by Mears (1984). Following the 1999 avalanche disaster, a comprehensive package of measures was introduced at Galtür, Austria. The exclusion zone for buildings, previously drawn up for a 1 in 150 year event, was extended and revised regulations required all new buildings to be reinforced against avalanche pressures. In addition, snow rakes were installed for the first time in the

Table 8.3 Vegetation in avalanche tracks as a rough indicator of avalanche frequency

Minimum frequency (years)	Vegetation clues
1–2	Bare patches, willows and shrubs; no trees higher than 1–2 m; broken timber
2–10	Few trees higher than 1–2 m; immature trees or pioneer species; broken timber
10–25	Mainly pioneer species; young trees of local climax species; increment core data
25–100	Mature trees of pioneer species; young trees of local climax species; increment core data
>100	Increment core data

Source: After Perla and Martinelli (1976)

Table 8.4 The Swiss avalanche zoning system

1 *High-hazard (red zone)*
   Any avalanche with a return interval <30 years. Avalanches with impact
      pressures of 3 t m^{-2} or more and with a return interval up to 300 years.
   No buildings or winter parking lots allowed. Special bunkers needed for
      equipment.

2 *Potential hazard (blue zone)*
   Avalanches with impact pressures <3 t m^{-2} and with return intervals of
      30–300 years.
   Public buildings that encourage gatherings of people should not be erected.
      Private houses may be erected if they are strengthened to withstand impact
      forces. The area may be closed during periods of hazard.

3 *No-hazard (white zone)*
   Very rarely may be affected by small airblast pressures up to 0.1 t m^{-2}.
   No building restrictions.

Source: After Perla and Martinelli (1976)

starting zone and an avalanche dam was constructed across part of the runout zone on the valley floor.

Avalanche hazard assessment in the United States has been provided for a number of towns in the Rockies, as shown by Ives *et al.* (1976). Particular problems have been created by the rapid expansion of winter sports resorts, such as the town of Vail, Colorado, which was built mainly during the 1960s before adequate assessments were available for avalanche and debris flow hazards (Oaks and Dexter, 1987). By the time the local authority had adopted a three-fold, Swiss-type zoning ordinance, several large construction projects were underway, including that of King Arthur's Court (Figure 8.9). Part of this project was located within the high-hazard zone at the base of an avalanche track. Construction loans were eventually withdrawn from the scheme because the developer made no adjustments to take account of the avalanche threat. The project was then abandoned and the

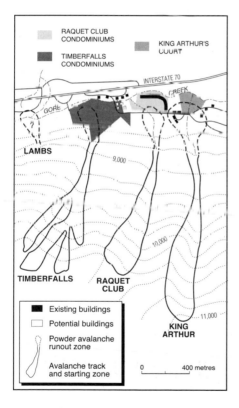

Figure 8.9   The planned King Arthur's Court development at Vail, Colorado, in relation to avalanche tracks. Several pre-existing structures were razed as a result of the hazard mapping project and much of the property has since been developed as a park.
Source: After Oaks and Dexter (1987)

site was eventually purchased by the town of Vail for development as a greenbelt park. Such site-specific assessments have helped to reduce avalanche hazards partly because they have alerted adjacent property owners to the dangers. Thus, the King Arthur's Court study identified an existing site within the moderate-risk zone where a retaining wall was subsequently built to deflect avalanches away from the property.

9   Severe storm hazards

ATMOSPIIERIC HAZARDS

Most environmental hazards are atmospheric in origin. Whilst only a proportion of the world's population lives near active faults or on unstable slopes,
all are exposed to the natural variability of weather and climate. The extremes
of most individual atmospheric elements can constitute a hazard. But, when
these hazardous elements combine, the threats increase. Table 9.1 shows
that, although all severe storms have some features in common, each type
of storm creates its own synergistic mix of damaging conditions. During
the 1980s, windstorm-related hazards caused an average of some 30,000
deaths and US$2.3 billion in property damage worldwide each year
(Housner, 1987). Tropical cyclones are responsible for about half of all these
windstorm losses. Outside the tropics, conflict between contrasting air masses
or strong surface convection can produce other violent storms. Tornadoes
and hailstorms are mainly mid-latitude threats which occur during the
warm season. Other hazards, notably storm-force winds, ice and snow are
posed by severe winter storms.

Atmospheric-related hazards arise from a variety of causes and occur on
a large range of scales. For example, tornadoes are micro-scale phenomena
with life-cycles measured in hours whilst tropical cyclones are meso-scale
events which are embedded in the global atmospheric circulation and
last for days. Large-scale hazards, like drought or excessive seasonal rainfall, can last for many months and they reflect major variations in the
atmospheric circulation. Some variations, like the monsoon rains and
their associated floods, are predictable annual events but other large-scale
fluctuations are less expected and, therefore, potentially more hazardous.
The prime example of such climatic variability is the *El Niño Southern
Oscillation* (ENSO) phenomenon which involves hemispheric-scale interactions between the atmosphere and the ocean over the Pacific basin. Such
events normally occur every 2–7 years and tend to appear around the
Christmas season, thereby resulting in the name 'El Niño' (the Christ
Child). They start with the local incursion of abnormally warm surface
water southward along the Peruvian coast but, in association with changed

Table 9.1 Severe storms as compound hazards showing some major characteristics and their associated impacts. All these storms are also responsible for death and injury to humans.

Tropical storms	Mid-latitude storms			
Tropical cyclones	Tornadoes	Hailstorms	Cyclones	Snowstorms
Wind	Wind	Hail	Wind	Snow
Rain	Pressure drop	Wind	Rain	Ice
Storm surge and waves	Updraughts	Lightning	Flooding	Glaze
Coastal erosion	Building damage	Agricultural losses	Landslides	Wind
Flooding	Agricultural losses		Coastal erosion	Blizzards
Landslides			Building damage	Transport disruption
Saline intrusion			Agricultural losses	Building damage
Building damage				Agricultural losses
Agricultural losses				
Transport disruption				

airflow patterns across the Pacific, El Niño can spread much more widely and last for over a year.

The term *Southern Oscillation* refers to the cycle of varying strength in the atmospheric pressure gradient between the Indo-Australian low- and the South Pacific high-pressure cell. Under normal conditions, shown in Figure 9.1a, this pressure gradient produces the regular *Walker cell circulation* which is characterised by the flow of the south-east trade winds across the ocean leading to a convergence of low-level air in the western Pacific. The resulting vigorous uplift of moist air brings seasonally heavy rainfall to eastern Australia and much of south-east Asia. A westerly return flow aloft contributes to subsidence in the eastern Pacific, thereby completing the Walker cell (Bigg, 1990). The offshore winds from Peru blow across upwelling cold water which tends to be at least 5°C colder than waters in the western Pacific. These cool sea surface temperatures and the stable descending air maintain the dry conditions along the South American coast while the upwelling cold water, rich in nutrients, supports an important fishing industry in Peru.

When El Niño happens, outbreaks of warm water occur off the Peruvian coast and the upwelling of cold water ceases. The atmospheric pressure

(a)

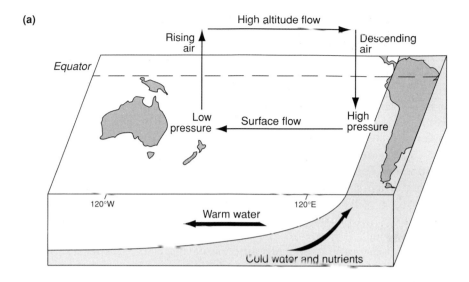

(b)

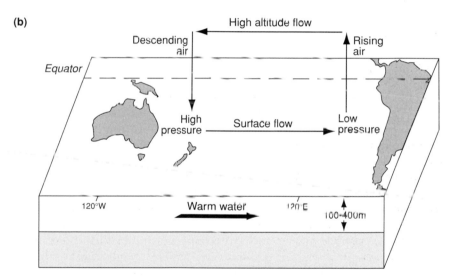

Figure 9.1 Idealised depiction of the two phases of the Walker atmospheric circulation cell in relation to the Southern Oscillation pressure variation: (a) represents the normal (Walker cell) pattern and (b) represents the El Niño phase. When the normal circulation becomes exceptionally strong, a La Niña phase of the cycle is recognised.

gradient across the Pacific changes so that the Walker cell is weakened and even reversed (Figure 9.1b). The anomalously warm water spreads throughout the Pacific basin and contributes to a low-level onshore flow of moist, unstable air along the coast of South America which brings heavy rainfall and floods, sometimes accompanied by disease epidemics, to the normally arid areas of Peru and Ecuador. The fishing industry in Peru collapses. At the same time, negative anomalies in sea surface temperature in the western Pacific ocean lead to the displacement of the convection zone usually centred over northern Australia and Indonesia. Drought replaces seasonal rains in these countries as well as on some Pacific islands such as New Zealand and Fiji. Staple agricultural crops fail and forest fires become common as the wet season is replaced by hot, dry conditions. Under the clear skies, highland areas of Papua New Guinea even suffer crop losses from frost. Weather patterns become disrupted well beyond the tropical Pacific ocean (Trenberth, 2000). During a severe El Niño episode, drought may extend much further westwards than Australia to reach Africa and even cross the Atlantic to north-east Brazil, while the coastal areas of California experience greater winter storminess and above-average rainfall.

The El Niño cycle is not fully understood but the frequency and intensity of events seem to be increasing, possibly as a result of global warming. During the second half of the twentieth century twelve El Niño events were recorded (1951, 1953, 1957–8, 1963, 1965, 1969, 1972, 1976–7, 1982–3, 1986–7, 1990–5, 1997–8) but the four strongest, and also the longest, were recorded after 1980. According to some estimates, the floods, droughts, wildfires and diseases attributed to the 1997–8 event claimed over 21,000 lives worldwide and the damage costs may exceed US$8 billion (IFRCRCS, 1999). The most severe economic losses were in South America. For example, floods in Peru made 500,000 people homeless and destroyed US$2.6 billion worth of public utilities, equivalent to nearly 5 per cent of national GDP. The Peruvian fishing industry declined by 96 per cent in the first three months of 1998 compared to the same period of the previous year. Across the Pacific basin forest fires burned out of control in Indonesia as the country was gripped by drought.

When exceptionally strong Walker cell conditions occur, so-called *La Niña* ('the girl') events arise. The cold surface water off the coast of South America then spreads further north to occupy a latitudinal band 1–2° wide around the equator where it produces sea surface temperatures as low as 20°C. Weather effects are now reversed, with exceptionally heavy rains over the western Pacific and south-east Asia. In 1988–90 there was a strong La Niña event, which greatly enhanced the strength of the easterly trade wind ciculation and brought considerable flooding to the Sudan and Bangladesh (see Chapter 11), as well as to eastern Australia. Although La Niña phases of the Southern Oscillation occur less frequently, and have attracted less attention than El Niño, they are increasingly being linked to high sea surface temperatures in the Caribbean and to more intense hurricane seasons in the Atlantic basin.

TROPICAL CYCLONE HAZARDS

About 15 per cent of the world's population is thought to be at risk from tropical cyclones, and Smith (1997) placed them second only to floods in terms of the number of hazard-related deaths. Most of these deaths are due to drowning in the storm surge. An estimate of the global loss from such storms was about US$10 billion annually at 1995 values (Pielke and Pielke, 1997). Some tropical cyclones bring benefits as well as losses. For example, there is a tendency for cyclonic rain to fall on drought-stricken areas in Australia and elsewhere.

The term 'tropical cyclone' is used in the Indian ocean, Bay of Bengal and Australian waters, whilst the same storms are called 'hurricanes' in the Caribbean, Gulf of Mexico and the Atlantic ocean. In the region of greatest frequency, which is the north-west Pacific in the vicinity of the Philippines and Japan, they are known as 'typhoons'. According to Landsea (2000), in an average year, about eighty-six *tropical storms* (winds at least 18 m s^{-1}), forty-seven *hurricane-force tropical cyclones* (winds at least 33 m s^{-1}) and twenty *intense hurricane-force tropical cyclones* (winds at least 50 m s^{-1}) are recorded worldwide. The greatest hazard exists for three landscape settings:

1 *Densely populated deltas* Bangladesh is the most vulnerable nation, with some 20 million people exposed to the cyclone hazard, mainly in heavily populated rural communities along the fertile delta at the head of the Bay of Bengal. About 10 per cent of all tropical cyclones form in the Bay of Bengal and this area averages over five storms per year with about three reaching hurricane intensity. This area has little topographic relief; in November 1970 up to 300,000 people died and damage of US$75 million occurred, when windspeeds around 65 m s^{-1} created a storm surge some 3–9 m in depth. In the absence of an effective warning, and no reliable evacuation plan, most of the survivors saved their lives by climbing into trees. On 29 April 1991, in the early pre-monsoon part of the cyclone season, the south-east coast of Bangladesh was struck by one of the strongest tropical cyclones recorded in the twentieth century in the Bay of Bengal. An estimated 200,000 people were killed by the 6 m high storm surge and up to 10 million were made homeless as the storm surge washed away the poorest houses made of mud, bamboo and straw. The greatest devastation occurred on the many off-shore islands of silt near the head of the bay. On Sandwip island, where 300,000 people lived, 80 per cent of the houses were destroyed.
2 *Isolated island groups* The Japanese, Philippine and Caribbean island groups are all at risk from tropical cyclones. The Caribbean lies in the path of most Atlantic hurricanes. Quite apart from the resulting fatal-ities, the agricultural sector of these islands is particularly vulnerable to damage from the defoliation of banana and other tree crops by strong winds and the washing away of food crops in heavy rain (Hammerton

et al., 1984). Future harvests can be affected by salt contamination of the soil from storm surge. Commercial crops, grown for vital foreign exchange, appear to be especially at risk. For example, in 1979 hurricane 'Allen' destroyed the following percentages of banana plantations: St Lucia 97 per cent, St Vincent 95 per cent, Dominica 75 per cent and Grenada 40 per cent.

3   *Highly urbanised coasts* The greatest property losses occur in the MDCs. A high damage potential exists along the Gulf of Mexico and Atlantic coastline of the USA, a country where more of the population is at risk from hurricanes than from earthquakes. The worst natural disaster death toll in US history occurred in September 1900 when more than 6,000 people were killed by a storm surge in Galveston, Texas, with the regional death toll probably exceeding 12,000 (Hughes, 1979). A total of 2,636 houses, nearly half of all the dwellings in the city, were destroyed largely because, at the time, Galveston's highest point was less than 3 m above sea level. The greatest threat today exists in Florida, as revealed by hurricane 'Andrew' which hit the south-east coast of Florida in 1992 with maximum sustained winds over 60 m s^{-1} and a storm surge of about 4.5 m (Rappaport, 1994).

The hazard impact of tropical cyclones is closely related to storm intensity. For example, although intense tropical cyclones (winds at least 50 m s^{-1}) account for only one-fifth of all hurricanes which make landfall in the USA, these severe storms account for over 80 per cent of all hurricane-related damage. 'Andrew' was a Category 4 hurricane on the Saffir–Simpson scale (see Table 9.2) and was the costliest US natural disaster since the 1906 San Francisco earthquake. In Florida as a whole, it destroyed about 28,000 residential structures, including some 5,000 mobile homes. The storm killed about sixty-five people and made 250,000 homeless. The east coast of the USA is also at risk, as in 1972 when hurricane 'Agnes' moved north from the Florida panhandle causing at least 118 deaths and more than US$3 billion in damages (Bradley, 1972). A similar high-risk situation exists in Australia's premier holiday resort area along the Queensland Gold Coast.

Table 9.2 The Saffir–Simpson hurricane scale

Scale number	Central pressure (mb)	Windspeed (m s^{-1})	Surge (m)	Damage
1	>980	33–42	1.2–1.6	Minimal
2	965–979	43–49	1.7–2.5	Moderate
3	945–964	50–58	2.6–3.8	Extensive
4	920–944	59–69	3.9–5.5	Extreme
5	<920	>69	>5.5	Catastrophic

Despite the high monetary losses in the MDCs, tropical cyclones hit hardest at poor countries. When hurricane 'Fifi' struck Honduras in 1974 it produced major landslides on the steep hills where most of the peasants had relocated after being forced off the more fertile valley land. Several thousand lives were lost. More recently, hurricane 'Mitch', a Category 5 storm, devastated much of central America in October 1998. It was the fourth strongest hurricane ever recorded in the Atlantic basin with sustained windspeeds of 80 m s^{-1} accompanied by intense rainfall that created many floods and landslides. Deaths were estimated at more than 9,000, with 13,000 injured, 80,000 homeless and 2.5 million people temporarily dependent on aid. Material losses were calculated at around US$6 billion with two-thirds of the loss concentrated in the primary economic sector of agriculture, forestry and fishing. Once again, Honduras – the second poorest country in the western hemisphere – was badly affected. In the mountains 100–150 cm of rain fell within 48 hours and created over one million landslips and mudflows. Six thousand Hondurans were killed, many in the capital, Tegucigalpa, which is built on a series of floodplains and surrounding hillsides. About 60 per cent of all bridges, 25 per cent of schools and 50 per cent of the agricultural base, mainly in the cash-crop sector of bananas and coffee, were destroyed. The economic losses totalled about 60 per cent of the annual GDP (IFRCRCS, 1999) and a US$625 million aid package was released from the World Bank to aid reconstruction.

Attention was first drawn to the hazard consequences of the human invasion of the coastal zone in the USA by Burton and Kates (1964b). A decade later it was estimated that six million Americans lived in areas exposed to hurricane flood surges which occurred at least once per century (Brinkmann, 1975). Since then the demand for homes located as close as possible to the shore has led to further developments. In the coastal counties of Florida alone, the population has grown from fewer than 500,000 in 1900 to over 10 million today (Figure 9.2) and about 75 per cent of the American population is now concentrated within 100 km of the coast; 40 per cent of these people live in zones where hurricanes have a return interval of 1 in 25 years. Most of the recent population increase of these 'Sun Belt' coasts has been in the over-65 age group, mainly located in mobile homes or in expensive apartments near the water's edge. In Queensland, Australia, a mixture of canal estates, high-rise apartments and beach-front residences extends into the fore-dune area (Hobbs and Lawson, 1982).

As a result of such population movements and growing concerns about the consequences of climate change, there is considerable interest in the identification of any trends in severe storm frequency. For example, Changnon and Changnon (1992) examined weather disasters in the USA over a 40-year period and found that the incidence was related to spells of cyclonic activity. In respect of tropical cyclones, much has been made of the long-term trend in the USA towards rising hurricane-related losses as a reflection of a growing tendency towards a higher incidence of severe

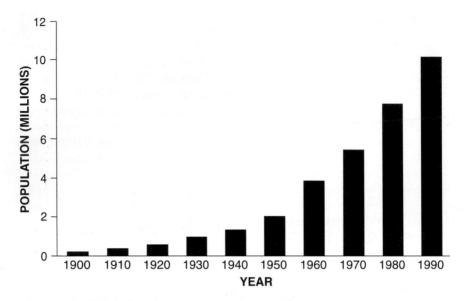

Figure 9.2   The growth in the coastal population of Florida from less than 1 million in 1900 to over 10 million in 1990.
Source: After Pielke and Landsea (1998)

storms. In particular, as shown in Figure 9.3a, there was an apparent rapid rise in economic losses during the final decades of the twentieth century (see also Chapter 2 and Figure 2.8). But Pielke and Landsea (1998) have argued that, if the loss data are normalised for increases in coastal population and exposed wealth, this trend disappears and the 1970s and 1980s actually show smaller damages than some earlier decades (Figure 9.3b). This suggests that – in real terms – the 'Great Miami' storm of 1926 was a costlier event than hurricane 'Andrew' and that the reported increases in material losses reflect social trends rather than any changes in hurricane climatology. If this is so, then the USA has actually been rather fortunate in having a relatively low number of damaging storms in recent decades.

NATURE OF TROPICAL CYCLONES

The term 'tropical cyclone' is a rather general one. The minimum mean speed threshold for *hurricane* winds is set at 33 m s^{-1}. Such winds blow around very low pressure centres with strong isobaric gradients. As the storm evolves from the initial closed circulation with only moderate depth and heavy showers to a fully developed hurricane, many authorities recognise the intermediate stages of a *tropical depression*, characterised by maximum

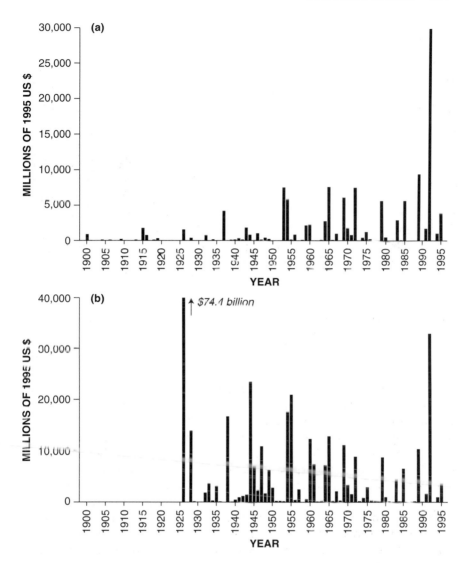

Figure 9.3   Annual hurricane damage in the USA: (a) shows the unadjusted values
1900–95; (b) shows the normalised values 1925–95. After adjustment
for changes in hazard exposure, the 1926 storm is more costly than
hurricane 'Andrew' in 1992.
Source: After Pielke and Landsea (1998)

mean wind speeds below 18 m s^{-1}, and a *tropical storm*, with maximum mean
wind speeds from 18 m s^{-1} to 32 m s^{-1}. The hurricane's severity can be classified according to either the central pressure, windspeed or ocean surge on
the Saffir–Simpson scale (Table 9.2). With scale 4 or 5, the system releases
more power in one day than the USA uses in a year. Category 5 storms

are comparatively rare, the only recent ones to affect the Caribbean being 'Camille' in 1969, 'Gilbert' in 1988 and 'Mitch' in 1998.

Since tropical cyclones depend for their existence on heat and moisture, they form over warm oceans with sea surface temperatures of at least 26°C. In fact, tropical cyclones originate mainly over the western parts of the main ocean basins, where no cold currents exist. Figure 9.4 illustrates the main areas of formation and the land areas most affected by the storm tracks. Tropical cyclones do not form in the eastern South Pacific ocean nor do they occur in the South Atlantic ocean because of low temperatures and unfavourable upper winds. Having said this, there is a high degree of variability in the occurrence of these storms and their occurrence can also be linked to major perturbations in the tropical ocean-atmosphere system, such as El Niño Southern Oscillation (ENSO) events (see also below, pp. 234–7 and Chapter 11). Most storm systems decay rapidly over land areas, although some cyclones remain dangerous for thousands of kilometres with sufficient energy to cross mid-latitude oceans and threaten higher-latitude coasts.

The meteorological development often begins with a small, low-pressure disturbance, perhaps a vortex near the Inter-Tropical Convergence Zone. If surface pressure continues to fall, by 25–30 mb, this may create a circular area with a radius of perhaps only 30 km and with strong inblowing winds. This disturbance can develop into a self-sustaining hurricane if four environmental conditions are satisfied:

1   The rising air, convected over a wide area, must be warmer than the surrounding air masses up to 10–12 km above sea level. This warmth comes from latent heat taken up by evaporation from the ocean and liberated by condensation in bands of cloud spiralling around the low-pressure centre. There must also be high atmospheric humidity up to about 6 km. If the rising air has insufficient moisture for the release of latent heat, or if it is too cool in the first place, the chain reaction will never start. In practice, this means that cyclones form only over tropical oceans with surface temperatures of 26°C or more.

2   Hurricanes need vorticity to give the low-pressure system initial rotation. Therefore, they do not develop within 5° latitude of the equator where the Coriolis force is almost zero and inflowing air will quickly fill up even a strong surface low. But between 5–12° north and south of the equator the airflow converging on a low is deflected to produce a favourable spiral structure.

3   The broad air current in which the cyclone is formed should have only weak vertical wind shear because wind shear inhibits vortex development. Vertical shear of the horizontal wind of less than 8 m s^{-1} allows the main area of convection to remain over the centre of lowest pressure in the cyclone. Although this is not a difficult condition to satisfy in the tropics, it explains why no cyclones develop in the strong, vertically-sheared current of the Asian summer monsoon. Most cyclones

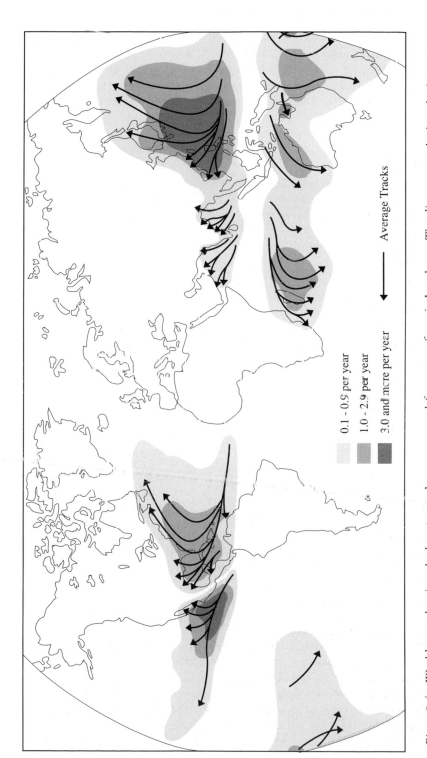

Figure 9.4 World map showing the location and average annual frequency of tropical cyclones. The diagram emphasises the importance of the western North Pacific region and the way in which the storm tracks curve polewards to threaten populated coastal areas.
Source: After Berz (1990)

0.1 – 0.9 per year

1.0 – 2.9 per year

3.0 and more per year

Average Tracks

occur after the monsoon season, in late summer and autumn when sea surface temperatures are at their highest level.

4   In combination with the developing surface low, an area of relatively high pressure should exist above the growing storm. As this happens only rarely, few tropical disturbances develop into cyclones. If high pressure exists aloft, it maintains a strong divergence or outflow of air in the upper troposphere. Crudely stated, this acts like a suction pump, drawing away rising air and strengthening the sea level convergence.

Once established, cyclones have efficient thermodynamic mechanisms for maintaining themselves. A ring-like wall of towering cumulus cloud rises to 10–12 km. Most of the rising air flows outward near the top of the troposphere, as shown in Figure 9.5 (vertical section), and acts as the main 'exhaust area' for the storm. The release of rain and latent heat encourages even more air to rise and violent spiralling produces strong winds and heavy rain. A small proportion of the air sinks towards the centre to be compressed and warmed in the 'eye' of the storm. The warm core also maintains the system because it exerts less surface pressure, thus maintaining the low-pressure heart of the storm.

Several features of the tropical cyclone contribute to its hazard potential. During the initial phases of development, as a tropical depression with a diameter of only 30–50 km, the cyclone can be difficult to detect, even with improved satellite imagery. Although all these storm systems move westward, driven by the upper air easterlies at about 4–8 m s^{-1}, they eventually recurve erratically towards the pole, as shown in Figure 9.4. Forecasting their detailed path, including the exact coastal landfall, with a sufficient warning lead-time is a major problem for the national weather services responsible for tracking their movements.

There are several hazards associated with the characteristics of tropical cyclones, all of which need to be accurately forecast:

1   *Strong winds* cause the most structural damage, as in the case of hurricane 'Andrew'. The atmospheric pressure at the storm centre often falls to 950 mb with winds of 50 m s^{-1}. One of the deepest lows ever recorded was 870 mb, when typhoon 'Tip' hit the Pacific island of Guam in October 1979. Windspeeds here reached 85 m s^{-1}. The inertial force of the wind, experienced when a structure is perpendicular to the moving air mass, is proportional to the windspeed, so the damage potential increases rapidly with storm severity. As shown in Figure 9.6, the destructive energy of a Category 5 hurricane, with windspeeds around 70 m s^{-1}, can be up to about fifteen times greater than the damage potential of a tropical storm with windspeeds around 20 m s^{-1}.

2   *Heavy rainfall* creates freshwater flooding and landslides, as in the case of hurricane 'Mitch'. At any one station the total rainfall during the passage of a tropical cyclone may exceed 250 mm, all of which may

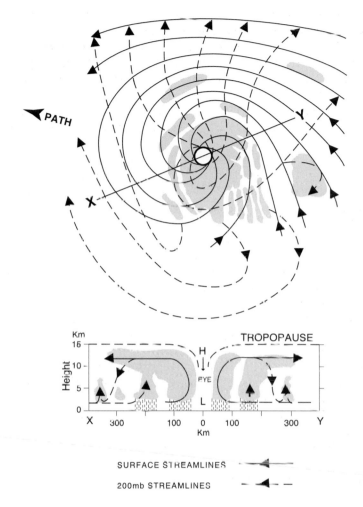

Figure 9.5 A model of the areal (above) and vertical (below) structure of a tropical cyclone. The spiral bands of cloud are shaded and areas of rainfall
are indicated in the vertical section X–Y across the system. The stream-
lines symbols refer to the upper diagram.
Source: After Barry and Chorley (1987)

fall in a period as short as 12 hours. Higher falls are likely if there are
mountains near the coast; Baguio in the Philippines received 1,170 mm
on a single day in 1911. Even in low-lying Texas, a hurricane in 1931
unloaded 587 mm in 24 hours.

3 *Storm surge* is often the feature that causes most deaths and also the salt
contamination of agricultural land. The height of the storm surge
depends on the intensity of the tropical cyclone, its forward speed of
movement, the angle of approach to the coast, the submarine contours

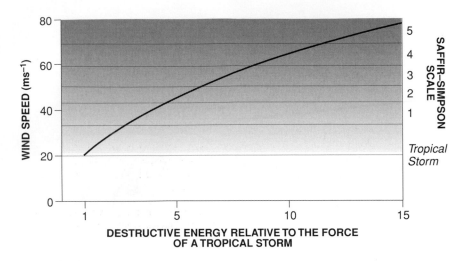

Figure 9.6   The relationship between hurricane windspeeds and their destructive
force compared with a tropical storm. Because the damage potential is
proportional to wind energy, winds of 65 m s^{-1}, typical of a Category
4 hurricane, have about ten times the destructive power of the airflow
in a tropical storm.
Source: Modified after Pielke and Pielke (1997)

of the coast and the phase of the tide. Wind-driven waves pile up water
along shallow coasts, with the greatest effects in bays. Swell waves move
outward from the storm, perhaps three to four times faster than the
storm itself, and can act as a warning of its approach to coastlines
1,000–1,500 km distant. In confined locations such as the Gulf of
Mexico and Bay of Bengal the total sea-level rise may exceed 3 m. In
addition, there will be a further increase in level due to low atmospheric
pressure and there is a 260 mm rise for every 30 mb fall in air pressure.

SEVERE SUMMER STORMS

Tornadoes

A tornado is a violently rotating and narrow column of air, averaging about
100 m in diameter, which extends to the ground. Visually, it is marked
by a funnel-shaped cloud that appears to hang from the cloud base. Although
most tornadoes are associated with cumulonimbus clouds, they may also
extend from lesser clouds that flank the thunderhead. The greatest hazard
exists when the funnel cloud touches the ground and creates some of the
strongest horizontal pressure gradients seen in nature. A scale of tornado
intensity was devised by Fujita (1973) and is shown in Table 9.3. It is

Table 9.3 The Fujita scale of tornado intensity

Category	Damage	Windspeed (m s⁻¹)	Typical impact
F-1	Light	18–32	Damage to trees, free-standing signs, some chimneys
F-2	Moderate	33–50	Roofs damaged, mobile homes dislodged, cars overturned
F-3	Severe	51–70	Large trees uprooted, roofs removed, mobile homes demolished, damage from flying debris
F 4	Devastating	71–92	Masonry buildings damaged, cars become airborne, extensive damage from large missiles
F-5	Disastrous	93–142	Wood-frame buildings lifted from foundations and disintegrate, cars airborne for more than 100 m
F-6 and over		>142	Currently thought not to exist

believed that about one-third of all tornadoes exceed F-2 and attain windspeeds greater than 50 m s⁻¹. The forward speed of the tornado is much lower, perhaps only 5–15 m s⁻¹. Most tornadoes are of short duration and have a limited destructive path, rarely more than 0.5 km wide and 25 km long. However, in May 1917 a tornado travelled about 500 km across the Midwest of the USA and existed for well over seven hours.

Tornadoes are highly localised events and often break out in groups in association with thunderstorms. Most tornadoes form in warm, moist air just ahead of a strong cold front when the contrast in air masses produces latent heating and the creation of a low pressure area near the surface. Their presence can often be detected because they exist beneath 'parent' thunderstorm clouds, sometimes associated with hail and strong surface winds, that can persist for several hours. The tornado hazard is highly concentrated in time and space. Over half of all tornadoes in the USA develop in the April to July period and there is a marked decrease after the summer solstice. The United States leads the world in tornado hazard, particularly along 'Tornado Alley': the area running from Texas through Kansas and Oklahoma and on into Canada. The maximum tornado frequency is located in central Oklahoma, which experiences approximately ten tornadoes per year, although the most destructive storms tend to occur north and east of this area.

When adjustment is made for size of the country, it is likely that the tornado incidence in Italy and some other mid-latitude countries is similar to that of the USA (Fujita, 1973). One difference seems to be that the USA suffers from severe storms with a high disaster potential, although data presented by Golden (2000) suggests that about 85 per cent of reported tornadoes are weak, 15 per cent strong and less than 1 per cent violent. In practice, perhaps only 3 per cent of the 900 or so tornadoes that occur in the USA each year are responsible for human deaths, up to half of which are suffered by occupants of mobile homes. The greatest tornado disaster recorded in the USA was the so-called 'Tri-State Tornado' of March 1925. Losses in the three Mid-Western states hit by this storm included 695 people dead, over 2,000 injured and damages equal to US$40 million at 1964 prices (Changnon and Semonin, 1966). These losses resulted from a combination of physical factors, such as the high ground speed, which approached 30 m s^{-1}, coupled with the long track and wide path, and human failings which included the lack of warning and inadequate shelter provision, since many homes were without storm shelters or basements. Following this event, tornado warnings giving clear advice to seek shelter have contributed to a significant declining trend in deaths. However, fatalities continue to occur in the most severe outbreaks. Schmidlin *et al.* (1998) reported forty-two killed in Florida in February 1998 whilst a Category F-5 storm in May 1999 killed forty-five people and created material losses of up to US$1 billion in Oklahoma City. Most of these fatalities occurred amongst people unprotected by strong buildings; for example, all the 1998 Florida deaths occurred in mobile homes or recreational vehicles.

Hazardous tornadoes are also found in Bangladesh. Here they exert a much higher death toll, as illustrated by a storm in May 1989 which killed between 800 and 1,300 people. In 1996, a tornado in the Tangail area killed about 700 people and destroyed approximately 17,000 homes (Paul, 1997). These high death rates have been attributed to a variety of factors including high population densities, weak building construction, an absence of preparedness and warning, poor medical facilities and a lack of rapid transport of injured victims to hospitals.

Hailstorms and lightning

Hail consists of ice particles which fall from clouds to the ground. The intensity of hailfall depends on the number of particles and the surface windspeed which drives them but the degree of hazard is also related to the size of the particles. The most destructive hailstones tend to exceed 20 mm in diameter. Large hail has been known to result in human deaths but the main damage is to property, especially standing crops. Most hail is produced by thunderstorms in which strong vertical motions are present. These motions give rise to towering cumulonimbus clouds which are often associated with thunder and lightning. Hailstorms result from strong surface heating and are warm-season features, although some are produced by cold fronts which wedge under warm, moist air to produce clouds and intense storms. Isolated hailfalls, very often of the greatest intensity, occur in and near to mountain ranges.

Few mid-latitude areas are immune from hail but most of the damage is concentrated in the continental interiors close to mountains. Notable regions include the upper Great Plains of the USA and Canada, many of the alpine countries of Europe, the Caucasus region of Asia and the Mendoza area of Argentina. Hail can also be a problem at high altitude in the tropics where, for example, the tea crop in Kenya is at risk. Comparisons of hail hazard are difficult because the key characteristic of hail is its enormous variability in size, time and space. According to Changnon (2000), significant hail damage in the USA occurs during the most severe 5–10 per cent of all storms. In an average year, hail in the USA causes about US$1.3 billion in crop losses and a further US$1–1.5 billion in property damage (at 1996 dollar prices). Most places in the United States experience only two or three hailstorms per year. However, in the lee of the central Rocky Mountains, between six and twelve hail days are recorded each year and between April and October crop-damaging hail falls somewhere in the eastern two-thirds of the USA almost every day. In temperate, oceanic climates, such as that of Britain, damaging hail is rare and is confined to the areas, such as south-east England, which experience the greatest convective activity in summer (Elsom and Webb, 1993).

Lightning is often associated with rain, hail and the powerful upcurrents of air within the clouds of summer storms. It occurs when a large positive electrical charge builds up in the upper, often frozen, layers of a cloud and a large negative charge – together with a smaller positive area – forms in the lower cloud. Since the cloud base is negatively charged, there is attraction towards the normally positive earth and the first (leader) stage of the flash brings down negative charge towards the ground. The return stroke is a positive discharge from the ground to the cloud and is seen as lightning. The extreme heating and expansion of air immediately round the lightning path sets up the sound waves heard as thunder. According to Elsom (1993), there was an average of about four deaths per year in England and Wales due to lightning in the 1980s. Despite the growth in population, this was a considerable reduction on the twenty deaths per year recorded over 100 years earlier, a decrease attributed to a decline in the number of outdoor workers and the greater degree of physical protection offered by present-day urban areas.

SEVERE WINTER STORMS

Extratropical cyclones tend to bring the strongest winds during the winter season when they may be accompanied by snow and ice hazards. Some attempt will be made to distinguish between the windstorm threat and the snowstorm threat, although many winter storms present a combined hazard.

Severe windstorms

Severe windstorms, frequently accompanied by heavy rain, are associated with deep mid-latitude depressions. The greatest damage is often sustained in coastal areas where wind-driven waves encroach upon the shoreline, eroding sea defences and other structures. For example, in the winter of 1977–8 a combination of strong waves, local storm surges and high tides caused losses of US$18 million along the coast of California (Pappas, 1978). Even greater losses were recorded in the El Niño winter of 1982–3 when beaches were eroded with consequent losses to shoreline properties. The southern North Sea coastline is vulnerable to this form of marine flooding, as happened with the tidal surge of 31 January 1953. In this event, a deep depression in the North Sea produced a strong northerly gale, which combined with a tidal surge of 2.5–3.0 m to cause extensive coastal flooding. In Holland 1,835 people were killed, 3,000 houses were destroyed, 72,000 people were evacuated and 9 per cent of agricultural land was flooded. Some major world cities, such as Venice and London, are subject to increasing storm-surge hazard because of long-term local subsidence and rising sea levels.

Some mid-latitude cyclones create a hazard because they develop very quickly and are called *rapidly deepening depressions*. The most favoured breeding grounds for these depressions lie off the east coast of continents and correspond to the areas of warm ocean currents. Depressions in this category appear to be most frequently found in the north Atlantic ocean and can pose a significant hazard in western Europe, notably to Britain which lies directly in their eastward path. Rapidly deepening depressions are difficult to forecast accurately because the standard models used by weather forecasters often under-predict the rate of deepening (Sanders and Gyakum, 1980). In addition, local processes may be involved. The break-up of deep clouds can produce turbulent eddies which can contribute to the strength of gusts at ground level to such an extent that it resulted in the highest gust speeds in the notorious windstorms of October 1987, January 1990 and January 1993.

In areas exposed to the frequent passage of winter depressions the ongoing economic loss can be high. For Britain, Buller (1986) estimated that an average of about 200,000 buildings are damaged by windstorms each year. Nearly 85 per cent of the damage, which is suffered mainly by domestic properties, occurs between November and March. In October 1987, a small depression deepened very rapidly in the Bay of Biscay and then moved over western Europe. Heavy losses were sustained in southern England but, because the storm came at night, only nineteen direct fatalities occurred, followed by eleven other storm-related deaths over the next few days. Casualties in other countries raised the total death toll to about fifty. Forestry suffered badly; in England alone more than 15 million trees were lost. The storm struck when most trees were still in full leaf and their grip in the ground was weakened by the waterlogged soil. Much of the damage to infrastructure was due to trees falling on to power lines, houses, roads and railways. The total insurance bill was estimated at £1–2 billion.

Between January and March 1990 four severe storms caused more damage than any other recorded natural disaster over western Europe; 230 people died and the insurance loss was estimated at more than US$8 billion. These storms produced high gust speeds over a wide area, mostly during daylight. Although the storms were forecast, adequate warnings advising people to seek protection, for example, by sheltering in cellars, were not issued. On the contrary, many people continued to drive their cars or went for walks, even in forests. Most deaths were caused by falling trees crashing on to road vehicles. In Britain alone, an estimated 3.5 million trees were lost in this event, fewer than in 1987 because deciduous trees were not in leaf and also because the largest and most vulnerable trees had been lost in the earlier storm. A similar storm devastated France and several other western European countries in December 1999 when eighty-eight people died.

Severe snow and ice storms

Following Rooney (1967), there is now a large body of literature on the urban snow hazard in North America. Approximately 60 million persons in the United States live in urban areas of the northern states with a high risk of snow storms. In March 1993 a severe snow storm affected much of the east coast of the USA and Canada, breaking several weather records. It killed over 240 people – including forty-eight missing at sea – and caused economic losses of around $3 billion (Brugge, 1994). This is around three times the combined death toll for hurricanes 'Hugo' and 'Andrew'. The storm was initiated over the warm waters of the Gulf of Mexico and moved north along the Atlantic seaboard as a rapidly intensifying surface low. In New York City, 22 m s^{-1} winds were experienced and over half a metre of snow fell in Manhattan. At one time, over 3 million people were without electrical power due to high winds and fallen trees.

Ice or _glaze storms_ are an important winter hazard in North America, especially in the Great Lakes region, where they can extend to areas over 10,000 km². The problem arises from thick accretions of clear ice on exposed surfaces. Ice accretes on structures whenever there is liquid precipitation, or cloud droplets, and the temperature of both the air and the object are below freezing point. Electric power transmission lines and forests are at greatest direct risk because the weight of ice may well be sufficient to bring such objects down. Chaine (1973) has described the effects of an ice storm in the area of Montreal, Canada, where 20,000 families remained without electrical power three days after the event. Occasionally, such storms sweep far to the south in the USA, where their unexpected arrival causes major losses. In 1951, for example, a glaze storm struck Tennessee (Harlin, 1952). It led to twenty-five deaths and almost US$100,000 damages (at 1951 prices), over half of which were to forests.

LOSS-SHARING ADJUSTMENTS

Disaster aid

Tropical cyclones create such comprehensive losses in developing countries that international disaster aid is usually necessary. Apart from the loss of life and structural collapse of buildings, it is likely that standing crops, water supply systems and telecommunications will also suffer. The rehabilitation of the infrastructure often requires technical support well beyond the emergency relief phase. After cyclone 'Namu' swept through the Solomon Islands in the Pacific in May 1985, the government declared a state of national emergency (Britton, 1987). Australia and New Zealand responded with aircraft, food, tents, water containers and medical supplies and also provided US$2.2 million which was supplemented by aid from other countries. Within twelve days of the disaster, the immediate needs for food and medicines were satisfied. The state of emergency lasted for little more than three weeks but several years were required to rebuild the economy.

Apart from cash donations, which enable emergency purchases to be made within the disaster region, specialist external help is often important. Small tropical islands are vulnerable to failures of electricity supplies and telephone systems following the destruction of overhead power lines in a cyclone. The lack of electricity, especially for refrigeration, poses a public health risk while the lack of communications greatly hampers relief efforts. For example, when hurricane 'Gilbert' struck Jamaica in September 1988 approximately 40 per cent of the electricity transmission system and 60 per cent of the distribution system was made unserviceable. After a reconnaissance visit, a team of British engineers was in the field within two weeks of the disaster, followed by teams from the USA and Canada (Chappelow, 1989). The British team was also activated almost one year later when hurricane 'Hugo' devastated the Leeward Islands of the eastern Caribbean in September 1989.

Aid is also necessary in the MDCs after major events, especially in multi-ethnic communities where problems may be created by poverty, gender and minority status. For example, after hurricane 'Andrew' only about 20 per cent of the population in Florida City (which is mainly black) applied for aid even though over 80 per cent of the homes had been destroyed. In Homestead, a mainly white area, 90 per cent of the population applied for aid and 80 per cent of such applications were successful (Peacock *et al.*, 1997).

Insurance

This measure is largely confined to the MDCs. For example, it has been estimated that only 2 per cent of the losses imposed on Central America by hurricane 'Mitch' were covered by insurance. Some countries have *all-risk policies* on building structures and contents as well as on motor vehicles; they contain cover for a variety of 'storm' losses including wind, hail, lightning and fire. Because of these package policies, it is difficult to identify the role of individual weather perils, although property damage arising from Atlantic storms in winter is believed to be the most expensive impact in the UK. Farmers in the MDCs often insure crops against storm-related losses. Hail insurance is fairly common in North America and about 25 per cent of the total crop value in the USA is covered by insurance (Changnon, 2000).

Major storms, especially hurricanes in the USA, can cause great strain within the insurance industry. Most of the structural damage to buildings is wind-related. Typically, insurers paid out US$2.6 billion for wind-related damage resulting from hurricane 'Hugo' in 1989, whilst only 10 per cent of the total insured loss was flood-related (Anonymous, 1995). On the other hand, for a few hurricanes, such as 'Opal' which affected Florida and Alabama in 1995, the major losses come from the storm-surge. In this case, Bush *et al.* (1996) demonstrated how the property damage was increased in certain areas by the prior removal of dunes compared to the lower losses suffered by houses elevated on pilings or set back behind sea walls. Hurricane 'Andrew' was the most costly storm ever recorded, with insured losses of US$17 billion. These costs are well beyond the estimates of a decade before when it was believed that individual storms causing insured losses of up to US$7 billion were possible (AIRAC, 1986).

According to Pielke and Pielke (1997), hurricane 'Andrew' acted as a 'wake-up' call to the insurance industry. Prior to that event insurers had kept death and damage statistics only and had taken on increasing property risks in order to maintain their market share (Mittler, 1997). Since then, private companies have become interested in hurricane climatology and have paid attention to storms that do not make landfall or those that hit sparsely populated areas. As a result, some companies became reluctant to underwrite further cover in parts of Florida, Hawaii and coastal South Carolina, and state catastrophe funds were required to resolve the crisis of residents unable to buy policies on the private market.

In the longer term, it is possible that the availability of storm-hazard insurance has increased the demand for coastal homes and even raised property values in some hazard zones. This is because the presence of insurance encourages either the perception that storm events are very rare or the more cynical judgement that the certainty of financial recompense more than balances out any risks for a property owner. Ever since 1968, shoreline construction in the USA has been endorsed by the federal government, which sells insurance to waterfront property owners through the National Flood Insurance Program (NFIP). Cross (1985) noted that both home-buyers and estate agents in the hurricane-prone Lower Florida Keys believed that the availability of flood insurance made residents more willing to locate in low-lying property or along the waterfront and also made it easier to sell property within these areas. One result is an increasing number of repeat claims. About 40 per cent of all claims under the NFIP have been for properties flooded at least once before and half of such repeat claims were in respect of buildings in the coastal zone (Anonymous, 1995). In this situation, commercial insurers will tend to withdraw from the residential market or lobby for the adoption and enforcement of more stringent building codes.

EVENT MODIFICATION ADJUSTMENTS

Environmental control

In the past, many countries have conducted weather modification experiments, especially during the period of scientific optimism that prevailed in the 1950s and 1960s. Enthusiasm has since waned and, at the present time, there is no established severe storm suppression technology.

Hurricane modification is the most attractive of all severe storm suppression goals. As already shown, the destructive power of a tropical cyclone increases rapidly with the maximum windspeed and it has been estimated that a 10 per cent reduction in windspeed would produce an approximate 30 per cent reduction in damage. As shown by Willoughby *et al.* (1985), attempts at weather modification in the United States started in 1947 but it was the creation of Project STORMFURY in 1962 that was mainly responsible for raising hopes for this mitigation measure.

The theory was that the introduction of freezing nuclei into the ring of clouds around the storm centre (eye) would stimulate the additional release of latent heat within the clouds, which were thought to contain abundant supercooled water droplets. It was believed that this would lower the maximum horizontal temperature gradients within the storm system, thus weakening pressure gradients and the damaging winds. Early computer models of hurricanes suggested that a 15 per cent reduction in maximum windspeed was possible. But clear results were slow to emerge and only hurricane 'Debbie' in August 1969 produced successful results (Gentry, 1970). Doubts continued to be expressed about the scientific basis of the experiments and Project STORMFURY was finally discontinued in 1983.

Hail suppression has been attempted in the alpine countries of Europe since medieval times. The firing of cannon and the ringing of church bells was thought to be an effective measure and, in the late 1940s, farmers in northern Italy began to fire explosive cardboard rockets into thunderstorm clouds. There is no real scientific basis for such methods, although it has been suggested that explosions may propagate pressure waves in the air which are sufficient to crack and weaken the ice making up the hailstone. By this means, the growth of large hailstones may be prevented.

Most hail suppression technology has relied, like hurricane modification, on cloud seeding with ice nucleants. The principal theory is that the introduction of artificial ice nuclei, such as silver iodide, will introduce competition for the supercooled water droplets which hailstone embryos feed on. The expected result is that, although the total number of ice particles will increase, all individual hailstones will grow to a smaller size. These smaller stones will then do less damage when they reach the ground. Hail clouds have been seeded with silver iodide by a variety of methods including ground-based generators, over-flying aircraft which drop pyrotechnic flare devices, and artillery shells.

Weather modification is unlikely to become a successful means of suppressing severe storms, until it can satisfy the following criteria:

1   *Scientific feasibility* This requires a more complete understanding of the microphysical processes within clouds. At the present time, it is thought that hurricane clouds contain too little supercooled water and too much natural ice for artificial nucleation to be effective. It is also believed that major hailfalls could be reduced only by very high applications of seeding agents, perhaps up to 100 times more silver iodide than was used in previous experiments.

2   *Statistical feasibility* Not enough sample storms are available for treatment in any area to provide the statistical proof that experimental results differ from naturally occurring changes in factors such as windspeed or hailfall intensity. For both hurricanes and hailstorms, fully randomised experiments have to be designed and evaluated in order to separate results into two categories of 'seeding' and 'non-seeding' days. All too often, the sample size is insufficient to permit a valid identification of weather modification effects.

3   *Environmental feasibility* Quite apart from the moral issue of interfering with atmospheric processes while in an incomplete state of knowledge, most seeding agents create a pollution threat. This is especially so for silver iodide. Little is known of its toxic effects, either on ecosystems, as a result of atmospheric fallout, or on the seeding operators.

4   *Legal feasibility* Several hail suppression programmes in the USA have been the subject of lawsuits. The most common charge by complainants is that their right to natural precipitation has been diminished by a reduction in rainfall which they attribute to hailstorm cloud seeding. In other cases lawsuits have claimed that the seeding has increased storm damage, either directly by intensifying hailfall or indirectly by contributing to storm runoff.

5   *Economic and social feasibility* Highly favourable projected cost:benefit ratios of the potential savings achieved from the successful modification of the most severe storms have been a major factor in releasing funds for cloud seeding experiments. Unfortunately, it has not proved possible to deliver the promised savings.

Hazard-resistant design

There are two prime purposes of hazard-resistant design against severe storms. The first is to protect the lives which are threatened by the storm surge generated by tropical cyclones. To some extent, this goal can be achieved by coastal defences. Almost 25 per cent of the US coast affected by hurricanes has some obstacle to storm surge in the form of man-made breakwaters, sea walls or the use of dunes and beach stabilisation measures to limit coastal erosion. Such schemes cannot always protect against severe storm-wave attack. Large sea waves frequently result in the severe scouring of beaches with the inevitable undermining of adjacent roads and buildings. In some tourist areas, such as parts of Florida, natural defences such as mangrove trees have been removed because they are unsightly. In Bangladesh, the low-lying deltaic topography offers few safe refuges from storm surge. Some special-purpose concrete shelters have been constructed and there are proposals for more large raised mounds, or escape platforms, for the most vulnerable villages. The mounds would be of excavated earth, with a top surface area of some 1,000 m² raised between 2–9 m above ground. For an average mound, each village would have to give up about 1 ha of agricultural land, but the side slopes could be used for crops whilst the area excavated around the base of the mound could be used for fish ponds.

Plate 8   A concrete shelter in Bangladesh specially constructed to provide protection from the storm surge associated with a tropical cyclone in the Bay of Bengal. There are few other refuges suitable for emergency evacuation in this low-lying terrain. (Photo: International Federation)

The second aim of hazard-resistant design is to reduce the property damage from tropical cyclones. The causes of wind-induced building failure are well-known (Key, 1995). For example, most losses in the USA occur as a result of structural damage to roofs. Often shingles and other roofing materials are disturbed by hurricane-force winds, a process which then allows rain to penetrate the building and cause additional damage to furnishings and fixtures. Most of these losses could be avoided if a few hundred extra dollars were spent at the house construction phase. Better waterproofing of the roof, the use of hurricane clips, rather than staples, for fastening roof cladding and roof sheathing, and the fitting of storm shutters to resist damage from wind-borne debris would do a great deal to reduce damage (Ayscue, 1996). In some other countries the gradual switch from wooden to masonry buildings has helped to reduce hurricane damage but there is still a need for careful design to resist the most severe storms.

The key to the mitigation of storm damage lies in adequate, and properly enforced, building codes. A comparative study of hurricanes in Texas and North Carolina showed that nearly 70 per cent of the damage to residential property was due to poor enforcement of the building codes, and it has been claimed that 25–40 per cent of the insured losses associated with hurricane 'Andrew' were avoidable (Mulady, 1994). Problems of uneven code enforcement usually arise due to lack of funds and qualified staff plus the poor training of site inspectors. However, there are signs of changing attitudes in the United States. For example, following hurricane 'Hugo' in 1989, Surfside Beach became the first community in South Carolina to adopt the high-wind design standard in the Southern building code. After hurricane 'Andrew' the South Florida building code was also strengthened. All new buildings are intended to be constructed with permanent storm shutters and to be protected from wind-borne debris. Exterior windows or shutters must pass a missile impact test with a 4 kg piece of timber striking at a speed of 15 m s^{-1}, and shingle and tiles must be tested as a system at 49 m s^{-1}. In addition, the Florida state legislature mandated that all new educational facilities in the state be designed to serve as public hurricane shelters.

VULNERABILITY MODIFICATION ADJUSTMENTS

Community preparedness

An effective public response to severe storm warnings depends on good community preparedness. This means knowing what specific actions to take when a warning is received. For example, Emergency Preparedness Canada has distributed self-help leaflets which address a variety of situations. The *Severe Storms* edition contains general advice on pre-planning for all wind-storm emergencies, including the preparation of an emergency pack, the trimming of dead or rotting trees around the home, the prior choice of a shelter (such as a basement or place beneath the stairs) plus the designation of a rendezvous point where separated members of a family could meet. In the case of tornadoes, it stresses the need to shelter indoors within inner rooms well away from windows and to get as close to the floor as possible. If caught outdoors, people are advised to leave their cars and seek shelter in a ditch or other depression. Within the towns in some tornado-prone areas, such as the Midwest of the USA, substantial public buildings are clearly identified as tornado shelters.

Hazard preparedness has great potential for reducing cyclone disasters within the LDCs. One example is the Pan-Caribbean Disaster Preparedness and Prevention Project (PCDPPP) which was the first regional scheme to be established, in 1980 (Toulmin, 1987). The PCDPPP project concentrated on technical assistance, the training of island nationals in emergency health and water supply provisions, and the preparation of training materials. These plans were given a practical test with hurricanes 'Gilbert' in 1988 and 'Hugo' in 1989. On the island of Jamaica fewer deaths were experienced than might have been expected. Thus, hurricane 'Charlie' in 1951 caused 152 deaths, compared with 45 from 'Gilbert' in 1988, despite the fact that the latter's damaging winds lasted longer and affected more of the island, which also had a much higher population in 1988 than in 1951. In Bangladesh the workshops organised through the national Cyclone Preparedness Programme (CPP) and the Cyclone Education Project have been credited with saving many thousands of lives in the cyclones experienced during the 1990s (Southern, 2000).

Forecasting and warning

Forecasting and warning systems are in existence for most storm hazards. Several national meteorological agencies operate a tiered pattern of public information which distinguishes between storm 'watches' and 'warnings'. In Australia, tropical cyclone warning progresses through a *watch phase*, initiated 48 hours before storm force winds are expected to reach the coast, to a *warning phase*, when storm winds are expected within 24 hours, and finally a *flash message phase* if any significant changes occur. Near landfall, warnings may be issued hourly and will contain information on storm surge and flood rainfall for the specific coastal zones under threat. In the USA warnings and watches are issued by the National Weather Service for hurricanes, floods, tornadoes and severe thunderstorms. These are distributed to a range of federal, state and local authorities such as FEMA, Corps of Engineers, state and local emergency services and the media. They are also released via the Internet. Depending on the type of severe storm, forecasts are available on a variety of time scales: long range (more than 10 days), intermediate range (3–10 days), short range (1–3 days), very short range (a few hours) and 'nowcasts' (events in progress). In the USA a *hurricane watch* is issued to advise a specified coastal sector that it has at least a 50 per cent chance of experiencing a direct strike from a tropical cyclone of hurricane force with 36 hours (Southern, 2000). A *hurricane warning* provides similar advice about a strike expected within 24 hours (minimum 12–18 hours). Simultaneous public bulletins advise about evacuation procedures or on measures to protect property.

For longer-term preparedness, reliable forecasts of the level of storm activity one season ahead would be valuable. The traditional 'forecasting' method is to rely on statistical climatology as a guide but attempts are being made to forecast Atlantic hurricane activity several months ahead (Gray and Landsea, 1992; Landsea *et al.*, 1994). Such methods rely on deriving functional relationships, or *teleconnections*, between global atmospheric and oceanic processes. Since the late 1960s there has been a marked decrease in the number of intense Atlantic hurricanes, a feature which has been associated with the severe 20-year drought in Africa (see Chapter 12). Such features may be linked with changes in the large-scale thermohaline ocean circulation, as described by Broecker (1991). This 'conveyor belt' is a global transport of water associated with the sinking of the cold, salty water in the North Atlantic ocean (Figure 9.7). The deep water then flows

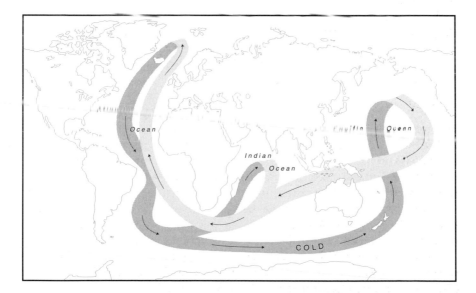

Figure 9.7   A conceptual view of the oceanic thermohaline circulation or 'conveyor belt' associated with the sinking of cold, salty water in the North Atlantic ocean. Variations in the strength of this circulation affect sea surface temperatures and may influence major climatic hazards, such as the frequency of Atlantic hurricanes and the duration of drought in sub-Saharan Africa.
Source: After Broecker (1991)

southwards and eastwards through the Indian ocean and wells up in the western Pacific. It then completes the circuit by returning as a surface flow westward through the Indian ocean before turning northwards to reach the North Atlantic again. It has been hypothesised by Street-Perrott and Perrott (1990) that this circulation has slowed down over the last 20–30 years. As a result, less heat has been transported across the equator in the Atlantic, leading to lower sea surface temperatures and a reduced frequency of Atlantic hurricanes. If the circulation speeds up again in the coming years, increased hurricane activity may well be found in the Caribbean and along the east coast of the USA.

Early and accurate storm detection is a prime requirement of all warning systems. Tornado watch programmes, linked to Doppler radar systems, now allow communities a short time, perhaps 2–3 hours, to prepare for the event. The progressive reduction in tornado deaths implies quite a high skill in the short-range forecasting of severe weather, coupled with increasing preparedness through tornado drills, spotter groups and other activities. Satellite sensing enables a tropical cyclone to be located with an average accuracy of about 100 km, although intense storms with well-developed 'eyes' can often be detected to within 30–50 km. When a cyclone has moved within some 250 km of the coast, weather radar permits a more accurate fix on the position, usually to within about 10 km. The National Hurricane Center in Miami maintains continuous real-time monitoring of such storms, often by flying specially equipped aircraft through the system, in order to forecast the subsequent path and ultimate landfall. Different computer models, some based on current synoptic conditions and some based on previous storm tracks, are used to predict the landfall site up to three days ahead.

Until recently, much less progress had been achieved in the LDCs. For example, the 1970 Bangladesh cyclone was capable of being forecast by an onshore radar station and satellite imagery but the warnings issued were inadequate and the coastal evacuation procedures, designed to take advantage of the limited elevation provided by flood embankments and some community centres, were never operated. Since then, preparedness and emergency management systems have been improved. For example, when the 1991 cyclone struck, two days' warning was given to the areas at risk and over 20,000 trained volunteers went into the field to encourage evacuation to the concrete cyclone shelters which have been built since 1970. As with all hazard forecasting and warning schemes, it is important to check the likely effectiveness of the arrangements before the event occurs, and Parker (1999) has described a methodology for evaluating a tropical cyclone warning system using Mauritius as an example.

A major purpose of severe storm warnings is to save lives by evacuating people from areas at risk. Evacuation efficiency in the USA has been greatly improved by the powers available to some city managers, despite the fact that the authority to enforce hurricane evacuation orders remains variable from state to state. For example, Baker (2000) cited maximum evacuation participation rates of over 90 per cent for communities living on vulnerable beaches and barrier islands whenever officials have been successful in reaching the residents with evacuation notices ahead of a severe storm. Evacuation from the path of hurricane 'Andrew', where the orders were disseminated by the media and directly by the police driving through evacuation zones with loudspeakers, was judged a success, mainly because of the low loss of life. However, according to Peacock *et al.* (1997), only 54 per cent of the threatened households evacuated entirely, although this proportion did rise to over 70 per cent in the lowest-lying coastal zone. It is clear that people consider a wide variety of factors when responding to emergencies and tend to make their own assessment of what is suitable behaviour. For example, recent experience of 'false' hurricane warnings and 'unnecessary' evacuation orders appeared to have little influence on the likely future evacuation behaviour of residents in South Carolina (Dow and Cutter, 1997).

With the increasing congregation of population in coastal areas, limitations of hurricane warning include the availability of shelters and the lead-time necessary for certain communities to achieve total evacuation. Although comparatively few evacuees in the USA seek refuge in public shelters, the 'shadow' evacuation of up to 10–20 per cent of additional households not under direct threat adds to the pressure on relief agencies and the evacuation routes. With a 12-hour lead-time in hurricane warning, it is doubtful if all the inhabitants on Key Biscayne and Virginia Key, Florida, could be successfully evacuated. Since a large proportion of the population are relative newcomers who have never experienced a severe hurricane, the warning response rate could be low. As much as six hours before hurricane landfall, the storm surge may start to flood low points on the highways, whilst the highway network itself would struggle with the peak traffic flows. The Lower Florida Keys lie over 100 km from the closest mainland and it has been estimated that it could take more than 30 hours to evacuate this area. Although many residents of the Florida Keys are absent in the summer, a late season storm could prove disastrous as it would be less expected and would involve the evacuation of large numbers of people along US Highway 1.

Land use planning

Land use is closely tied to the planning process and restrictions on new near-shore development are an important tool in reducing losses from hurricanes. This is because much of the structural damage along the coast is caused by the wind-driven waves which expend their energy on the foreshore dunes or the houses within 100 m of the shore. But rapid changes in coastal land use, together with the perceived desirability of waterfront locations, continue to expose more people and buildings to storm surges.

Land use planning, operated through zoning ordinances, has operated in the vulnerable Lower Florida Keys area, USA, since at least 1960. According to Cross (1985), it has had comparatively little influence on residential development because the home-seeker's primary requirement is to secure a water-front property with access to deep-water boating facilities. These aspirations conflict with the hazard from storm flooding because virtually all the land lies in designated flood zones less than 1.5 m above sea level. Most of the recent home-buyers were informed by estate agents about the hazard but, because of the declared desire of purchasers to be as close to the sea as possible, only 12 per cent felt that the information made any difference to their home selection.

Since 1975 all newly constructed houses have had to have floors at least 2.4 m above mean sea level (the 1:100-year flood level) in order to comply with the National Flood Insurance Program. Although 90 per cent of new residential construction is on elevated stilts, the value of this mitigation adoption has been limited by the building of enclosed garages and recreation rooms in the space below the property. Moreover, ground-level houses remain an attractive option, especially for more elderly residents, and still account for approximately one-third of all homes sold. The determination to colonise such hazard zones was confirmed by the fact that more than nine out of ten residents indicated that they intended to rebuild their homes at precisely the same location if the property were ever destroyed by a hurricane.

Without the political will to create better maps of coastal risk and enforce both building codes and zoning bye-laws, the hurricane threat will remain strong. As with building codes, there is now some evidence that hurricane-prone communities in the USA have become willing to slow the rate of population growth, especially in areas which need to evacuate in advance of a landfall. According to Baker (2000), the island of Sanibel, Florida, accepted restrictions on the annual number of new housing units as early as 1977, citing the potential need for emergency evacuation as the prime reason. Since then the state of Florida has regulated some of the residential and commercial developments which would have adverse effects on either emergency shelter demands or evacuation clearance times. So far, there is little evidence of the beneficial effects of such measures on the human response to hurricane hazards.

10 Biophysical hazards

Temperature extremes, epidemics and wildfires

BIOPHYSICAL HAZARDS

The term 'biophysical hazards' covers a wide spectrum of environmental risk which exists because of interactions between the geophysical environment and biological organisms, including humans. In some cases, variations in the physical environment create the hazard directly, as when periods of unusually hot or cold weather threaten human life through physiological stress. The average human body is most efficient at a core temperature of 37°C and, compared with the natural variations of air temperature, physiological comfort and safety can be maintained within only a narrow thermal range. If the heat balance of the body is modified beyond this critical zone, physiological stress results. Irreversible deterioration and death often occurs if the internal body temperature falls below 26°C or rises above 40°C.

In other instances, the hazard arises from the biological end of the spectrum. Disease is a very common environmental hazard and can lead to disaster when a pathogen (virus, bacterium or parasite) creates a disease outbreak amongst a human population which lacks immunity. For some diseases, vectors are necessary to carry the infection from a carrier to the potential victim, for example, the mosquito for malaria or lice for typhus. Major epidemics, called *pandemics*, have occurred throughout history. The Black Death pandemic is reputed to have killed more than 50 million people, up to one-third of the population of Europe, during the fourteenth century. More recently, the influenza pandemic of 1918–19 swept the world. At the most conservative estimate, 21 million people died, more than twice the number of people killed during World War I. Lung damage was the major cause of death at a time when antibiotics were unavailable. About 16 million people died in India, while in New York more than one in every hundred persons was killed. Sometimes epidemics occur as a result of natural disasters, as when floods contaminate drinking water supplies or destroy sewerage systems. For the most part, however, disease is endemic in many poor countries and an epidemic can be triggered by quite minor events.

Other biophysical hazards, often triggered by atmospheric events, disrupt the assumptions of stability upon which agricultural output depends. For example, the economic well-being of some fruit farmers may be threatened by temperature extremes, such as frost. More seriously, a rapid upsurge in pests and diseases may threaten food security on a wider scale. An outbreak of potato blight disease (*Phytophthora infestans*) in mid-nineteenth-century Ireland was triggered by unusually warm and wet weather in 1845 and led to at least 1.5 million famine-related deaths over the following three years and the forced emigration of a further 1 million people.

Like most organisms, pests produce more offspring than required for replacement of the adults when they die, and the stability of populations is usually maintained through high juvenile mortality. If the environmental factors controlling juvenile mortality are eased, many species have the capacity to attain plague proportions very quickly. This is most likely to occur in arid and semi-arid areas which, although normally harsh, low-productivity environments, can be transformed by rains. The desert locust (*Schistocerca gregaria*) responds quickly to climatic conditions which enable it to switch from a solitary phase to a swarming phase. Rain encourages the female to lay her eggs in wet ground and provides vegetative growth to feed the immature, wingless locusts after they have hatched. Once they become adult, winged locusts migrate between areas of recent rainfall, some-times over thousands of kilometres in a matter of weeks, and can consume vast quantities of crops. Together with grasshoppers, locusts also consume the natural flora. If plant regeneration is hampered by moisture stress, this can lead to a prolonged loss of vegetation cover which, in turn, can accel-erate environmental degradation. Present locust control strategies in Africa, based on early-warning surveys followed by large-scale insecticide spraying, have been estimated to cost more than US$10 million per year (Krall, 1995). Most of the funds come from overseas aid and, given the difficulty of proving a positive cost–benefit relationship for the low-value crops protected, it is doubtful if the current approach is successful.

A particular combination of vegetative fuel and weather conditions will result in *wildfires*. They may be triggered either by natural events, such as lightning strikes, or by human actions, such as sparks from a campfire. Some wildfires start in comparatively remote areas and may be confined to forest and brushland; but others reflect the increasing recreational and devel-opment pressures at the interface between wildland and urban areas. All wildfires develop because material is sufficiently dry to burn and prevailing weather conditions encourage the fire to spread. Unusually hot, dry weather is the main common factor.

PHYSIOLOGICAL HAZARDS

Physiological cold stress to the human body is created by combinations of low temperature and high windspeeds. *Windchill* or *cold stress* hazard is a normal feature of high latitude and high altitude areas. The greatest threat is associated with the unexpected outbursts of cold air into the mid-latitudes during severe winter spells. The development of a winter ridge of high pressure over north-western North America often encourages arctic air to penetrate through the Midwest and may bring frost conditions as far south as Florida. Three days of cold during December 1983 produced at least 150 weather-related deaths throughout the USA, although the total number may well have been nearer 400 (Mogil *et al.*, 1984). Some of these deaths were from hypothermia and cold-aggravated illness while many others were caused indirectly by snow shovelling, exposure, house fires resulting from the use of emergency heaters and automobile accidents. In warmer climates surges of polar air can create health risks at much higher temperatures.

Abnormally high temperatures can also create a directly life-threatening hazard by imposing severe *heat stress* on the human body. The amount by which the temperature exceeds the mean is more important than the absolute value of temperature. Under conditions of relatively high temperature and humidity, physical discomfort soon turns into disease and mortality. After several days of excessive heat, the mortality rate tends to rise to two and three times the normal seasonal rate. About 20,000 people died in the USA in the 40-year period 1936–75 from prolonged heat and humidity (Quayle and Doehring, 1981). This hazard is likely to increase as a direct result of global warming. For example, over 5,000 excess deaths attributed to heat stress were recorded in the USA in 1979–80, a greater mortality than that from all other weather-related disasters combined.

Extreme heat waves can appear in about half the United States, mainly affecting the elderly (over 65 years of age) and those with existing heart disease. Other high-risk groups comprise the urban poor, especially people who lack domestic air conditioning or who are dependent on alcohol or drugs. Mortality tends to be highest in the first heat wave of the season before acclimatisation has been established. One example occurred in Illinois in July 1966 where in some areas a 36 per cent increase in deaths was recorded over a five-day period (Bridger and Helfand, 1968). In September 1955 a heat wave associated with dry south-easterly winds created 946 excess deaths in Los Angeles, more than twice the mortality recorded in the 1906 San Francisco earthquake and fire (Oechsli and Buechly, 1970). Urban areas pose the greatest risks of thermal stress, partly because of the enhanced heat-island effect and partly because of endemic socio-economic problems in the inner cities. In an attempt to quantify these effects, Beechley *et al.* (1972) studied a July 1966 heatwave in the New York–New Jersey metropolitan area and claimed that 150–200 extra deaths occurred in the city core over the number of fatalities that would have occurred under suburban heat levels and living conditions.

As far as thermal stress on humans is concerned, the most common hazard adjustment is insulation through suitable clothing and housing. In the MDCs, central heating and air conditioning are widespread responses to cold and heat respectively. Where this level of protection is not available, the shading of houses by trees against direct solar radiation may be an adequate response to heat stress. The increased frequency of heat-stress episodes is likely to be one of the most direct consequences of global warming and there may be a doubling of heat-related deaths worldwide by 2020. The wider introduction of domestic air conditioning could offset some of this projected increase in mortality but the widespread lack of such facilities for most inhabitants of Third World countries suggests that they will suffer disproportionately from this aspect of climate change.

FROST HAZARDS IN AGRICULTURE

Frost is a hazard to many crops in the middle latitudes. The major threat exists from late spring frosts which occur when orchard trees are in blossom and high-value horticultural crops are almost ready for early marketing. The majority of frosts are *radiation frosts*, which occur on still and cloud-less nights when the ground surface cools rapidly to produce a strong temperature inversion in the lowest atmospheric layers. This causes marked temperature gradients both above and within the canopy of many plants. *Wind frost* is generally much rarer, occurring when air has been cooled else-where and then advected across the area at risk by a cold airstream. Although less frequent, such events are often more damaging owing to their wide-spread nature.

The risk of frost is specific to individual crops and to critical periods in their development. Orchard and citrus fruits are most vulnerable at the early flowering period and also at the budding stage since, if either the blossom or the bud is destroyed, the tree will not produce fruit that year. For soft fruits, such as blackcurrants and strawberries, damage can even occur after fruiting if temperatures fall sufficiently low to freeze water solutions within the plant cells and cause structural damage. Some 'light' frosts may be beneficial. For example, the fruit grower may welcome a short frost down to $-2.0°C$ which destroys a little fruit but ensures that the remainder has space to develop fully.

Once the critical temperature which first causes frost damage to a specific crop is determined, estimates of the long-term risk can be made based on the minimum temperatures which have occurred during the sensitive growth phase, as shown in Table 10.1. In practice, experienced farmers rely on individual hazard perception, which often shows a good relationship to damaging frost thresholds. Horticultural crops can justify a high invest-ment in semi-permanent frost protection. The shelter afforded by plastic or glass structures creates an artificial climate, thus protecting against both wind frosts and radiation frosts. However, many farmers in frost-prone areas make use of weather forecasts before taking emergency frost-reducing measures. Most meteorological agencies offer a specialised frost warning

Table 10.1 Agricultural frost risk classification

Criteria	Frequency (%)	Rating	Class
Frost very rare	0–2	0	No risk
Frost rare (once or twice per generation)	2–8	1	Slight risk
Frost once or twice in 10 years	8–20	2	Medium risk
Frost very frequent	20–50	3	High risk
Frost nearly every year	50–100	4	Very high risk

service which is often available from mid-day onwards, giving the farmer 12 hours or more lead-time to take action.

Overhead sprinkler irrigation can be used to prevent frost damage in orchards and in intensive vegetable plots. The latent heat liberated when a fine spray of water is applied to the plant surfaces prevents the temperature of buds, flowers and small fruits from falling below freezing point. A continuous spray ensures a mixture of ice and water with a temperature no lower than 0°C, but the sprinklers must be kept going until all the ice has melted. This method can be used in both radiation and wind frosts but a heavy ice load may build up on orchard trees and cause limb breakage. If sprinkling is required several nights in a row, the ground may become waterlogged.

Direct orchard heating is the oldest and most common method employed in the California citrus groves. Oil was traditionally burned in cans and pails holding from 10–20 l of oil, but the burners were objectionable in built-up areas due to smoke pollution. Increasingly, small stack heaters have been used to provide more complete combustion of the oil, or users have turned to other fuels. Large heat outputs are required to combat moderate radiation frost, and the cost of heating a small isolated orchard is much higher per unit area than that required for a larger orchard under the same frost conditions. Better results are obtained with a large number of small fires than with a small number of large fires, and generally about 125 heaters per hectare will be necessary to achieve a net warming of 5–6°C in the tree canopy.

Wind machines have been used since the late 1930s and are widespread in both the USA and Canada. This method works only for radiation frosts, with strong thermal inversions giving layers of warmer air above the orchard level. Under these conditions, large fans pull the warm air down and mix it with the colder air below. Such machines operate best with inversions of 5–7°C between screen height and 20 m above the surface. Several experiments have been conducted and, depending on frost severity, a single machine can provide some protection for an area ranging from 2 to 4 ha.

One of the best responses is hazard avoidance through the choice of frost-free sites. In general, the best sites are on open, gently sloping ground, especially near large bodies of water. Large lakes release heat during cold nights, and in Florida many orange groves are on the south side of inland lakes to gain protection from wind frosts arriving from the north. Site selection is easiest against radiation frosts, since a slope of 2° is needed to promote katabatic drainage. Major frost risk zones can be identified from topographic maps. Very often, however, topographic factors operate on much smaller scales, as for example with the pooling of cool air on mid-slope valley benches. Such a problem occurs in the Okanagan valley of British Columbia, Canada, where the benches offer relatively flat, good land for orchards. Sites which should be avoided are those where cold air ponds up on slopes behind ridges or trees, nor should orchard trees be planted in dense rows across the slope, encouraging the stagnation of cold air.

EPIDEMICS

The World Health Organization has defined an epidemic as:

> the occurrence of a number of cases of a disease, known or suspected to be of infectious or parasitic origin, that is unusually large or unexpected for the given place and time. An epidemic often evolves rapidly so that a quick response is required.

From this definition it is clear that epidemics share many characteristics of rapid-onset environmental hazards. Bacterial, viral and parasitic infections are still capable of causing large-scale disasters, especially in the poorest countries of the world where the control of cholera, malaria, meningitis and yellow fever is far from satisfactory.

Over millions of years, biological adaptation has enabled *Homo sapiens* to evolve through a process of natural selection which has helped the species to resist disease. For example, many Africans possess a sickle-cell trait in their blood which provides some immunity to malaria and which has developed as a result of prolonged exposure to the disease (Burton *et al.*, 1993). More generally, migrations and inter-marriage have produced genetic diversity which promotes resistance because it reduces the chance of a new, virulent organism being introduced into a susceptible community. In more recent times, the spread of knowledge about public health, with an emphasis on clean water supplies and effective sanitation, has reduced the incidence of many diseases.

Despite these developments, epidemics resulting from well-established pathogens, including cholera, malaria and meningitis, kill about 10,000 people per year and affect a further 300,000. In the LDCs, especially the poorest countries of Asia and Africa, the illness associated with diseases such as measles and tuberculosis, which have been largely eradicated from the MDCs, has a continuing debilitating effect on people. It reduces their ability to produce food or earn a living and thereby makes them more vulnerable to other hazards. Other diseases pose a quite different threat. The AIDS pandemic started to spread in the mid-1970s and by 1993 around 13 million people were infected with the Human Immunodeficiency Virus (HIV). This represented about one in every 250 adults worldwide (IFRCRCS, 1993). On average, the time from infection by the HIV to the development of AIDS is 10 years and approximately 90 per cent of those with the disease die. By the late 1990s, AIDS had become the fourth biggest killer worldwide and over 2.5 million people per year were dying as the epidemic spread from Africa to Asia as a result of poor health, poverty and drug abuse.

The conditions necessary for epidemics often exist before the outbreak occurs and the disease may well be endemic in the region. In all cases, a pathogen needs to be introduced into a suitable environment, followed by an incubation period. The outbreak itself is often triggered by other events, which may include natural disasters, since these provoke large population movements and the subsequent overcrowding and poor sanitation found in refugee camps (Morris *et al.*, 1982). The migrants themselves may bring in new pathogens or may move into a contaminated area and catch the disease because of their lower resistance. But the underlying predisposing factor is poverty. Infants and children in the LDCs are several hundred times more likely to die from diarrhoea, pneumonia and measles than those in Europe or North America (Cairncross *et al.*, 1990). Poor housing, malnutrition, lack of hygiene to protect against vectors, inadequate clean water supplies and restricted access to health care facilities all play a part. For example, in Peru the poor suffer from more than twenty water-borne diseases and a major cholera outbreak in the early 1990s was attributed to a combination of contaminated water supplies and poor hygiene (Witt and Reiff, 1991). Many urban centres lack a safe sewerage system and render their populations vulnerable to many diseases. As with other disasters, the main impact of epidemics is on the young, the old and the disadvantaged.

Despite the above comments, it appears that major outbreaks of communicable diseases after natural disasters are comparatively rare, although considerable potential exists. According to Seaman *et al.* (1984), such a disaster-related disease outbreak could occur as a result of six factors:

1   the diseases present in the population before the event;
2   ecological changes resulting from a disaster;
3   population movements;
4   damage to public utilities;
5   the disruption of disease control programmes;
6   altered individual resistance to disease.

Such influences often combine. For example, vector-borne diseases, especially malaria and yellow fever, commonly increase after floods in tropical areas due to the increase in mosquito (and other insect vector) breeding sites and the greater exposure of the population caused by the loss of housing. In the worst cases, malarial epidemics may spread to urban areas previously free of such diseases. Other sources of epidemic disease are rats, which act as reservoirs of plague and often emerge from sewers after floods, and abandoned dogs infected with rabies which are more likely to bite humans following a breakdown in living conditions. One of the best-documented examples of a vector-borne epidemic following a natural disaster is still that of the malarial outbreak after hurricane 'Flora' crossed the southern peninsula of Haiti in October 1963 (Mason and Cavalie, 1965). Most of the houses in the area were destroyed, leaving about 200,000 people homeless, and it was estimated that some 75,000 cases of malaria occurred there over the following six months. The epidemic was attributed to a combination of an incomplete malaria eradication programme, the removal of insecticide from houses by heavy rain, a rapid increase of mosquito breeding due to flooding and a lack of shelter for the local population, which became more mobile in their search for food, construction materials and employment.

The physical disruption of water supply and sewage disposal systems, especially in areas where sanitation levels are already minimal, can be an important factor. In Bangladesh about four-fifths of the population rely on tube wells for drinking water and use surface sources – such as shallow ponds – for bathing, washing and cooking. After the 1991 cyclone, about 40 per cent of the tube wells were damaged and the surface sources became highly contaminated with sewage and salt (Hoque *et al.*, 1993). As a result, water became extremely scarce and there was a large increase in diarrhoeal diseases.

Climate change is likely to encourage the latitudinal spread of some vector-borne diseases presently restricted to the tropics. Malaria is the major mosquito-borne disease, and about 40 per cent of the world's population is currently at risk of infection. The transfer of malaria from mosquitoes to humans is highly dependent on temperature and is most effective within the range 15–32°C with a relative humidity of 50–60 per cent (Weihe and Mertens, 1991). The largest changes in potential increase are expected to occur at the latitudinal and altitudinal limits of the present risk areas. As a result of global warming and an extension of warm, humid seasons into higher altitudes within the tropics, seasonal epidemics of malaria could spread into new areas where the population has little or no immunity to the disease and where health care is limited (Martens *et al.*, 1998). The accuracy of these estimates depends on the performance of the climate models, assumptions about future greenhouse gas emissions and population changes. Figure 10.1 shows the potential spread of *P. falciparum* malaria, which is clinically more dangerous than the more widely distributed *vivax* form of the disease, into the temperate zones and beyond the approximate current limits of latitude 50° north and south. Under the worst-case scenario, an additional 290 million people worldwide could be at risk by the 2080s,

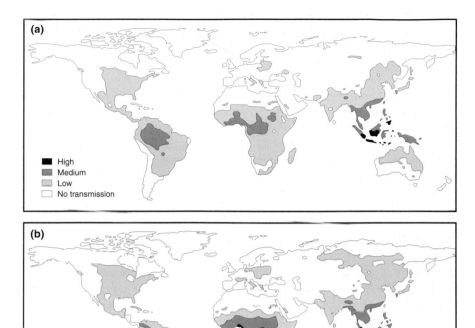

Figure 10.1   The potential spread of high-, medium and low-risk areas for malaria
(*P. falciparum*) from (a) the baseline climatic conditions of 1961–90
to (b) the climate change scenario estimated for the 2050s
Source: After Martens *et al.* (1998)

with the greatest increases in risk in China and central Asia as well as the
eastern USA and Europe. Although malaria is unlikely to take hold in
developed countries with effective health services, some estimates indicate
that, as early as 2100, 60 per cent of the world population will be at risk
unless precautionary health programmes are adopted. Any major increase
in the spread of infectious disease or the incidence of heat-stress deaths
within the MDCs could affect the insurance industry through the provision of life assurance and pensions policies.

Epidemic control depends in the short term on comprehensive immunisation and pesticide application programmes. Global immunisation
programmes in the 1960s and 1970s largely eradicated small pox, and the
World Health Organization (WHO) is now taking epidemics seriously. The
WHO reorganised its emergency division during 1993 in an attempt to
improve disaster response. If a disease is well known, like cholera, it may

be quickly recognised and action can be taken to control the epidemic. Mass vaccinations could similarly save millions of lives each year from common diseases such as tuberculosis, measles, whooping cough, tetanus and diphtheria. In the longer term, better domestic water supply and sanitation are necessary, especially to combat diseases like cholera which are spread by contaminated food and water. Such improvements must be reinforced by primary health care programmes, including local pharmacies and reference laboratories and good community health practice through regional workshops for emergency health managers and local officials. Pesticide sprays are still used to control some epidemics. Although liquid pesticides are effective in controlling diseases like malaria, the longer-term ecological effects of large-scale applications are not well understood.

WILDFIRE HAZARDS

Wildfire is a generic term for uncontrolled fires. In the past, these fires have been a rural hazard but, increasingly, they break out on the rural–urban fringe with a potential to threaten major population centres. In Australia the term *bushfire* is used to denote any fire in a rural area. In North America a *brushfire* is a conflagration whose chief fuel is low, scrubby vegetation. Apart from Antarctica, no continent is entirely free from the combinations of ignition source, fuel and weather conditions necessary for a wildfire hazard. In general, high temperatures and drought following an active period of vegetation growth provide the most dangerous conditions. This means that the most hazardous zones tend to have either a Mediterranean or a continental climate with either *xerophytic* or *sclerophyllous* vegetation. In the former, most of the rain which supports the natural cover of forest and grassland, as well as the agricultural crops, falls in the winter so that the vegetation is dry during the annual summer drought. Many of these areas, such as California or southern France, are popular holiday destinations in the summer with the added risk of fire started by tourists. In both Corsica and the Var region of France, which includes resorts such as St Tropez, some 5,000 ha of forest is burned annually. In addition, most continental interiors experience dry air for much of the year and have a long fire season.

It is likely that the fires, which burned some 1,700,000 ha in Wisconsin and Michigan in October 1871 and claimed the deaths of about 1,500 people, created the world's greatest wildfire disaster. These fires, which broke out on 8 October, the same night that a fire in Chicago killed 250 people, were preceded by a drought in the Midwest that had lasted for fourteen weeks. Many small fire outbreaks in the forests surrounding Peshtigo and other small townships were not considered a real threat until strong winds whipped up the flames and created uncontrollable spot fires within the firebreaks that had been constructed around the settlements. The disastrous 'Ash Wednesday' fires which affected large parts of Victoria and

South Australia in February 1983 were caused by a classic 'fire weather' situation with air temperatures up to 40°C combined with windspeeds over 20 m s^{-1}. Seventy-six people died, 8,000 were made homeless and the estimated direct losses were put at A\$200 million (Bardsley *et al.*, 1983). Bushfires in New South Wales between December 1993 and January 1994 affected more than 1 million ha and destroyed 200 buildings but killed only four people. At their height, more than 300 fires were burning along almost the entire 1,100 km coastline, fanned by strong winds caused by unseasonally deep lows off eastern Australia.

Australia is the most fire-prone country in the world. Fires have been started by lightning strikes for at least 100 million years and most native vegetation is adapted to regular burning. But lightning is now responsible for less than 10 per cent of the 2,000 or so wildfires which occur each year, many of which are started illegally. In Australia, as in North America, a major fire can extend over 100,000 ha. During the 1974–5 season, an estimated 15 per cent of the continent was burned, although this was largely in remote, arid land and the level of damage was relatively low. The major characteristic of Australian fires is the speed with which they spread. According to Mercer (1971), Australian wildfires can engulf up to 400 ha of forest in 30 minutes compared with as little as 0.5 ha over the same period in the slower-burning coniferous forests of the northern hemisphere. The general unavailability of large water sources in Australia means that many fires end only with the arrival of rainfall.

Rural wildfires damage ecosystems. After a major event, timber and forage may be destroyed, animal habitats disrupted, soil nutrient stores depleted and amenity value greatly reduced for many years. When the burned areas consist of steep canyons, debris flows, rill erosion and floods are likely to follow. These fires also adversely influence timber production, outdoor recreation, water supplies and other resources. During a five-day period in the late summer of 1987 over 4,500 known lightning strikes ignited nearly 2,000 fires in the western USA, mainly in northern California and southwestern Oregon. The US Department of Agriculture estimated that its emergency fire suppression actions cost US\$100 million and that a further US\$4–5 million was required for land rehabilitation measures which included the seeding of steep slopes, clearing of sediment from stream channels and replacement of bridges and culverts.

Figure 10.2 illustrates that, between the mid-1980s and the mid-1990s, both the number of large wildfires and the area burned in all national forests in the USA increased by about four times. The greatest threat exists in the dry, inland part of the western United States, where over 15 million ha of forests are at risk of catastrophic wildfires (Hill, 1998). Since 1990, over 90 per cent of all large (>400 ha) forest fires and over 95 per cent of the area burned in the USA have been in this region. In 1988 nearly 300,000 ha of the Yellowstone National Park was burned out, despite the efforts of more than 9,000 fire fighters; this raised important issues about fire

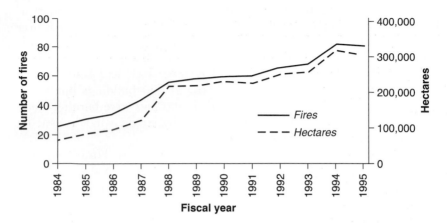

Figure 10.2   The cumulative increase in the number of wildfires, together with the total area burned, in all national forests in the USA, 1984–95. Source: After Hill (1998)

management strategies in rural areas with a high-profile, heritage status (Romme and Despain, 1989). Rural wildfires pose a particular threat to fire-fighters. In the South Canyon, Colorado, fire of 1994, fourteen fire-fighters were killed when a dry cold front moved into the burning area, which was covered mainly by the pinyon-juniper fuel type. Under strong winds up to 21 m s^{-1}, the fire spotted back across a canyon floor and moved on to very steep slopes covered by highly flammable Gambel oak immediately below the fire-fighters. Within a few seconds the fire, with flames up to 90 m high, spread up the slope at a speed which was impossible to out-run.

The spread of human activities into areas of predominantly natural vegetation has increased the number of wildfires and led to more losses to life and property. In 1990 about one-quarter of the population of the USA lived in rural communities of less than 2,500 people and over 25 per cent of the fires attended by public fire departments occurred in timber, brush or grass in these areas (Rose, 1994). It has been estimated that people in such rural communities are almost twice as likely to die in a fire as people living in communities of 10–100,000 people (Karter, 1992). There is a particular problem in California which has approximately 8 million ha of brush land that is highly flammable. Timber, brush and rangeland fires have become more frequent as development has created a greater urban/wildland 'intermix' and, during the past 14 years, over three times more structures have been consumed by fire than over the previous 25 years (Hazard Mitigation Team, 1994). As shown in Table 10.2, of the twelve fires creating the greatest loss of built structures in California's history, five have occurred since 1990.

Today, wildfires threaten the suburban areas of some of the world's largest cities. The attractions of a rural environment, together with effective

Table 10.2 Twelve highest structure losses in California fires

Date: location	Structures[a]	Casualties
October 1991: Oakland/Berkeley Hills	2,900	25
June 1990: Santa Barbara County	641	1
August 1992: Shasta County	636	0
September 1923: Berkeley	584	0
November 1961: Bel Air	484	0
September 1970: Cleveland National Forest, San Diego County	382	5
October 1993: Laguna Beach	366	0
November 1980: San Bernardino County	325	4
November 1993: Malibu Area	323	3
September 1988: Nevada County	312	0
July 1977: Santa Barbara County	234	0
October 1978: Malibu Area	224	0

Source: California Department of Forestry and Fire Protection; after Hazard Mitigation Team (1994)

Note

a Includes all buildings – homes, outbuildings, etc.

commuting facilities, has encouraged the expansion into natural bushland of low-density suburbs in Sydney, Melbourne and Adelaide in Australia, and Los Angeles and the San Francisco Bay communities in the USA. In Canberra, the Australian capital, a series of semi natural ridges, which are used for open space recreation and nature conservation, run through the city which, in many areas, backs directly on to rural areas without any transitional land uses (Lucas-Smith and McRae, 1993). In 1991 a wildfire in the East Bay Hills area of San Francisco killed twenty-five people, injured more than 150 and made over 5,000 homeless (Platt, 1999). With estimated losses of US$1.5 billion, it was the third most costly urban fire in US history. The fire started under classic conditions of high temperatures, low air humidity and strong winds and spread rapidly aided by a dry vegetation cover. Fire-fighters were hampered by narrow and congested access roads, plus a critical loss of water pressure, and some 60 years of urban development in this area was destroyed, leaving only the building foundations. Handmer (1999) described the wildfire which affected Sydney, Australia, in January 1994. Once again, this was caused by a combination of high fuel loads and hot, dry weather. In this event four deaths occurred and 200 houses were destroyed despite the efforts of over 20,000 fire-fighters mobilised from all over Australia.

Communities threatened by wildfire often have little awareness of the hazard. Without appropriate guidance, home-owners build with highly flammable materials, such as weatherboard or wood shingle roofs, deliberately retain thick vegetation too close to their property and disregard the adequacy of fire-fighting equipment. This is despite the existence of legislation, such as the 1987 Public Resources Code 4291 operative in California, which requires property owners in State Responsibility Areas to remove flammable vegetation for a distance of at least 10 m from a structure or to the property line, whichever is closer. When disaster strikes, the residents are ill-prepared to cope. McKay (1983) has shown that newspaper reports presented the victims of the 'Ash Wednesday' fires in south-east Australia as completely helpless and gave little prominence to reports containing warning or response information, either before or during the event.

Characteristics of wildfires

Of the fires which threaten life and residential areas, the majority – perhaps 80–90 per cent – are due to human actions. The main ignition sources are probably agricultural fires, tourists who discard cigarette ends or let campfires get out of control, and arson. Assuming that a source of ignition exists, the occurrence and severity of wildfires are determined by two interdependent factors: fuel and weather.

Fuel influences both the intensity of the fire (heat energy output) and the rate of spread. Thus, grassland fires rarely produce the intensity of burn and the degree of threat associated with forest land. Apart from its quantity, the moisture content of the fuel is important and this depends on the *weather*. These relationships lead to a marked seasonal procession of risk, illustrated for Australia in Figure 10.3, which is climatically driven by the sequence of rains. According to Cunningham (1984), south-eastern Australia is the most hazardous wildfire region on earth. One major reason is the nature of the fuel in this area. Most Australian forests accumulate a great deal of fuel on the forest floor, mainly from bark shedding. In dry sclerophyll forests litter can accumulate at the rate of 7 t ha yr^{-1}. This will progressively decay but fuel in the *stringybark* forests of the south-east can still build up to 50 t ha after 15 fire-free years. These forests, like many others in Australia, are dominated by the genus *Eucalyptus*, which is itself fire-promoting. This is because eucalypts contain highly volatile oils within the leaves which greatly aid the combustion process. At ambient fire temperatures of around 2,000°C these oils can create a spontaneous gas explosion.

In addition, bark shedding creates a special problem of rapid fire spread known as *spotting*. This occurs when ignited fuel is blown ahead of an advancing fire front by strong winds to create 'spot' fires. Australian eucalypts have the longest spotting distances in the world. The main reason for this is bark shedding by the *stringybark* and *candlebark* species, which produce loose, fibrous tapers easily torn away by strong winds and convection currents. Spotting distances of 30 km or more have been authenticated, at least twice the distance recorded in the deciduous hardwood and coniferous forest fires of North America.

Weather conditions are crucial for wildfires. Drought periods followed by hot, dry winds blowing from arid continental interiors over a period of days

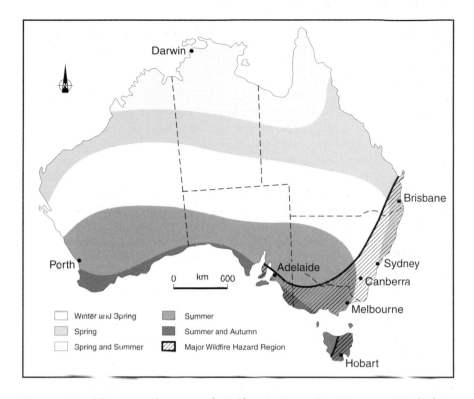

Figure 10.3 The seasonal pattern of wildfires in Australia. The central half of the continent is sparsely vegetated and populated so the major hazard region is confined to the southern and eastern parts of the country shown on the map.
Source: Modified from R.H. Luke and A.G McArthur, *Bushfires in Australia*, Forestry and Timber Bureau, Division of Forest Research, Commonwealth Scientific and Industrial Research Organisation, Canberra: Australian Government Printing Service, 1978

to give a cumulative heating and drying effect on vegetation are particular culprits. These are also the atmospheric conditions which promote dry lightning storms, which are a frequent ignition source. Very exceptional weather conditions can create very extensive fires. During the 1997–8 El Niño event, parts of south-east Asia suffered the worst drought for about 50 years. The rain forests became very dry and, aided by uncontrolled slash-and-burn agricultural activities, widespread fires broke out. Over 5 million ha of forest was burned out in Kalimantan and Sumatra and the associated smoke pall covered large areas of south-east Asia. Fire-related deaths were estimated at 1,000 and about 40,000 people were treated in hospital for the effects of smoke inhalation (IFRCRCS, 1999). Some airports were closed because of poor visibility, and in Kuala Lumpur, Malaysia, reported cases of asthma rose seven-fold above normal levels. Such fires have clear regional-scale impacts but, arguably, are of global significance since the forest clearances are driven by worldwide economic forces and the fires are major sources of greenhouse gas emissions which contribute to global warming.

Once ignited, the rate of wildfire spread is closely related to the surface wind strength. Brotak (1980) compared extreme fire hazard situations in the eastern USA and south-eastern Australia and found many synoptic similarities. Most fire outbreaks occur near surface fronts, particularly in warm, dry conditions ahead of a well-developed cold front with unstable temperature lapse rates and strong winds at low levels. In California, easterly Santa Ana winds, which occur mainly in September and October – the driest and warmest months in the Bay Area – create an extreme hazard in the fall season which may last right through to November (Monteverdi, 1973). Strong north-easterly Santa Ana-type winds developed in late July 1977 and led to a disastrous wildfire which began in the hills and advanced to within a mile of the downtown area of the city of Santa Barbara. Over 230 homes were destroyed (Graham, 1977). In October and November 1993, twenty-one major wildfires developed in six Southern Californian counties fanned by hot, dry Santa Ana winds. Three people were killed, 1,171 structures were destroyed and some 80,000 ha were burned. The combined property loss was estimated at US$1 billion.

Cheney (1979) indicated that most wildfire damage, including loss of life, occurs during a relatively short period of time – usually a few hours – compared with the total duration of the fire. These high-loss episodes are associated with extreme fire risk weather, often involving high winds which shift in direction and cause the fire to accelerate in an unexpected direction. Such fire acceleration can be greatly aided by the topography. For a fire driven upslope, wind and slope acting together increase the propagating heat flux by exposing the vegetation ahead of the fire to additional convective and radiant heat. The combined effect of wind and slope is to position the advancing flames in an acute angle so that, once the slope exceeds 15–20°, the flame front is effectively a sheet moving parallel to the slope. Data from experimental fires in eucalypt and grassland areas in Australia have shown that the rate of forward progress of a fire on level ground doubles on a 10° slope and increases nearly four times when travelling up a 20° slope (Luke and McArthur, 1978).

The combined effect of wind and topography was very evident in the Clare, South Australia, bushfire of 21 February 1965 (Figure 10.4). Starting at 1335 h, the fire first swept south through orchards and vineyards driven by a north-west wind. At 1700 h there was an abrupt change to a south-westerly wind. The fire became uncontrollable along the eastern flank as it raced through pasture and stubble, partly because it was travelling with the grain of the hills and valleys rather than against it. The long tongues of burnt country on Figure 10.4 reflect the nature of the hill–valley topography between Mintaro and Clare.

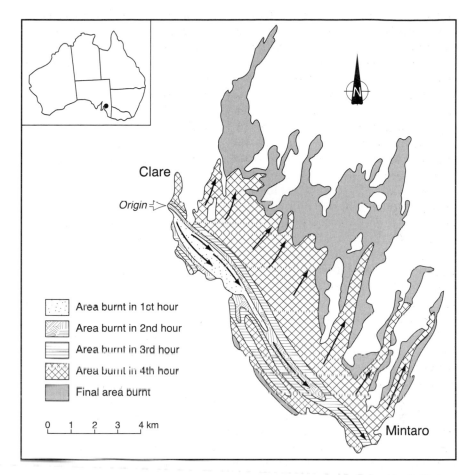

Figure 10.4 The spread of the Clare, South Australia, wildfire on 21 February
1965. The fire was aided by classic high-risk weather conditions,
including an air temperature of 40°C, relative humidity of 10 per
cent and an initial wind from the north-west gusting at speeds of
8–13 m s⁻¹.
Source: After Butler (1976)

Wildfire hazard reduction

Disaster aid has a role when disaster strikes. The 1983 'Ash Wednesday'
fires in Victoria and South Australia raised a total donation of some A$12
million which was channelled through an appeal fund administered by the
Department of Community Welfare (Healey *et al.*, 1985). About three-
quarters of this sum originated within Australia itself. Other assistance
came from the federal government under the National Disaster Relief

Arrangements. A large part of the assistance was in the form of interest-free repayable loans rather than direct grants.

Disaster appeals always raise the issue of whether people with insurance who incur property losses should be compensated from appeal funds to the same extent as those who did not have private insurance. Clearly, if all are treated alike by disaster compensation, there is little incentive for a hazard-zone occupant to take out insurance. Current fire insurance arrangements tend to rely on the private sector. For example, about two-thirds of all the home-owners affected by the 1991 East Bay Hills fire had replacement-cost insurance cover. This was a major factor in defraying the recovery costs to the federal government and also ensured that recovery and rebuilding went ahead quickly (Platt, 1999). However, there is usually no real differentiation in private policy premiums according to risk. At best, the standard residential policy considers only the presence or absence of adequate fire-fighting services when premiums are set. In the future, there is scope for premiums to vary in response to the effectiveness of the community in enforcing fire-safe building codes and vegetation management. For example, roofing materials have long been recognised as a risk factor that can increase the chances of a structure igniting, so a premium reduction could be offered for fire-resistant roofing materials.

Arson is the suspected cause of many fires. In California, it is estimated that about one-quarter of all wildfires are due to arson but that only about 10 per cent of arson investigations lead to an arrest. Consequently, after wildfire disasters, it is usual to see increased demands for fire ban legislation. This is unlikely to be effective unless it can be properly enforced. Fire bans on specific days of high risk are often necessary but legislation must also match public expectations. In Australia the procedures for Total Fire Bans, when no fires may be lit in the open, ensure that legitimate exceptions can be made. Total fire bans may increase the risk of a major event in the future due to the availability of a fuel supply which has been allowed to build up over time. The recognition of this relationship has led to the increasing use of low-intensity fires to reduce fuel accumulation. The main advantages of prescribed burning under controlled conditions lie in reducing the intensity of wildfires and lowering the spotting potential of fibrous-barked trees. On the other hand, so-called *controlled burns* often lead to wildfires, cause air pollution and have controversial effects on local ecosystems (Handmer, 1999). There is an increasing view that, in order to suppress forest fire hazards in the western USA, fuel must be removed by mechanical means – including commercial timber harvesting – as well as controlled burning.

Community preparedness, including plans for the early detection and suppression of wildfires, is a key element in disaster reduction (Britton, 1984). In most countries, rural fire-fighting groups are the first line of defence in coping with wildfires. But such groups are composed of volunteers who are often taken for granted by state and federal governments. For example, in the USA the value to the nation of rural fire-fighting services has been estimated to exceed US$36 billion each year but the fire-fighters feel they can neither influence policy nor obtain the resources needed to work effectively (Rural Fire Protection in America, 1994). Rural fire services need adequate training and access to some specialised equipment. Because piped water supplies are not always available in rural areas, fire teams need methods to deliver and use water more efficiently; this could mean dedicated items such as tankers for transporting water or access to aircraft. There is also a need for more general tools such as earth-moving plant to construct access tracks and firebreaks.

In the United States the emphasis is now on the process of integrating all rural fire and emergency response activities under a common incident management system. Most major wildfires cross local government boundaries of land managed by private landowners and state and federal agencies, and affect *inter-mix* areas. A comprehensive fuel modification plan should be agreed to reduce fire intensity, including prescribed burns and vegetation thinning by any means, including machinery, hand crews or even herds of goats. It is also necessary to have an overall evaluation of fire fighting infrastructure, including water supply and equipment. This approach was tried in California after the Oakland–Berkeley Hills firestorm of 1991 when the cities of Oakland and Berkeley formed a consortium with other major 'inter-mix' landowners to develop a coordinated hazard reduction plan. Similar bushfire management committees, representative of local interest groups, exist in Australia.

Forecasting and warning play an important role in wildfire hazard reduction. At appropriate times of the year, fire danger ratings are issued as a specialised meteorological service (Haines *et al.*, 1983). In populated areas, lookout points may be sufficient for early detection but, in more remote regions, regular surveys by aircraft or other remote sensing means, such as satellite imagery, may be necessary. If soil water is in short supply, plants reduce the amount of evapotranspiration from their leaves with a consequent increase in the surface temperature of large vegetation stands, such as forests. These changes in temperature, which reflect the drying weather, can be detected on satellite images, and the derivation of an appropriate 'vegetation stress index' can be used as an indication of where wildfire outbreaks are most likely to occur (Patel, 1995). Given such information, it is possible either to intensify ground surveillance in these areas or to exclude the public until the fire risk starts to fall.

The need for early fire detection varies according to the source of fires and the length of the fire season. In the western states of the USA, about 90 per cent of the fires are caused by summer lightning, so investment in automated lightning detection systems, with a follow-up aerial survey, is a prudent strategy. Florida, on the other hand, is subject to a year-round fire season because of the fuel types and weather patterns, and most of the fires are human-caused. Here early fire detection is based on a fixed-position, passive infra-red system (Greene, 1994). This comprises an integrated set of computer-controlled, infrared sensors, weather sensors and video cameras located on towers at remote observation points. At each site, an infrared sensor and camera scan the horizon and mountainsides for thermal variations. Each infrared sensor can detect thermal variations in an area of land about 20 m^2 at a distance of up to 20 km. Thus, the remote site coverage can be as much as 100,000 ha per installation, depending on the type of terrain. Each remote sensing point provides the control centre with fire location, estimated size of the fire, weather conditions, video picture of the area scanned and sensor status.

Much of the present threat exists because local governments have not factored wildfire hazards into the urban development control system. Therefore, land use planning and public education have an essential role to play in future hazard reduction. The first step should be a map showing severe wildfire hazard areas. The need is for neighbouring jurisdictions to work together on vegetation management and to develop a common database for use as a planning tool in the vital 'inter-mix' zones. *Fire breaks* are an integral part of rural land use planning to reduce wildfire hazards. They may be provided along natural boundaries to exclude fire or to isolate crops, timber plantations or other high-risk areas. It also follows that public fire prevention education deserves more attention. The persuasive approach can be reinforced in a variety of ways. For example, the provision of barbecue places set up by local authorities in safe clearings alongside roads tends to discourage indiscriminate fire lighting. In addition, an increased understanding of the benefits of prescribed burns could help officials to obtain the cooperation of landowners in carrying out such fuel management practices. Finally, when areas have been burned out, consideration should be given to the government acquisition of some land for public open space and the rebuilding of properties at lower densities on larger plots.

11 Hydrological hazards

Floods

FLOOD HAZARDS

Flooding is the most common of all environmental hazards. Every year, floods claim over 20,000 lives and adversely affect around 75 million people worldwide. The reason lies in the widespread geographical distribution of river floodplains and low-lying coasts, together with their long-standing attractions for human settlement. Bangladesh is by far the most flood-prone country in the world, accounting for nearly three-quarters of the global loss of life from both river and coastal floods. China also suffers badly and some 5 million lost their lives in floods between 1860 and 1960, despite the fact that the flood defence of cities goes back over 4,000 years (Wu, 1989). Large losses continue. In the 1998 Chinese floods, over 3,000 people died, some 15 million were made homeless and the direct property damage was estimated at US$20 billion. Investment in flood control and better disaster preparedness, combined with improved sanitation and control of post-flood diseases – such as cholera and typhoid – has reduced the mortality in Asia but large numbers of people are still affected by floods. Flood impacts are not restricted to the LDCs. The historic Midwest floods of spring and summer 1993 on the Mississippi and Missouri rivers affected nine states or more than 15 per cent of the contiguous United States. More than 50,000 homes were damaged or destroyed, some 54,000 persons were evacuated from flooded areas, over 4 million ha of farmland were flooded, the national soybean and corn yields were down 17 per cent and 30 per cent respectively below the 1992 level. Total losses ranged between US$15–20 billion (US Dept of Commerce, 1994) although fewer than fifty deaths were reported.

Physical damage to property, especially in urban areas, is the major cause of tangible loss. There are also secondary losses associated with a decline in house values after the event. Such declines in property prices appear to be largely temporary (Montz, 1992), although repeated flooding can have continuing effects (Tobin and Montz, 1997). Damage to crops, livestock and the agricultural infrastructure can also be high in intensively cultivated rural areas. In India, for example, almost 75 per cent of the direct flood damage has been attributed to crop losses (Ramachandran and Thakur, 1974). In addition to deaths from drowning, there is mortality as a result of illness after floods due to disease epidemics. Gastrointestinal diseases regularly break out in the LDCs where sanitation standards are low or when sewerage systems are damaged. In tropical countries the incidence of some other water-related diseases, such as typhoid or malaria, may more than double above the endemic rate. In the MDCs flood survivors have been found to suffer mental illness and 18 months after the Buffalo Creek, West Virginia, disaster in 1972, over 90 per cent of the survivors were suffering from mental disorders (Newman, 1976). Other losses on which it is diffi-cult to place a monetary value include the environmental changes brought by floods, such as water pollution. It has been estimated that river-bank erosion of farmland and villages in Bangladesh renders up to 1 million people landless and homeless every year (Zaman, 1991).

More than any other environmental hazard, floods bring benefits as well as losses (Smith and Ward, 1998). The seasonal 'flood pulse' is a vital part of most river ecosystems where the annual flow regime helps to maintain a range of wetland habitats. After the initial ecological disturbance associ-ated with major floods, there is often a burst of biological productivity (Allen, 1993). Floods maintain the fertility of soils by depositing layers of silt and flushing salts from the surface layers. Although silt-laden flood water regularly reaches only a small area of Bangladesh, the new alluvium

Plate 9   Regular floods bring benefits as well as losses. Productive farmland, supporting crops of sugar cane growing on enriched alluvial soils, is a feature of parts of the Richmond river floodplain in northern New South Wales, Australia. Further downstream the same flooding causes losses in urban areas such as Lismore. (Photo: K. Smith)

enriches the phosphorous and potash content of the soil. Floods provide water for natural irrigation and for fisheries, which are a major source of protein in many LDCs. Flood-retreat agriculture, where the moist soil left after flood recession is planted with food crops, is widely undertaken in the tropics. The seasonal inundation of large floodplains in semi-arid West Africa is of crucial ecological and economic importance and is responsible for a larger agricultural output than that associated with formal, highly capitalised irrigation systems (Adams, 1993). Along low-lying coasts and estuaries, regular inundations by salt water and marine sediments help to maintain saltmarshes and mudflats, which are often rich in wildlife, as well as specialised vegetation such as mangrove forest. In a normal year, floods may be expected to bring all these benefits and it is only the rare, high-magnitude events which create disaster.

The nature and scale of the flood risk varies greatly. In most countries rivers are the greatest hazard, as in the United States where river flooding accounted for nearly two-thirds of all federally declared disasters between 1965 and 1985. But in Britain rivers represent only about one-third of the total flood risk, for two reasons. First, storm rainfall maxima are only about a quarter of recorded world extremes, thus creating less aggressive rivers. Second, virtually all buildings are constructed of brick or stone so that they are not as easily washed away as timber or adobe structures. On the other hand, sea flooding is a serious threat caused by the coastal configuration of eastern and southern England combined with long-term land subsidence, rising sea levels and under-investment in sea defences over many years. In February 1953 over 300 people died in eastern England when these defences were overtopped.

In a comparison of flood risk in selected MDCs, Handmer (1987a) concluded that less than 2 per cent of the population of England and Wales and of Australia was exposed to flooding compared with almost 10 per cent of people in the USA who live within the 1:100 year floodplain. Exposure to risk shows little relationship to overall population density because flood risk in the MDCs is concentrated in urban areas. Although New Zealand has a low population density, nearly 70 per cent of the towns and cities with populations in excess of 20,000 have a river flood problem (Ericksen, 1986). Direct comparisons with those LDCs which have dense rural populations are more difficult. The worst river floods occur in parts of Asia where the monsoonal climate ensures that 70–90 per cent of the total annual rainfall occurs in the May–October wet season. In China vast alluvial river plains cover over 1 million km^2 and contain half the total population. In some countries, such as Bangladesh and Vietnam, there is a combined threat from river, delta and sea floods. Thus, in Vietnam the low-lying river deltas and coastal lands have been intensively exploited for wet-rice agriculture to the extent that over 70 per cent of the population is at risk from flooding (Department of Humanitarian Affairs, 1994).

The most vulnerable landscape settings for floods are listed below.

Low-lying parts of floodplains

In their natural state, these settings will suffer the most frequent inundation. For example, Smith *et al.* (1979) have shown that the town of Lismore, Australia, on the floodplain of the Richmond river, suffered a damaging flood at about three-yearly intervals during the 1945–75 period. Because of the high frequency of events, such areas in the MDCs are often given some protection by engineering works and are also subject to planning controls.

Within the LDCs the risk of disaster is much greater. In Bangladesh over 110 million people are relatively unprotected on the floodplain of southern Asia's most flood-prone river system. As shown in Figure 11.1, the area of the Ganges–Brahmaputra–Megna basin extends over more than 1,750,000 km². In an average year it receives about four times the annual rainfall of the Mississippi basin in the USA. Over the flat, deltaic country of Bangladesh, monsoon-generated flooding regularly covers an estimated 20 per cent of the total land area and very high floods may cover half of the country (Rogers *et al.*, 1989). For example, the 1988 flood inundated 46 per cent of the land area and killed an estimated 1,500 people.

Low-lying coasts and deltas

Estuarine areas are often exposed to a combined threat from river floods and high tides, as in the case of the Thames in London, England. Such areas can be submerged when river floods are prevented from reaching the sea, perhaps as a result of high-tide conditions, and a mixture of fresh and marine water spills over the land. More direct marine flooding occurs when salt water is driven onshore by wind-generated waves or storm surges (see Chapter 9). Storm surges are responsible for most of the worldwide loss of life from coastal floods. Other, much rarer marine invasions can result from tsunami waves, created by earthquakes out at sea, that move into shallower coastal waters and submerge beaches and bays (see Chapter 6).

At the close of the twentieth century, seventeen out of the top twenty-five cities, defined as those with more than 9 million people, were coastal cities (Timmerman and White, 1997). These cities, which are often surrounded by heavily populated rural areas, tend to be in countries which lack effective coastal zone management and development planning controls. A marine flood risk exists for cities as varied as Venice, Alexandria and Shanghai, plus their surrounding areas, especially during severe spells of weather. For example, El Niño storms in late 1997 carried a sea surge 15 km inland and flooded the main square in the coastal city of Trujillo, Peru (IFRCRCS, 1999).

Small basins subject to flash floods

Flash floods are found mainly in the arid and semi-arid zones where there is a combination of steep topography, little vegetation and high-intensity,

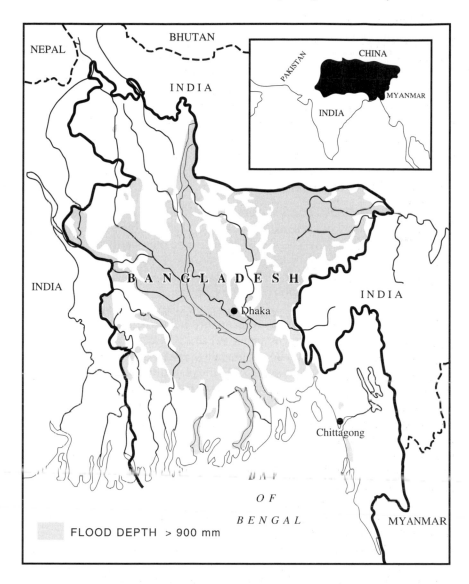

Figure 11.1 Areas of Bangladesh that are subject to flooding to depths greater than 90 cm in a normal year in relation to the major rivers. The inset map shows the location and area of the Ganges–Brahmaputra–Megna drainage basin.
Source: Modified from Rogers *et al.* (1989)

short-duration convective rain storms. They can also occur in narrow valleys and heavily developed urban settings. Warning times are invariably limited and flash floods are now the main cause of weather-related deaths. A typical event was the Big Thompson Canyon, Colorado, flood of July 1976 when 139 people were drowned and there was millions of dollars damage after a thunderstorm produced 300 mm of rain in less than six hours. Many of the dead were recreationalists with little awareness of flash flood dangers and the need to escape from the canyon floor. Estimates suggest that in tropical countries some 90 per cent of the lives lost through drowning are the result of intense rainfall on small steep catchments upstream of poorly drained urban areas. For example, the growing city of Kuala Lumpur, Malaysia, is situated at the foot of a relatively steep, fan-shaped basin which has almost perfect hydrological conditions for generating flash floods (Sehmi, 1989).

Areas below unsafe or inadequate dams

There are around 30,000 sizeable dams in the USA alone and more than 2,000 communities have been identified as being at risk from dams which are believed to be unsafe. As with flash floods, there may be little opportunity for warning and evacuation. For example, 421 people died when the foundations of the Malpasset dam, France, failed in 1959. Even dams which are structurally sound may be overtopped by surges of water induced by earth movements. In 1963 a landslide created a major flood surge behind the Vaiont dam in Italy. Although the structure held, the subsequent wave of water killed 3,000 people downstream. When a poorly maintained dam burst in 1972 in the coal mining valley of Buffalo Creek, West Virginia, there was no warning and 125 people were killed and 4–5,000 were made homeless. According to D.I. Smith (1989), few countries have prepared inundation maps or made emergency plans for such events even though the likelihood of a dam failure killing 1,000 or more people is statistically significant when compared to the risk from certain other environmental hazards.

Low-lying inland shorelines

These extend for thousands of kilometres and involve much property, as around the Great Lakes and the Great Salt Lake in North America. Fluctuating lake levels from high river inputs is the main problem. Lake levels rise to damaging heights only after a period of wet years, as shown by Changnon (1987) but the erosion of barrier islands, sand dunes or bluffs removes any natural protection from wind-driven wave attack on buildings and other shoreline facilities.

Alluvial fans

These environments create a special type of flash flood threat, especially in semi-arid areas where the fans often support urban development. It has been estimated that 15–25 per cent of the arid American West is covered by alluvial fans, which often provide attractive development sites due to their commanding views and good local drainage (FEMA, 1989). The hazard is underestimated because of the prevailing dry conditions, which lead to long intervals between successive floods and the absence of well-defined surface watercourses. The braided drainage channels meander unpredictably across the steep slopes, bringing speeds of 5–10 m s^{-1} and high sediment loads produce hydrodynamic forces capable of destroying built structures.

There is a high degree of inter-annual variability in flood losses but certain broad trends can be detected. Figure 11.2 shows annual deaths and economic losses in the USA 1925–94 (Smith and Ward, 1998). In this 70-year period, flood damage has increased by about thirty times, from US$100 million to US$3,000 million, even when allowance is made for cost inflation. A much less steep increase exists for flood deaths, probably associated with the continuing problem of flash floods. Flood damages rose throughout most of the twentieth century at about 4 per cent per year in real dollar costs, despite the fact that since 1936 the United States government spent more than US$15 billion on structural flood control works. Other

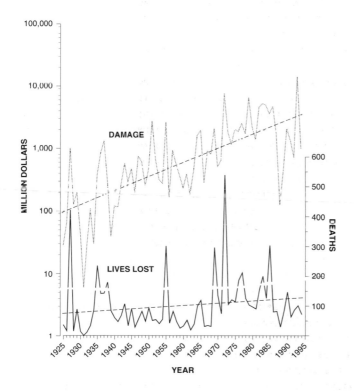

Figure 11.2 The long-term annual trend in deaths and economic losses caused by flooding in the USA, 1925–94. Damages on the logarithmic scale are in millions of US$ adjusted to 1990 values. The high cost of the 1993 Midwest floods is clear.
Source: After Smith and Ward (1998)

countries show a similar picture (Ericksen, 1986). It should be noted that the number of deaths is higher in the LDCs than in the industrialised countries, partly because flood warning and emergency evacuation is less effective.

The fact that flood losses continue to increase, despite growing investment in alleviation schemes, could be attributed to a variety of factors. But, assuming that allowances are made for continuing cost inflation and the increased availability of flood-loss data, there are really only two plausible explanations:

1   *a physically-driven increase in the frequency and magnitude of flood events*;
2   *a human-driven increase in disaster vulnerability caused by greater floodplain occupancy.*

In a few instances, changing hydrometeorological factors may be implicated. Thus, an increase in annual rainfall totals, together with greater storm activity, in eastern Australia since 1945 may account for upward hazard trends on the Richmond river at Lismore (Figure 11.3). This shows that damaging floods, which start when the river height reaches 10 m, occurred on average every 2–3 years between 1945 and 1975 compared to about once every 5 years over the entire 1875–1975 period of record. The hydrological shift also implies that the 1:100 year flood would also occur much more often. But the consensus is that continued floodplain invasion accounts for most of the upward trend in flood losses. Despite the fact that floodplains are one of the most topographically obvious of all hazard-prone environments, widespread invasion has occurred as a result of countless individual decisions rooted in the belief that the locational benefits outweighed the risks. An appreciation of these attitudes is as important as flood hydrology in understanding flood hazard.

Over the past 50 years there have been important changes in flood hazard mitigation policy. The emphasis has moved away from physical control and structural measures towards reducing human vulnerability through non-structural approaches. Platt and McMullen (1980) identified three overlapping stages:

1   *Structural era, 1930s-1960s* A period with almost exclusive reliance on engineering structures (reservoirs, levees, sea walls) designed to control floods physically.
2   *Unified floodplain management era, 1960s–1980s* A period with a mix of mitigation measures but increasingly characterised by non-structural approaches (flood warning, land use planning, insurance) designed to reduce vulnerability to floods.
3   *Post-flood hazard mitigation era, 1980–?* A period which is still emerging and typified by measures such as property acquisition and stricter land use controls under-pinned by a growing realisation of the ecological benefits of floods and the need to live with them in greater harmony.

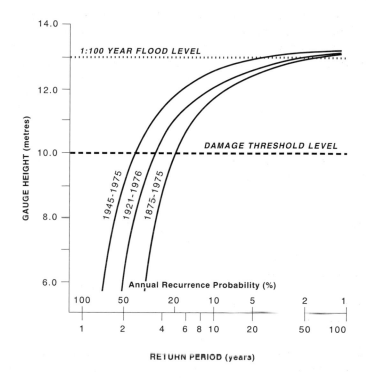

Figure 11.3 Changes in flood frequency for the Richmond river at Lismore, Australia. Damaging floods occur when the river stage rises above 10 m. The increased frequency since 1945 applies to several towns along the northern coast of New South Wales and has been associated with a rise in mean annual rainfall.
Source: After D. I. Smith *et al.* (1979)

CAUSES OF FLOODS

Physical causes

River floods

Physically, a river flood is a high flow of water that overtops either the natural or the artificial banks of a river. However, such an event is not a hazard unless it threatens human life and property. For the hydrologist, flood magnitude is best expressed in terms of instantaneous peak river flow (*discharge*) whilst the hazard potential will relate more to the maximum height (*stage*) that the water reaches. Smith and Ward (1998) distinguished between the primary causes of floods, which mainly result from external climatological forces, and secondary flood-intensifying conditions, which tend to be more drainage basin-specific. Another way is to relate the physical causes of floods to other types of environmental hazard (Figure 11.4).

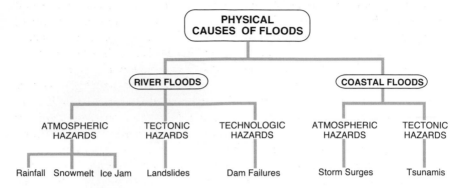

Figure 11.4   The physical causes of floods in relation to other environmental hazards. Atmospheric hazards resulting in large amounts of rainfall are the most frequent cause of floods.

Atmospheric hazards, especially excessive rainfalls, are the most important cause of floods. These vary from the semi-predictable seasonal rains over wide geographic areas, which give rise to the annual floods in tropical areas, to almost random convectional storms giving flash floods over small basins. The more intense the precipitation is beyond a flood-producing threshold, the lesser is its duration and areal coverage. Therefore, the smaller the drainage basin, the greater the unit-depth of flood runoff and the more rapid the flood flow concentration into the channel is likely to be.

Some floods are associated with atmospheric and oceanic processes on a large scale. The 1993 USA floods have been associated with an El Niño Southern Oscillation (ENSO) episode (Lott, 1994) whilst Pearce (1988) linked severe flooding in Sudan and Bangladesh in 1988 to a La Niña event (see Chapter 9). Over the Mississippi river watershed, precipitation was the greatest since 1895 for the April–July period. Prolonged rainfall over large drainage basins is also associated with tropical cyclones or the intense depressions of mid-latitudes. These have a preferred seasonal incidence. Most floods in Britain are associated with deep frontal storms between October and March and over 75 per cent of peak flows on British rivers occur in this six-month period. In other climates, flood concentrations also depend on seasonal concentrations of rainfall. For example, about 70 per cent of India's rainfall comes during the 100 days of the summer south-west monsoon.

The highest-intensity rainfall is associated with more localised storms. If the intense convectional cells coincide with small drainage basins, then catastrophic flash floods can result. These floods occur mainly in the summer season, especially in continental interiors. They produce large volumes of water, rapidly concentrated in both time and space, with great damage potential. In June 1972, Rapid City (South Dakota) was devastated by a flood partly caused by a dam collapse. There were 238 deaths, the highest recorded loss of life from a single flood in the United States.

Melting snow is responsible for widespread flooding in the continental interiors of both North America and Asia in late spring and early summer. Many disasters have resulted. In May 1948, after heavy winter snow and a rapid temperature rise, the Fraser river in British Columbia, Canada, produced a major flood. More than 2,000 people were made homeless and compensation payments of more than Can$20 million (at 1948 prices) were made. The most dangerous melt conditions often arise from rain falling on snow to give a combined flow. This occurred in the Romanian floods of May 1970, when the Transylvanian basin was devastated by heavy rain from a deep depression plus snowmelt from the Carpathian mountains. Meltwater floods may be greatly compounded by ice jam flooding. This occurs when an accumulation of large chunks of floating ice, resulting from the spring break-up, causes the temporary damming of a river. The floating ice lodges at bridges and other constrictions in the channel or at shallows where the channel freezes solid. The largest ice masses can destroy buildings and shear off trees above the water level. Near lake shorelines, pressure ridges in the ice can rise several metres above the banks and dislodge houses from their foundations.

Flood-intensifying conditions cover a range of factors which increase the drainage basin response to a given precipitation input. Most of these factors, such as those relating to the hydraulic geometry of the basin or the effect of frozen soils in reducing infiltration, are entirely natural. Together with the precipitation characteristics, these factors will determine key features of a flood event such as the magnitude of the flood, the speed of onset, the flow velocity, the sediment load of the river and the duration of the event.

The damage potential of flood waters increases exponentially with velocity. Velocities as low as 0.5 m s^{-1} are capable of sweeping victims off their feet and at speeds above 2 m s^{-1} floodwaters can undermine the foundations of buildings. The physical stresses on structures are raised further, probably by hundreds of times, when rapidly flowing water contains debris such as rock, sediment or ice. The ensuing collapse of sewerage systems and storage facilities for products such as oil or chemicals means that flood waters create pollution and other hazards. In November 1994 over 100 people were killed in Durunqa, Egypt, when floods destroyed a petroleum storage facility and carried burning oil into the heart of the town.

Other flood-intensifying conditions arise from changes in land use. Some changes may be semi-deliberate, such as the increase in agricultural land drainage designed to speed the runoff from productive fields. Thus, in England and Wales, the area of under-drained agricultural land rose from 10,000 ha in 1945 to more than 100,000 ha at the present time. On a world scale the chief effects are associated with more inadvertent land use changes, notably urbanisation and deforestation.

Urbanisation increases the magnitude and frequency of floods in at least four ways:

1   The creation of highly impermeable surfaces, such as roofs and roads, inhibits infiltration so that a higher proportion of storm rainfall appears as runoff. According to Hollis (1975), small floods may be increased up to ten times by urbanisation and the 1:100 year event may be doubled in size by a 30 per cent paving cover of the basin.
2   Hydraulically smooth urban surfaces, serviced with a dense network of surface drains and underground sewers, deliver water more rapidly to the channel. This increases the speed of flood onset, perhaps reducing the lag period between storm rainfall and peak flow by half.
3   The natural river channel is often constricted by the intrusion of bridge supports or riverside facilities, thus reducing its carrying capacity. This increases the frequency with which high flows overtop the river banks. For example, successive navigation works on the Mississippi river have reduced the capacity of the natural channel by one-third since 1837 (Belt, 1975). A major flood in 1973 was reported as a 1:200 year event in terms of water level, although the flow had an average recurrence interval of only 30 years.
4   Insufficient stormwater drainage following building development is a major cause of urban flooding. The design capacity of many urban stormwater drainage systems, even in the MDCs, is for storms with return periods as low as 1:10 to 1:20 years. Many sewerage systems cannot cope with the resulting peak flows and the excess water is delivered to the surface for storage. Some countries have a sewerage system in an advanced state of neglect. In Britain, there are about 5,000 sewer failures every year and the surcharging of storm drainage systems is a widespread concern in low-lying urban areas.

Deforestation appears to be a likely cause of increased flood runoff plus an associated decrease in channel capacity due to sediment deposition. In small basins more than four-fold increases in flood peak flows have been monitored together with suspended sediment concentrations as much as 100 times greater than those in rivers draining undisturbed forest land. Specifically, the flood which claimed thirty-three lives and damaged 1,400 works of art and 300,000 rare books in the city of Florence, Italy, in 1966 was partially attributed to long-term deforestation in the upper Arno basin.

Charoenphong (1991) and others have claimed similar effects for much larger basins, but direct cause and effect relationships between forest cutting in the headwaters, mainly for fuel wood, and increased floods far downstream are hard to find. In a review with special reference to the Himalayas, Hamilton (1987) conceded that forest cutting followed by abusive agricultural practices may aggravate flooding but cautioned against the misunderstanding of natural processes. Despite the availability of hydrological records for almost 100 years, no statistically reliable increase in physical flooding has been found in the plains and delta areas of the Ganges–Brahmaputra river system. Ives and Messerli (1989) concluded that there was no evidence to support any direct relationship between human-induced landscape changes in the Himalayas and changes in the hydrology and sediment transfer processes in the rivers of the plains. This is probably because the high monsoon rains in the Himalayas, combined with steep slopes, ensure rapid runoff and high sedimentation rates irrespective of the vegetation cover.

Coastal floods

Hazardous flooding of coasts and estuaries tends to occur when the height of the sea surface is raised above the normal fluctuations created by waves and tides. Such increases in height are likely to result from short-term factors, such as storm surges and tsunamis, or from very much longer-term processes, such as land subsidence and progressive sea level rise.

Storm surges are primarily dependent on atmospheric processes such as very strong onshore winds but certain coastal configurations can experience flooding from lesser windspeeds (see Chapter 9). For example, the semi-enclosed, low-lying coast of the North Sea is exposed to storm surges driven by northerly winds, especially towards the south where the narrowing of the sea forces the water to pile up. This combination of meteorological and geographical features has rendered necessary a complex system of barrages to protect extensive areas of the Netherlands and a line of sea walls throughout south-east England. Tsunamis are much rarer events and are created by earthquakes on the sea floor (see Chapter 6). Most damaging tsunamis are generated in the Pacific Basin, partly because of the tectonic instability of the area and partly because of the large expanse of ocean. At least twenty-two countries around the Pacific Rim are at risk.

Relative increases in sea level along low-lying coasts create much longer-term threats. During the last 100 years, there has been a *eustatic* (worldwide) increase in sea level of 0.10–0.20 m. This has been attributed to a combination of the thermal expansion of sea water and ice-cap melting following the end of the last ice age and now exacerbated by global warming. In addition, some coastal areas have experienced an additional *isostatic* (local) increase in sea level due to a lowering of the land surface. For example, the south-east corner of England, including London, is slowly sinking as the north-west of Britain rises progressively in response to the removal of the mass of ice which accumulated there more than 10,000 years ago. The city of Venice is sinking into the Adriatic due to local land subsidence brought about by the over-extraction of ground water. In the lowest-lying coastal zones, a combination of the increased volume of water in the ocean basins and local subsidence has resulted in a net rate of sea level rise of about 0.3 m per century. As a consequence, natural shore defences, such as salt marshes, beaches and dune systems, are suffering increased erosion and many of the 300 barrier islands along the coast of the United States are driven further landward with onshore storm winds. Where coasts are already protected by sea walls and other engineering works, sea level rise increases the frequency with which such defences will be overtopped by wind-driven waves or storm surges.

Human causes

In order to achieve immunity from the flood hazard, no intensive land use should occur on floodplains or along sea shores, although social processes have long made this an unrealistic ideal. Moreover, it should not be automatically assumed that floodplain development is uneconomic. A net economic benefit can occur if the additional benefits derived from locating on the floodplain (that is, the benefits over and above those available at the next best flood-free site) outweigh the average annual flood losses. In practice, it is virtually impossible to demonstrate such economic efficiency through floodplain use, not least because of the problems of assessing costs and benefits accurately at both local and national levels.

Human attitudes towards flood-prone areas have always been ambivalent. Some of the earliest settlers were aware of the dangers. In Europe major floodplain invasion often did not occur until the late nineteenth and early twentieth centuries when many urban areas expanded rapidly. Yet, in Australia, the coastal floodplains of the Hunter, Brisbane, Hawkesbury and Yarra rivers were quickly developed. In a study of twenty-six cities, Goddard (1976) found that one-sixth of all urban land in the United States lay within the 1:100 year floodplain and that more than half of the available floodplain land in the country was developed by 1974. Moreover, urban areas were expanding on to floodplains at the rate of about 2 per cent per year. At Rapid City, South Dakota, the initial site was laid out south of the floodplain (Rahn, 1984). A surge of building activity led to progressive floodplain invasion from 1940 onwards and by 1972, the year of the flash flood disaster, the entire floodplain within the city limits had been urbanised. Similar processes of urban development have affected coastal cities. It is now estimated that 21 per cent of the world's population lives within 30 km of the sea and that these populations are growing at twice the overall global rate (Nicholls, 1998).

Once floodplains become urbanised, there follows an almost inevitable demand from the local community for flood protection. This demand has been particularly strong in the USA where, as a result of the Flood Control Act of 1936, Congress initiated a major flood defence programme funded by the federal government. The policy was soon criticised, mainly because over-reliance on structural engineering works not only failed to control flood losses but actually encouraged some further floodplain development. This is because of the *levee effect* whereby the construction of flood embankments, or other physical controls, is erroneously perceived to render part of a floodplain safe for development. Land values rise and development tends to follow.

Despite a progressive shift towards more planning control of floodplain development, it has been difficult to shake off the massive structural legacy. In a re-examination of nine of the original United States cities investigated in 1958, Montz and Gruntfest (1986) found that structural controls still predominated and that floodplain encroachment had increased rather than decreased as a result of continued development pressures. These pressures were strongest for the cities experiencing high economic growth rates, which were difficult to control with local zoning laws. This process is illustrated for Datchet on the Thames floodplain, west of London, England, in Figure 11.5 (Neal and Parker, 1988). This small town of 6,000 people is in an area of active economic development. Despite the absence of flood protection works, planning controls failed to prevent the location of an additional 425 new houses on the floodplain in little more than a decade. Alternatively, cities undergoing economic recession are also prone to increased hazard since the local authorities are so desperate for investment that they are willing to attract floodplain development rather than no development at all.

The circular link between flood control works and floodplain encroachment (the levee effect) can be explained in terms of three factors:

1   The greater the amount of floodplain development and the greater the existing investment, the greater are the economic benefits to be obtained from flood control structures. Thus, flood protection schemes are more likely to be justified and implemented on cost–benefit grounds.

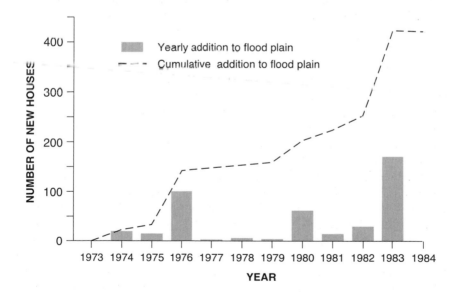

Figure 11.5   The encroachment of new residential housing on to the river Thames floodplain at Datchet, England, between 1974 and 1984.
Source: After Neal and Parker (1988)

2   The cost–benefit ratio also weighs in favour of construction when land can
    be protected from risk and freed for development. The higher land values
    in the 'protected' area then make further floodplain invasion more likely.

3   Above all, the real cost of protection (and encroachment) has not often
    been borne by the parties most directly involved. Most flood control
    structures have been financed by central government in an attempt to
    serve national economic efficiency. But local authorities pursue more
    local goals, which include the stimulation of development and employ-
    ment. Since private investment in the floodplain was protected by public
    money raised from taxes, it was perfectly rational for an individual to
    locate in the potentially hazardous area and for encroachment to occur.

LOSS-SHARING ADJUSTMENTS

Disaster aid

Calls for government relief are common following disastrous floods, but
there is increasing recognition that the taxpayer cannot be expected to fund
losses which should have been insured. This attitude is sometimes reinforced by legislation which limits aid to uninsurable losses. In these
circumstances, aid is restricted to charitable appeals. For many of the LDCs,
international aid, which includes technical assistance as well as immediate
disaster relief, is an important factor in flood mitigation. Following major
floods in Sudan in 1988, there was damage to agriculture, property and
social services totalling around US$1 billion. Some 200,000 homes were
either damaged or destroyed and about 2 million people were left homeless. After the emergency relief phase, the World Bank helped the Sudanese
government to prepare a US$408 million flood reconstruction programme
(Brown and Muhsin, 1991). Sectoral funding was allocated, as shown in
Table 11.1, and made up of both local and foreign investment aimed at
longer-term rehabilitation. The most successful efforts tend to be those that
depend on one or two donors only but this limits the scope for an ambitious programme, such as that in Sudan.

Insurance

Flood insurance is unobtainable in many parts of the world, but it is a key
loss-sharing strategy for many industrialised countries. In the UK householders buy flood insurance from private companies within standard
policies covering both buildings and their contents. Before 1961 cover was
typically excluded for buildings but, following major floods in the 1960s,
protection for structures was introduced. The main reason for this change
was a fear amongst insurance companies that the government might either
introduce a natural disaster insurance scheme or nationalise the industry.
As indicated by Arnell *et al.* (1984), this cause-and-effect type of policy

Table 11.1 Funding allocated for reconstruction aid after flooding in the Sudan in 1988

Sector	Local cost (US$ millions)	Foreign cost (US$ millions)	Total cost (US$ millions)
Agriculture	33.8	63.6	97.4
Rural water	6.6	17.4	24.0
Education	11.9	24.3	36.2
Health	5.9	32.7	38.6
Industry/construction	15.0	35.3	50.3
Power	5.9	29.0	34.9
Telecommunications	3.3	31.1	34.4
Transportation	7.9	25.6	33.5
Urban	31.3	25.0	56.3
Programme coordination and flood prevention	0.6	1.4	2.0
Total	122.2	285.4	407.6

Source: After Brown and Muhsin (1991)

evolution suggests a crisis response to flood hazard by the insurance industry rather than a well-considered strategy.

Few individuals in flood-prone areas purchase cover voluntarily. For example, significantly less than one quarter of all flood-prone households in the USA have insurance, according to Kusler and Larson (1993). Many households that do elect to take cover will be underinsured. Tenants, pensioners and lower social status householders are least likely to have adequate cover and least likely to recover financially after a flood. In Britain, flood insurance has not been used for hazard mitigation. Until recently, the premium paid by private householders took no account of individual hazard exposure and the lack of real competition between the companies has so far ruled out the introduction of significant premium incentives to policy-holders adopting damage mitigation measures, such as flood-proofing.

By contrast, in the USA the National Flood Insurance Program (NFIP) lies at the heart of floodplain management (Arnell, 1984). The NFIP was introduced by the federal government in 1968 because of the reluctance of private industry to continue selling cover owing to fears of catastrophic losses. Moreover, at this time there was a growing awareness of the part non-structural methods might play in flood reduction. Therefore, the scheme was designed to provide financial assistance to flood victims and to establish better land use regulations for floodplains. As shown in Figure 11.6, such planning regulations ideally aim to restrict development on the floodplain so that no construction is permitted in the floodway and only flood-proofed buildings are allowed on the 1:100 year floodplain (Hewitt, 1997). This base flood standard was adopted in recognition of the benefits, as well as the costs, associated with floodplain development.

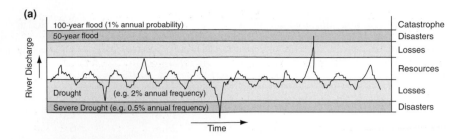

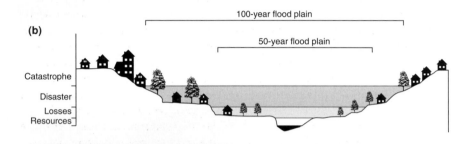

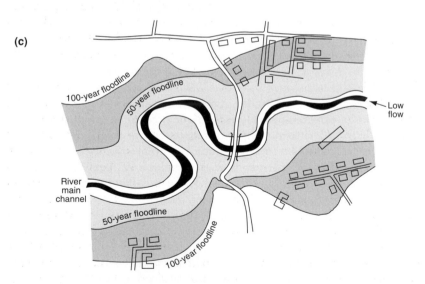

Figure 11.6   Schematic representations of river flow as a temporal and spatial hazard: (a) shows variations in river stage over time leading to floods and droughts with different return intervals; (b) shows the identification of flood hazard zones across a floodplain; (c) shows the corresponding map of flood risk which can be used to restrict further development onto the floodplain.
Source: After Hewitt (1997)

The scheme works by the Federal Emergency Management Agency (FEMA) first publishing a Flood Hazard Boundary Map outlining the approximate area at risk from flooding. In order to join the NFIP, a community must agree to adopt certain minimum land use regulations within this area. During the first phase (the 'Emergency Program') the community must effect minimal land use controls and flood insurance is then made available at nationwide subsidised rates. Any new development in the hazard area must be sanctioned by a building permit and be constructed so that it would suffer minimal damage.

If the hazard is judged to be serious, FEMA will then provide more detailed maps which delineate the 1:100 year floodplain and the floodway, defined collectively as the area within which the 100-year flow can be contained without raising the water surface at any point by more than 0.3 m (see Figure 11.6). For the USA, this amounts to a total area almost the size of California. A few communities also regulate some types of development in the 1:500 year floodplain. The community must then join the 'Regular Program' and implement more stringent controls. Further development in the floodway is theoretically prohibited and residential development in the rest of the floodplain (floodway fringe) must be elevated to at least the 1:100 year flood level. Non-residential development may be flood-proofed. The floodplain is divided into risk zones on the basis of a Flood Insurance Rate Map (FIRM) which is used to allocate variable insurance premium ratings to individual properties within these Special Flood Hazard Areas (SFHAs). About 10 per cent of the nation's 100 million households are located in designated SFHAs. All new property-holders within the 1:100 year floodplain must buy insurance at actuarial rates.

The NFIP has encouraged more than 17,000 flood-prone communities (out of a total of 21,000) to adopt at least some floodplain regulations and has provided maps of flood hazard areas to nearly 20,000 communities. It has also provided low cost insurance. But the scheme has been criticised for not providing sufficient incentives to local officials to promote active floodplain management. It has been asserted that the scheme is not financially sound, and Figure 11.7 shows that expenses outstripped income in most financial years between 1977 and 1997 (National Wildlife Federation, 1998). Typically, less than 30 per cent of all properties located in designated floodplains are within the program and 37 per cent of the properties insured receive subsidised rates. There is also evidence that the NFIP has failed to protect properties subject to frequent flooding and to discourage further development in the very high hazard areas. Over the 1977–97 period, repeat insurance claims, which arise from only 2 per cent of all insured properties, represented 25 per cent of the total losses paid by the NFIP. According to Rahn (1994), the biggest defect of FEMA's floodplain policy is the continuing reliance on the 100-year flood because an increasing percentage of the annual flood damage results from very large floods of low probability, such as the Mississippi floods of 1973 and 1993.

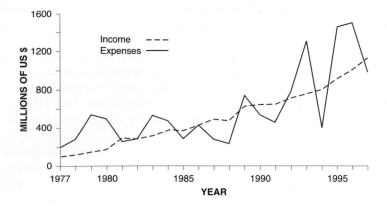

Figure 11.7   Annual income and expenditure (US$ million) under the US National
Flood Insurance Program, 1977–97. For most fiscal years, expendi-
ture exceeded income.
Source: After National Wildlife Federation (1998)

EVENT MODIFICATION ADJUSTMENTS

Environmental control

The physical control of floods depends on just two measures which may be
adopted singly or in combination: flood abatement and flood diversion.

Flood abatement measures

Flood abatement, or flood reduction, involves decreasing the amount of
runoff potentially able to create a flood peak in a drainage basin.
Theoretically, it can be achieved either by weather modification or by watershed treatment:

1   *Weather modification* This response does not exist on an operational
     basis for flood abatement because of the uncertainties highlighted in
     Chapter 9.
2   *Watershed treatment* This strategy aims to reduce flood peaks over a
     drainage basin following flood-producing rains. To be at all effective,
     the land use practices involved have to be adopted over most of the
     drainage basin. Typical strategies include reforestation or reseeding of
     sparsely vegetated areas to increase evaporative losses; mechanical land
     treatment of slopes, such as contour ploughing or terracing, to reduce
     the runoff coefficient; comprehensive protection of vegetation from wild-
     fires, overgrazing, clear-cutting of forest land or any other practices
     likely to increase flood discharges and sediment loads. In addition,

downstream decreases in peak flow can be achieved by the clearance of sediment and other debris from headwater streams, construction of small water- and sediment-holding areas (farm ponds) and the preservation of natural water detention zones such as sloughs, swamps and other wetland environments. Within urban areas some water storage can be achieved by the grading of building plots, detention ponds and the creation of parkland.

The theoretical advantages of dealing with flood problems through an integrated approach involving soil, vegetation and drainage processes was recognised quite early: for example, New Zealand set up a Soil Conservation and Rivers Control Council in 1941. Despite the variety of available methods, the experimental evidence for effectiveness is rather conflicting. Most flood reduction appears to have been achieved very locally, almost confined to on-site conditions in some cases, or is restricted to floodflows from comparatively small basins. For large basins the area to be treated is so large that it would take decades of reforestation and soil conservation to have any appreciable effect. In other words, forests will not prevent floods or sedimentation in the lower reaches of major rivers nor will they significantly reduce flood losses arising from major storm events.

Flood diversion measures

Once floodflows have been generated, they can be partially controlled and redirected away from specific areas at risk by a wide range of engineering structures. The widespread existence of flood controls reflects the fact that they are often wholly, or partly, funded by government departments or federal agencies. Few large-scale schemes go ahead without this financial support, which removes much of the direct cost of protection from the beneficiaries.

Levees

Levees (embankments, dykes or stopbanks) are the most common form of river control engineering (Starosolszky, 1994). They are designed to restrict flood waters to well-defined, low-value land on the floodplain. It is relatively cheap to construct earth banks and they offer protection up to the height or design limits of the particular flood. In China dykes built largely since 1949 now protect large alluvial plains from floods with a 10- to 20-year return period. Over 4,500 km of the Mississippi river, USA, is embanked in this way. Major cities, such as New Orleans, lie below river level and rely on such structures. During the flood of 1993 most of the levees performed as designed. Although 18 per cent of federal levees and 78 per cent of the non-federal levees failed or were overtopped (Dept of Commerce, 1994), the difference in the failure rate was due to the fact that most federal levees are designed for 1:100- to 1:500-year return intervals whilst non-federal levees, mainly protecting agricultural land, are designed to withstand floods with recurrence intervals of 50 years or less.

Levees fail for a variety of reasons. In the Mississippi floods failure rarely happened until the river stage reached a metre or more above the design level. Other levees behave differently. Gilvear *et al.* (1994) found that on the river Tay, Scotland, levees on the outside of bends and overlying old river courses are vulnerable, whilst a recent climatic trend towards wetter conditions is likely to increase the threats to these flood banks (Gilvear and Black, 1999). When floodbanks are breached, they increase total floodplain storage and water conveyance in relief channels behind the levee, thereby reducing flood stages downstream. Figure 11.8 shows the effects along a 150 km stretch of the Mississippi during July 1993. The levees above

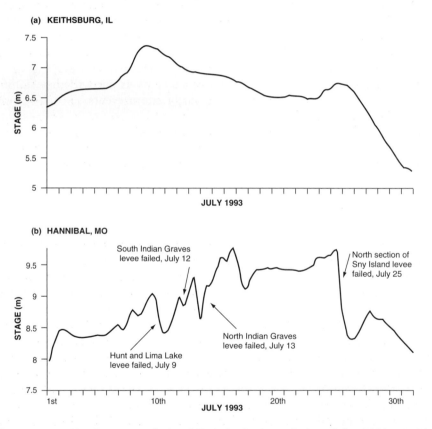

Figure 11.8   Flood stages of the Mississippi river during July 1993 at (a) Keithsburg and (b) Hannibal. The levees in the Rock Island area above Keithsburg held and the river stage shows smooth rises and falls compared to the sudden drops associated with local levee failures and widespread floodplain storage upstream of Hannibal.
Source: After Bhowmik (1994)

Keithsburg held, resulting in a smooth variation in river stage within the confined channel. But multiple levee failures just upstream of Hannibal led to large areas of floodplain storage which produced sudden drops in the river level at these sites as the flood wave was attenuated. Such upstream levee breaks, plus the security of the 15.9 m 'wall' protecting St Louis, which just managed to hold back the record 15.1 m flood crest on 1 August, were instrumental in saving that city from severe flooding.

Channel improvements

Channel improvements can be achieved in a variety of ways. Channel enlargement increases the carrying capacity of the river by increasing the cross-sectional area of the channel so that flood flows are contained within the banks. Dredging of the river Arno at Florence, Italy, followed the disastrous flooding of November 1966 with a view to lowering the river thresholds at two of the old bridges by 1 m in order to increase the channel discharge capacity from 2,900 m^3 s^{-1} to 3,200 m^3 s^{-1}. Although this is still below the estimated maximum discharge of 4,300 m^3 s^{-1} of the 1966 event, such bed lowering could increase the flood return period from 1:120 to 1:150 years. Flood relief channels can be used to provide extra overspill storage or can be used to divert water around an area of urban development. Such channelisation is increasingly criticised because, in addition to any visual intrusion, it isolates the river from its alluvial plain, with negative consequences for the riparian ecosystem and the functioning of the 'river corridor' as a whole.

Reservoirs

Reservoirs for flood reduction work on the principle of storing excess water in the upper drainage basin so that, by careful regulation, it can subsequently be released at a non-damaging rate (Harmancioglu, 1994) Surface storage reservoirs are a conventional technology which has been used for over 2,000 years but large dams are expensive and may well be subject to earthquakes and rapid siltation. In some countries, such as Bangladesh, the volumes of annual flood water are so large that it is impractical to retain them in storage reservoirs. The sixty-six flood control reservoirs in the upper Mississippi and Missouri basins worked well to reduce downstream river levels during the 1993 flood. Flood discharges were reduced by 30–70 per cent, despite the fact that the inflow to some of the reservoirs was several times their total storage capacity (Dept of Commerce, 1994). The maximum benefit was achieved on the Big Blue river, within the Kansas river basin, where Tuttle Creek Lake withheld a daily mean flow of 3,029 m^3 s^{-1} on 5 July (Figure 11.9), thus greatly reducing the instantaneous peak which would have caused far more damage than the 1,700 m^3 s^{-1} controlled release later in the month (Perry, 1994).

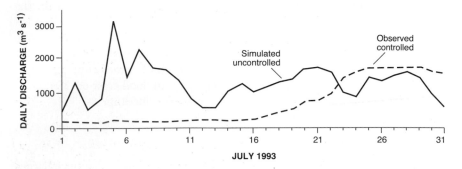

Figure 11.9   The controlling effect of reservoirs on river discharge during the upper
Mississippi floods of July 1993. Without the reservoir storage, the
Big Blue river near Manhattan, Kansas, would have overtopped the
federal levee and flooding downstream along the Kansas river would
have been much more severe.
Source: After Perry (1994)

Hazard-resistant design

So-called *flood-proofing* involves alterations to buildings, and their contents,
designed to reduce flood losses. Some of these changes can be temporary
and are activated on receipt of a flood warning. Temporary responses include
the blocking-up of certain entrances, the use of shields to seal doors and
windows and the use of sand bags to keep flood waters away from structures. Further measures include removing damageable goods to higher levels
and the greasing and covering of mechanical equipment.

Permanent modifications to buildings are more effective. The most
common structural alteration consists of raising the living spaces above the
likely flood level. Flood-proofing is increasingly implemented in combination with floodplain zoning and other local ordinances. Figure 11.10 shows
that property can be elevated above the prescribed design-flood level
(commonly the 1:100 year flood height) either by structural means (stilts)
or by raising the property on land fill. Usually a safety factor called the
freeboard, amounting to about 0.5 m, is added to the design flood level to
determine the flood construction level. This is the minimum elevation for
the underside of the floor system for habitable buildings. It also establishes
the top of any dike or river bank protection structure. Other precautionary
measures, such as setting the building back from any water body and the
water-proofing of any basement spaces, are also likely to be specified in
local planning regulations. Experience from the USA suggests that, unlike
the deployment of large-scale flood defences, which tend to raise land values,
regulations which require properties to be raised above the design-flood
level are likely to lower land values (Holway and Burby, 1993). This has
the advantage of making flood-prone land less desirable for development
although it is unlikely to halt the invasion of floodplains completely.

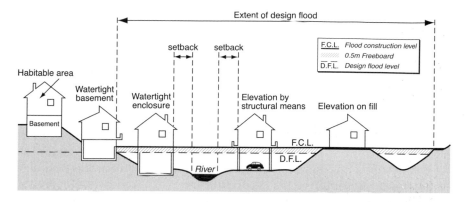

Figure 11.10 The location and design of flood-proofed residential buildings on an idealised floodplain. Habitable areas are raised above the flood construction level. In turn, the flood construction level allows 0.5 m of freeboard above the predicted maximum height of the design flood, for example, the 1:100 year event.
Source. Modified from Rapanos *et al.* (1981)

VULNERABILITY MODIFICATION ADJUSTMENTS

Community preparedness

Some countries rely on the routine civil emergency arrangements, including voluntary organisations and the armed forces, to combat flood losses. However, specialised flood preparedness programmes have increased in importance with the spread of forecasting and warning systems. The greatest need exists in flash flood events with short warning times, where even basic advice may be significant. For example, Handmer and Ord (1986) showed that, even in the absence of a formal warning, many lives can be saved in flash floods provided people run immediately to high ground, rather than continuing to drive in vehicles.

Preparedness has become a key flood mitigation factor in the LDCs. According to IFRCRCS (1999), there are now 30,000 Red Cross-trained flood volunteers in Bangladesh charged with a wide range of tasks, from raising hazard awareness, health and hygiene education and first aid techniques through to emergency response skills which include warning villages through loud hailers and evacuating people to refuges and higher ground.

Forecasting and warning

During the last 20 years, flood forecasting and warning schemes have become widely applied to flood hazard mitigation (Penning-Rowsell, 1986). Such schemes have proved most effective for large rivers, such as the Danube in

Europe and the Mississippi in the USA. The two main forces behind this trend have been the knowledge that improved warnings save lives coupled with the scientific and technical advances which have led to improved forecasts.

Flood hydrology has now developed to the point where storm rainfall and runoff conditions can be modelled to high levels of accuracy. Advances in hydrometeorological understanding have been complemented by great strides in flood monitoring and in real-time data handling and communication, which can reduce the lead time for warnings. For example, Clark (1994) has described the development of the flood forecasting system in the United States and shown how improved hydrologic networks and new technology, including the use of satellite and radar sequences, has been of benefit. Quantitative precipitation forecasting, using satellite and radar data, is a key element in current forecasting strategies for all areas where the watershed response time is shorter than that available for effective social mobilisation against floods. Flood waves moving down large rivers can be tracked by remote sensing. This could be especially useful in a country like Bangladesh where some 90 per cent of river flow originates in another country (Rasid and Pramanik, 1990).

Large sums of money have been spent on sophisticated flood forecasting systems for the MDCs. In Britain the combination of short rivers and highly urbanised basins ensures very limited warning times. As a result, a network of integrated weather radars was built during the 1980s, mainly for flood warning purposes. Despite this, over 50 per cent of the dwellings at risk in England and Wales have less than 6 hours flood lead time, which is regarded as a minimum period for effective warning in some other countries. In heavily urbanised basins the flood warning lead time may be as little as 30 minutes and the outstanding challenge for flood response worldwide lies in small basins prone to flash floods where there has so far been little systematic attempt to reduce risks. Flood estimation for such basins is notoriously prone to error, largely because the catchment characteristics used by the standard prediction methods may be either poorly represented or overridden by local factors, such as soil type or the degree of urbanisation.

Case studies within the MDCs have indicated that flood forecasts and warnings have the potential to reduce economic losses by up to one-third on the floodplains of large rivers. In practice, actual damage reduction may well be less than half of the pre-installation estimates, although even these savings can be impressive. There are many reasons why forecasting and warning schemes fail to perform well but difficulties often lie in the dissemination and response phases. Many people may not receive a warning. For example, 60 per cent of those who survived the Big Thompson Canyon, Colorado, flash flood in 1976 received no official warning. Often the response rate is poor. Penning-Rowsell *et al.* (1978) surveyed 160 people who had received flood warnings in Britain and found that only 54 per cent responded with damage-reducing actions. The remainder were either too old or infirm, or the flood took them by surprise despite the warning. Many were sceptical of the warning received because they had been given false warnings in the past. Emergency actions taken within short warning periods of less than four hours produced only limited savings, most of which were between one-quarter and one-half of the total potential damage.

Land use planning

During the past few decades, urban communities in the United States have adopted a regulatory approach whereby land use management policies have been used to limit further floodplain development. As with insurance, flood risk mapping, which may include water depth, flow velocity and flood duration, is an important planning tool. According to Marco (1994), flood mapping was first attempted in the USA and is less well developed in Europe, where the EU has shown little initiative in setting continent-wide standards. Burby *et al.* (1988) studied ten communities across the USA and concluded that land management had been effective in protecting new development from losses up to the 1:100 year flood event. The benefits, which far exceeded the costs to either individuals or government, were achieved mainly through influences on the decisions of builders and land developers. They also noted that floodplain development pressures could be reduced if, through a mixed policy of land annexation, service extension and zoning bye-laws, the community made available an adequate supply of flood-free land for development.

In the UK controls on floodplain development have traditionally been of a voluntary nature. After broad regional 'structure plans' have been approved by central government, detailed development of land is the responsibility of the local planning authority. This body has the power to refuse 'planning permission' on land zoned as prone to flooding but this designation is advisory only and can be overturned. It has been claimed that such essentially voluntary controls have been effective in limiting floodplain encroachment in the UK. Whilst it is true that there has not been encroachment on the North American scale, the reason probably has less to do with effective planning legislation than the relatively low rate of population growth. For example, in the 30-year period 1952–82, the population in England and Wales grew by only 12 per cent compared to the USA, 50 per cent, Canada, 73 per cent and Australia, 78 per cent. Where pressures for development have been great, as at Datchet (Figure 11.5), the regulatory system has clearly failed. Even with low rates of floodplain invasion, there can still be increased floodplain investment, as rising prosperity and property prices increase the value of houses and their contents, often at a level above general inflation.

As part of the emerging era of post-flood hazard reduction, Handmer (1987b) has highlighted the increasing use of public funds for the purchase of flood-prone land and property in both North America and the USA. The main motive for the public acquisition of these assets is public safety but other benefits may result, such as the creation of parkland, the preservation of wetland habitats and the improvement of waterfront access. Typically, the acquisition areas have a low socio-economic status with low property values. Australia has entirely voluntary schemes whereby the authorities offer to buy at an independently derived market price. This type of scheme can secure community acceptance because the relocation offers families an opportunity to better themselves. Buyouts are seen as a cost-effective use of public funds because, in return for the one-off purchase cost, the property becomes ineligible for any future financial support following disaster. In the USA, FEMA's Hazard Mitigation Program allows the voluntary buyout of floodplain properties on such terms. Following the introduction of the scheme in 1988, over 17,000 properties have been bought nation-wide but, since the remaining high-risk floodplain properties make repetitive-loss claims and account for a large fraction of the NFIP's costs, there is a view that much more use could be made of this strategy (National Wildlife Federation, 1998).

There are few examples where partial urban relocation has been properly assessed as an alternative to more conventional flood mitigation strategies. One is at the small settlement of Soldiers Grove sited near the Kickapoo river in south-western Wisconsin, USA (David and Mayer, 1984). When the town suffered a series of floods in the 1970s, the Army Corps of Engineers proposed to build two levees in conjunction with an upstream dam to protect the central business district, sited within an oxbow on the floodway, and other property (Figure 11.11a). The reservoir project was eventually shelved in 1977 but the residents had been considering other solutions for some time. Following a flood in 1978, they decided that a relocation of the entire business district would yield more benefits than just flood damage reduction for the large cost of the structural scheme. The plan involved public acquisition, evacuation and demolition of all structures in the floodway together with flood-proofing properties in the flood fringe (Figure 11.11b). The community has clearly gained through its non-structural option. Although the levees would have protected the village from most floods, they would not have provided other opportunities; for example, due to compensation payments, businesses could build improved premises. Relocation also gave scope for the use of energy conservation and solar energy at a more attractive site along the major highway which brings commerce into the area.

Figure 11.11   Adjustment to the flood hazard at Soldiers Grove, Wisconsin, USA:
(a) shows the floodway and the flood fringe, together with the location of two proposed levees; (b) shows the areas eventually flood proofed and abandoned, together with the relocation sites.
Source: After David and Mayer (1984). Reprinted by permission of the *Journal of the American Planning Association* 50: 22–35

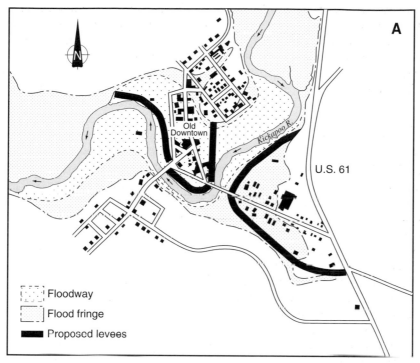

A

Old
Downtown

Kickapoo R.

U.S. 61

Floodway

Flood fringe

Proposed levees

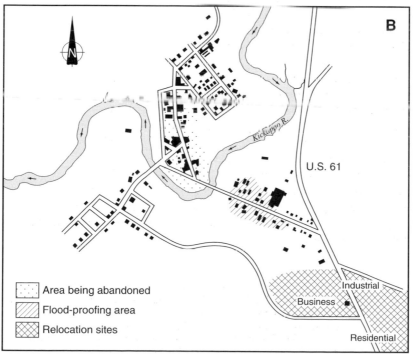

B

Kickapoo R.

U.S. 61

Area being abandoned

Flood-proofing area

Relocation sites

Industrial

Business

Residential

In the future, land use planning is likely to include a much greater element of what has been called the *living with floods* approach. Following disastrous 1:100-year flooding in Bangladesh in 1988, the United Nations Development Programme commissioned various flood studies in collaboration with the Bangladeshi government. The proposal was for increased reliance on embankments along the Brahmaputra and Ganges rivers together with special defences to protect the capital Dhaka and at least eighty other towns where floodwaters are eroding the foundations of buildings. This plan has not been adopted because there are alternative strategies for such LDCs which place increasing reliance on more traditional and sustainable flood responses (Cuny, 1991). These include village-level warning and evacuation schemes, organising working parties to repair levee breaches, developing plans to provide emergency supplies of food and fresh water, and stockpiling vital tools and equipment. Small-scale, self-help strategies which fit in with present land use practices and reduce the ecological impacts of engineering schemes are likely to assume increasing importance in the future. Within the MDCs, there is a similar movement towards *multi-objective river corridor management* which seeks to improve floodplain development so that these areas are better equipped to cope with the complex, and sometime conflicting, demands which are placed upon them (Kusler and Larson, 1993).

12 Hydrological hazards
Droughts

DROUGHT HAZARDS

Drought is different from other environmental hazards. First, it is a 'creeping' hazard because droughts develop slowly and have a prolonged existence, sometimes over many years. Second, droughts are not constrained to a particular tectonic or topographic setting and their impact can extend over very large regions. Thus, drought has similarities with long-term environmental degradation and it is often difficult to tell where drought ends and human-induced desertification begins. Third, the impact of drought varies greatly between the developed and less-developed countries. In the MDCs no one dies because of drought today but, in many LDCs, the effect of unusually low rainfall on already precarious food supplies creates a link between drought and famine-related death.

Famine is the most serious outcome of drought but the linkage is not always direct. This is because famine is also associated with social unrest and civil war so that, in many of the poorest countries of the world, it is difficult to distinguish the early effects of drought from the consequences of war and malnutrition. Malnutrition has been described as the most widespread disease in the world (Garcia and Escudero, 1982). Fully one-third of the population of the LDCs are malnourished. Because drought is so mixed up with other processes that make people prone to famine, the severity of drought impact in the LDCs is likely to be underestimated. There is a general under-registration of deaths in the LDCs, and in many countries reliable data on the causes of death do not exist. It has been said that where there are statistics there is no malnutrition and, more significantly, where there is malnutrition there are no statistics. Given these limitations, it is impossible to produce a reliable estimate of the average annual number of people either killed or affected by drought-related famine.

Most famine-related deaths occur in the semi-arid areas of sub-Saharan Africa, which include the 20 million people living in the countries of the Southern Africa Development Community (SADC), plus South Africa. The term *Sahel* derives from a local word meaning 'the edge of the desert' but the climate is only one of a number of factors responsible for drought disasters in this region, notably the African food crisis of the 1980s. In February 1985 the United Nations estimated that 150 million people living in twenty African countries were affected, of which 30 million were seriously affected and in urgent need of food aid. An estimated 10 million of these people abandoned their homes in search of food and water and 100,000 to 250,000 people died. There were also huge losses of cattle and sheep. But there have been many other drought-related disasters. As recently as 1959–61, perhaps 30 million peasants died from famine in northern China (Jowett, 1989).

Other Asian countries, such as India, are regularly affected by drought but have sought to avoid famine through policies which seek national self-sufficiency in the production of food grains (Mathur and Jayal, 1992). According to Paul (1995), droughts affect Bangladesh at least as frequently as major floods and cyclones with an average frequency of 2.5 years. In South America, the semi-arid area of north-eastern Brazil has also suffered from frequent droughts, notably in 1915, 1919, 1934, 1983 and 1994, and an estimated 2 million people died from starvation in the earlier 1877 disaster (Brooks, 1971). The population of this Nordeste region has doubled since the 1950s to make it the most densely populated area of northern Brazil. Children under five years of age constitute almost 20 per cent of the population, despite the area having the highest infant mortality rate in the country. Endemic malnutrition greatly increases the vulnerability to drought in this region. Drought is also a recurrent feature of Australia. No one dies here as a result but there are important economic consequences. The 1979–83 drought was one of the most severe in the country's history (Gibbs, 1984). One survey revealed that, by November 1982, over half of all the farms, accounting for more than 60 per cent of the nation's livestock, were affected. The ecological significance of such episodes, leading to the removal of vegetative cover and accelerated soil erosion, is also important.

The atmospheric processes that initiate drought are not well understood and drought is invariably defined in terms of effects rather than causes. The simplest definition is 'any unusual dry period which results in a shortage of water'. Rainfall deficiency is, therefore, always the 'trigger' but it is the shortage of *useful* water – in the soil, in rivers or reservoirs – which creates the hazard. Furthermore, it is important to view any water shortage in relative, rather than absolute, terms. It is especially simplistic to view a drought solely in terms of rainfall amounts. In other words, drought and aridity are not the same. This is because humans adapt their activities to the expected moisture environment: a yearly rainfall of 200 mm might be reasonable for a semi-arid pastoralist but could be a disastrous drought for a wheat farmer accustomed to an average of 500 mm per year. Droughts are not confined to areas of low rainfall any more than floods are confined to areas of high rainfall. Therefore, drought should be viewed in the context of a particular climatic regime where the impacts are a result of specific interactions between the supply and demand for water on a regional scale (Wilhite and Glantz, 1985).

Human adjustments to drought tend to rely on crisis management. Treating drought as a crisis can sometimes serve a political purpose by enabling a government to appear decisive in protecting the public from a hostile environment. Emergency methods focus on highly visible measures such as water rationing, cash or food aid. Longer-term adjustments often favour increasing the supply of water to meet anticipated demands, for example, by building more storage reservoirs. Much less attention has been paid to improving efficiency in water use and to promoting the management of water demand as well as supply. A demand-based approach means developing more sustainable responses to water shortages, such as water recycling in urban areas, better agricultural land use practices and the increasing selection of drought-resistant crops. Wilhite and Easterling (1987) were critical of the failure of governments in the MDCs to distinguish between the objectives of such differing drought policies. Such criticisms have even more force when applied to the LDCs, where drought related famine is a hazard. What is required everywhere is a smoother transition from an explicit drought policy, which incorporates long-term development strategies, to specific drought action-plans suitable for the country concerned.

TYPES OF DROUGHT

Figure 12.1 shows the four types of drought hazard commonly recognised and these are described below.

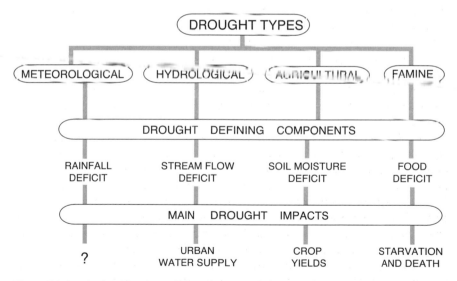

Figure 12.1   A classification of drought types based on their defining components and the main hazard impacts. Disaster potential increases from left to right across the diagram. Rainfall deficit alone does not always produce visible impacts.

Meteorological drought

Meteorological drought is the least severe form and occurs as a result of any unexpected shortfall of precipitation. Rainfall deficiency may not in itself create a hazard because the links between precipitation and the useful water which is available to meet a particular demand are often indirect. For example, strictly speaking, rainfall does not supply water to plants: the soil does this. Equally, rainfall does not supply water for irrigation or domestic use: this is done by rivers and ground water.

The concept of meteorological drought has led to a great variety of simple definitions based on rainfall data. One approach has been to define drought by the duration of a particular rain-free period, the total length of which has varied from 6 days (Bali), 30 days (southern Canada) up to 2 years (Libya). Other definitions depend on the rainfall amounts within a season or a year which fall either within the lower quartile or below 10 per cent of the long-term average. These definitions are entirely arbitrary. They pay no regard to the impacts on water resources or agriculture, not least because the consequences of an equivalent rainfall deficiency may vary greatly through the year. For example, the effect of a summer shortfall in rain, when crops are actively growing, will tend to be much more severe than the same rainfall deficit in winter.

A more complex method, widely applied in the USA, is the *Palmer Drought Severity Index* (PDSI). This approach, by Palmer (1965), is based on a soil moisture budgeting system that considers precipitation and temperature for a given area over a period of months or years. Thus, drought is defined in terms of a reduction of available moisture below that which is normally available and allows the degree of dryness to be rated according to expected climatic conditions. The severity of drought is considered to be a function of the length of period of abnormal moisture deficiency as well as the magnitude of this deficiency. The main advantage is that the PDSI represents a single all-purpose drought index that attempts to combine the impacts of soil moisture, groundwater shortage and low streamflow, although the Index values cannot be easily related to specific environmental or economic hazard impacts.

Hydrological drought

Hydrological drought occurs as a result of a marked reduction in natural streamflow or groundwater levels, plus the depletion of water stored in surface reservoirs and lakes for water supply. If these sources are used directly for supply purposes, hydrological drought can provide a measure of impact by relating the shortfall of supply to demand. Hydrological drought tends to be associated mainly with urban areas and the MDCs, although it can also be recognised elsewhere. For example, in the rural areas of north-eastern Brazil, there are no permanent rivers and water supplies are dependent on seasonal rains, which are stored in relatively shallow reservoirs and ponds prone to high rates of evaporation. After two or three years with below-average rains, this stored water dries up. Drought gives rural dwellers here even less access than usual to clean water supplies, with highly negative consequences for community health and mortality. During such conditions isolated communities may well have to rely on the extensive distribution of small quantities of water by road tankers.

The main impact of hydrological drought is on water resource systems. For example, legislation governing the abstraction of water from a river, or the release of effluent back to a river, will almost certainly be based on a minimum expected flow level under the current climatic regime. The 95 per centile value on the flow duration curve, as shown in Figure 12.2, is often taken as an appropriate minimum discharge for the setting of such legal consents. By definition, drought flows are river discharges below this level, when water abstractions have to be restricted or when effluent discharges have to be reduced, unless other sources can be found. For example, when drought affects the generation of hydropower the only options are to increase the water supply available or to reduce the demand for electricity. Streamflow drought may also be recognised by its effects on irrigation systems dependent on surface water and on river water quality, which is likely to deteriorate with reduced levels of dissolved oxygen and the discharge of sediment-laden 'bottom water' from reservoir storage.

During the mid-1970s, drought affected much of western Europe. Over England and Wales, the 16-month period from May 1975 to August 1976 inclusive had about 10 per cent less rain than had been recorded over any other period of this length since 1920 and was easily the driest May–August period since rainfall records began in 1727. The estimated recurrence interval for any 16-month period was 1:400 years and, for such a period ending in

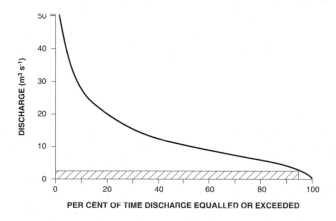

Figure 12.2 An idealised flow duration curve for a river showing the normal dry weather discharge based on the 95 per cent exceedance flow level.

August, the estimate was 1:1,000 years. Not surprisingly, a drought of this magnitude had major impacts on water resources. In England and Wales about one-third of the water used for public supply comes from groundwater and in the winter of 1975–6 recharge into the main aquifers was less than 30 per cent of average. For many rivers, the flow in 1976 was the lowest on record. In the worst-affected urban areas, water rationing was imposed.

The United States drought of 1988 was the most severe in many areas since 1936. Some of the worst impacts were recorded over the Mississippi basin and, by June 1988, over 80 per cent of the drainage system was experiencing severe drought. These conditions exacerbated low flow problems on the Mississippi, Missouri and Ohio rivers. Barge shipments on these rivers represent one of the nation's major means of hauling bulk commodities like grain, petroleum and coal. By early July 1988 blockages due to low discharge and sediment deposition were evident on the Ohio and Mississippi rivers and navigation could be maintained only by controlled releases of water from federal reservoirs and increased channel dredging. Even so, river traffic was down by 20 per cent at one point in the summer and most river ports experienced reduced shipments, especially of grain. The reduced riverflow caused hydropower generation to fall 25–40 per cent below average over large areas of the USA, with significant losses in revenue for the industry (Wilhite and Vanyarkho, 2000).

Agricultural drought

Agricultural drought is important because of the implications for food production. Such droughts tend to be recognised most clearly in those developed countries, such as Australia, which depend on agricultural output for their economic well-being, as well as in those LDCs where agriculture provides a direct livelihood for most of the population. All farmers, whether arable or pastoral, ultimately rely on the water available in the soil for plant growth. Therefore, an agricultural drought occurs when soil moisture is insufficient to maintain average crop growth and yields. Indices of agricultural drought should be based on soil moisture measurements but these are often assessed indirectly by water balance calculations such as the PDSI. Such droughts, which often cover wide areas, may also impact on other economic sectors apart from agriculture.

The first consequences of agricultural drought are in the reduced seasonal output of crops and other production. During 1982, national wheat production in Australia fell by 37 per cent compared with the average for the previous five years and was followed by livestock slaughtering. After such a severe event, it can take up to five years for the numbers of sheep and cattle to recover fully. In 1988 the USA experienced a costly agricultural drought because of the lack of precipitation and the extreme heat in the Midwest. The US Department of Agriculture reported that the 1988 average corn yield was 31 per cent below the progressive upward trend which is driven by improved technology. This was the largest drop since the mid-1930s (Figure 12.3). More than one-third of the American corn crops were destroyed, a loss put at US$4.7 billion (Donald, 1988). The soybean yield was 17 per cent below trend, the largest decline over the previous 60 years, at a cost of at least US$3.7 billion. The overall 1988 American grain harvest amounted to no more than 192 million tonnes, the smallest since 1970 and smaller than the Soviet harvest for the first time in decades. Agricultural drought on this scale within the MDCs can disrupt world trade in food commodities. For example, the 1988 event in the USA led to world grain stocks falling to 288 million tonnes. This represented a 63-day supply, the lowest since the mid-1970s.

Within the MDCs, agricultural droughts have severe financial impacts. Sweeney (1985) estimated that the 1984 drought on the Canadian prairies cost in excess of Can$2.5 billion. In some areas up to half the grain crop was lost due to drought in combination with heat, a severe grasshopper infestation and duststorms. The 1982–3 drought in Australia also created significant losses. Purtill (1983) reported that, given normal seasonal conditions, the yearly cash operating surplus on Australian farms would have averaged an estimated A$21,700, considerably more than the A$12,200 attained. Drought impacts were believed to have reduced receipts per farm by around 23 per cent, mainly because of reduced crop production, while costs were projected to be about 16 per cent lower due to deferred purchases.

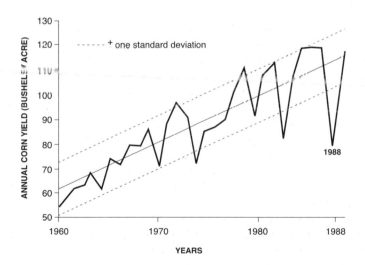

Figure 12.3   Average annual corn yields in the USA, 1960–89, showing the effect of the 1988 drought. In 1988 yields were more than 30 per cent below trend. This was the largest annual drop recorded since the mid-1930s.
Source: Updated after Donald (1988)

At the farm level, severe drought will disrupt normal activities and cause serious financial problems, including a diversion of capital from farm development to drought-reducing strategies, a fall in liquidity and a rise in debt.

In the poorest countries, drought disrupts the normal inadequacies of food supply and increases the seasonal hunger experienced in most subsistence societies. This happened during the 1990–2 drought in southern Africa. In general, the harvest failure was 30–80 per cent below normal and 86 million people were affected over an area of almost 7 million km^2. The drought caused severe hardship, although there was comparatively little loss of human life directly attributable to drought-related malnutrition. In Zimbabwe, for example, the volume of agricultural production fell by one-third and contributed only 8 per cent to GDP, compared to 16 per cent in normal years. By November 1992 half the population had registered for drought relief.

Conditions in Zambia were typical. Zambia's national food security has been repeatedly threatened by previous droughts in 1982–4, 1987 and 1990. In the Southern, Western and Eastern Provinces, yields of maize were down by 40–100 per cent and the drought affected some 2 million rural people (IFRCRCS, 1994). According to Kajoba (1992), part of the grain shortfall was due to the cultivation of hybrid maize under imported fertilizer regimes, rather than a reliance on more traditional drought-resistant crops like sorghum, millets and cassava. Food stocks were soon exhausted and many Zambians had their buying power eroded as a result of a high incidence of disease amongst livestock. Most of the worst-affected communities were in remote areas badly served by transport links and with limited access to health care and education. The drought impacts cascaded rapidly through these areas, leading to the closure of primary schools and a decline in tourism as wildlife camps became deserted. Due to low water levels, the Kariba, Kafue and Victoria Falls hydropower stations were working at only 30 per cent capacity and the government was forced to impose daily power stoppages. Many households resorted to charcoal for their domestic energy needs and the price of this fuel doubled in some towns.

Famine drought

Famine drought is often regarded as an extreme form of agricultural drought which destroys food security to the point where large numbers of people become unable to maintain an active healthy life. At worst, this is seen to result in mass deaths from starvation. But the relationship between drought and famine is rarely direct because drought is a geophysical hazard whereas famine is a cultural phenomenon. Very severe agricultural droughts occur in parts of the MDCs but no deaths result. On the other hand, famine drought appears almost endemic in some of the poor countries of sub-Saharan Africa.

Famine has been defined as 'a protracted total shortage of food, in a restricted geographical area, causing widespread disease and death from starvation' (Dando, 1980). Such disasters are often multi-causal and include an element of war or civil unrest. They have been recorded for at least 6,000 years and have only recently been eliminated from the developed world. Today, famine tends to be associated with semi-arid areas of subsistence, or near-subsistence, agriculture where crop failure has resulted from drought. However, evidence from the Darfur, Sudan, famine of 1984–5 and elsewhere has challenged the common concept of mass starvation arising directly from crop failure. According to de Waal (1989), the great majority of people survived in Darfur, despite the severe drought and a doubling of the overall mortality rate. The excess mortality was heavily concentrated on children and the elderly because people between 10 and 50 years old accounted for less than 10 per cent of the excess deaths. In practice, the deaths were caused by the transmission of disease arising from the crowding of refugees into centres where water supplies were poor and health care was inadequate. The majority of all the famine deaths were attributed to measles, diarrhoea and malaria rather than to starvation. This interpretation clearly casts doubt on the conventional indicators of famine drought, such as mass starvation, and on the assumed efficacy of conventional disaster reduction strategies, such as the supply of food aid.

Although food scarcity is the leading factor in famine drought, its effects are greatly increased by underlying health-related problems, such as limited access to potable water, a lack of modern sanitation and inadequate health care, especially for the very young. Another feature of the most severe droughts is that they undermine livelihoods and rural stability to create migration. After the 1985 drought in north east Brazil, as many as 1 million people, mainly men, abandoned their small farms in search of work. This produced a major wave of rural–urban migration which added significantly to the peri-urban *favellas* or shanty towns surrounding every Brazilian city.

CAUSES OF DROUGHT

Physical causes

Drought impacts tend to be worst in areas with drier climates for two reasons. First, a low mean annual rainfall is associated with high variability. It is the lack of *rainfall reliability* in these areas – from season to season or year to year – rather than the low rainfall totals that are responsible for drought hazards. Second, the duration of drought is longer in the drier lands. In humid areas, drought may be created by a rainfall deficit over a comparatively short period, for example, a few months or a single growing season. But in drier areas drought tends to build up more slowly over a number of years. For example, the 1975–6 drought over north-west Europe lasted no longer than 16 months whereas the more recent drought in the African Sahel region was created by persistently dry conditions over at least 16 years from 1968, leading to widespread famine in 1984–5.

There is, therefore, an inverse correlation between mean annual rainfall and the impact of drought. The arid Southwest and the semi-arid Great Plains are the regions of the USA most vulnerable to recurrent drought. Major droughts tend to occur on the Great Plains every 20 years, the worst being that of the 1930s – the Dust Bowl years. During the droughts of the 1890s and 1910s, widespread reports of deaths due to malnutrition were reported but little was done by the authorities to relieve distress. The effects of the major drought of the 1930s were compounded by poor farming techniques, poor market prices and a depressed economy. Afterwards, massive state and federal aid resulted in greater control of soil erosion and the development of irrigation. These measures, combined with improved farm management and crop insurance, meant that when drought struck again in the 1950s, its impact was less severe.

Under 'normal' climatic conditions the African Sahel is, at best, a semi-arid area. The mean annual rainfall ranges from 100–400 mm in the northern zone (on the Saharan edge) to 400–800 mm along the southern margins, although the most critical area receives some 200–400 mm. But these mean values are misleading. The annual rainfall patterns are characterised by both high seasonality and high variability from year to year. In the Sahel as much as 80 per cent of the annual rainfall occurs between July and August. As shown in Figure 12.4, the average annual variability, expressed by the

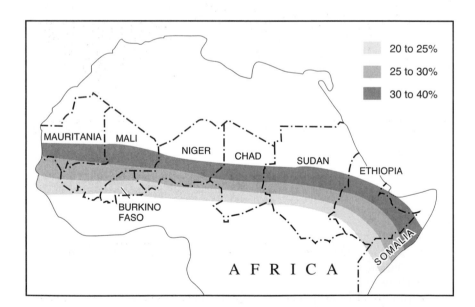

Figure 12.4   The countries of the Sahel zone of Africa showing relatively high mean annual departures from average rainfall. Where the variability of rainfall is high and the amount of rainfall is low, drought is likely to be a recurrent feature of the climate.

coefficient of variation, ranges from 25 per cent to 40 per cent, leading to a low reliability of yearly precipitation.

Persistent lack of rain has been the final 'trigger' mechanism in creating famine. Figure 12.5 shows an annual rainfall index derived for the African Sahel for the twentieth century up to 1998 (Parker and Horton, 1999). The graph indicates that the most recent drought started in 1968 and during the early 1980s built up to produce the lowest rainfall totals this century. Overall, the Sahel experienced a period of some 30 years with below-average rainfall conditions broken only by the widespread rains in the 1994 wet season. The agricultural impact of the drought was made worse by the good rains of the 1950s and 1960s, which encouraged rain-fed cropping into marginal lands and led to increased herd sizes.

Lack of rain, combined with pressure on land resources, has produced environmental deterioration in the Sahel. According to Grainger (1990), *desertification* is the degradation of lands in dry areas which is mainly due to poor land use practices but which can also be made worse by a shift in

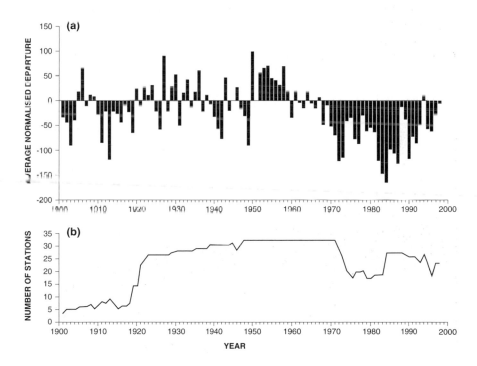

Figure 12.5   Rainfall variations over the Sahel region during the twentieth century: (a) shows annual rainfall anomalies (in percentage standardised units based on about 50 years of data ending in 1973); (b) shows the number of rainfall stations used. Rainfall has been consistently below average since 1968.
Source: After Parker and Horton (1999). Crown copyright

climate. The exact relationship between drought and desertification remains unclear but the chief effects are in reduced biological productivity. It is estimated that more than 80 per cent of the drylands of Africa now suffer from desertification and that the continent accounts for more than one-third of all such degraded land in the world.

The meteorological causes of drought can be traced to little-understood anomalies in the general atmospheric circulation. The most direct relationship is often with a displacement of the jet stream from its normal track which steers rain-bearing storms away from the stricken area. However, recent research has concentrated on the search for *teleconnections*, which are the linkages between climatic anomalies occurring at long distances from each other. In particular, there is growing evidence that large-scale interactions between the atmosphere and the oceans may be implicated. This view emphasises the importance of sea surface temperature anomalies (SSTAs) since it is known that these influence the flux of sensible heat and moisture at the ocean–atmosphere interface. Moisture conditions are at least as important as temperature because they influence both the subsequent latent heat release and the amount of precipitable water in the atmosphere.

Drought is most likely to be initiated by negative (relatively cold) SSTAs leading to more stable, anticyclonic weather as, for example, when El Niño events bring descending air to the countries of the western Pacific and south-east Asia. According to Dilley and Heyman (1995), worldwide drought disasters occur with twice the frequency recorded in all other years during the second year of an El Niño event. For example, during 1982–3, droughts in Africa, Australia, India, north-east Brazil and the United States coincided with a major El Niño phase. The 1988 North American drought was linked to a shift in the Southern Oscillation associated with a widespread decrease in Pacific sea surface temperatures. This led to a northward displacement of the Inter-tropical Convergence Zone southeast of Hawaii and the eventual appearance of a strong anticyclone at upper levels over the American Midwest (Trenberth *et al.*, 1988). The probable starting point for the drought that affected northwest Europe in 1975–6 was abnormally low sea surface temperatures over the Atlantic ocean north of 40°N. This SSTA typically causes near-surface stability in the atmosphere and a high frequency of blocking anticyclones over western Europe.

Related attempts have been made to link SSTAs in the tropical Atlantic to rainfall in the Sahel zone of Africa. It is known that there are recurring SSTA patterns and that these tend to differ around the globe depending on wet or dry conditions in Africa (Owen and Ward, 1989). As indicated by Gray (1990), a season-to-season link has been found between the frequency of Atlantic hurricanes and rainfall in the Sahel which is thought to be due to changes in the upper-tropospheric circulation that accompany changes in the monsoon structure and fluctuations in the strength of the easterly waves produced over north Africa. Other research has suggested that the underlying forcing agent might be the global transport of oceanic water which is dependent on the sinking of the cold, salty water in the North Atlantic ocean (see Chapter 9).

More local feedback mechanisms between the land surface and the lower atmosphere are thought to extend existing drought conditions. The European drought of 1975–6 was prolonged towards the end of the summer of 1976 by the excessively dry ground, which created a local feedback process and helped to maintain the rainless atmospheric state. During the late spring and early summer as much as 90 per cent of the incoming solar radiation over Britain was being used to heat the ground and the air whereas, in a normal year, more of the net energy would be used in evaporation (Ratcliffe, 1978). This is because dry soil has a higher albedo, or proportional reflectance of solar radiation, than moist soil. Similar effects were probably responsible for the high summer temperatures recorded during the US drought of 1988. Where prolonged dryness has reduced the vegetation cover and created greater dustiness, as in parts of Africa, it is possible that drought may become almost self-perpetuating.

Human causes

Sub-Saharan Africa suffers from many of the key factors found in disaster-prone LDCs (see Chapter 2). Despite the importance of agriculture, which accounts for more than 40 per cent of GDP in some countries, population growth is outstripping food production. In most countries the population is doubling every 20–30 years and many of the Sahelian countries have great difficulty in feeding themselves. Figure 12.6 reveals that, whilst most other developing regions of the world have improved their food supply position, there has been a progressive decline in Africa since 1969, coinciding with the onset of dry conditions. These long-term trends are creating environmental refugees. As many as 10 million people have migrated during recent years and there may be greater future redistributions of the African population in response to the deteriorating food situation.

According to UNEP data, two-thirds of Africa is dryland and over 70 per cent of its agricultural drylands are classified as degraded. The Sahel suffers from all the land use practices which give rise to desertification: over-cultivation of crop lands, over-grazing of rangelands, mismanagement of irrigated cropland and deforestation. About 90 per cent of pasture land and 85 per cent of crop lands in the countries closest to the Sahara have been affected. Deforestation is an important catalyst of land exhaustion and soil erosion. In Africa, more than 90 per cent of all wood is used for cooking and other energy needs, and the demand for fuel wood has grown considerably since the oil price rise in 1974. Since kerosine is expensive to buy, there is an 'urban shadow' of stripped land around most settlements. In effect, economic and social pressures – made worse by drought – have caused the breakdown of the traditional system of land use which was adapted to this fragile environment.

(a)

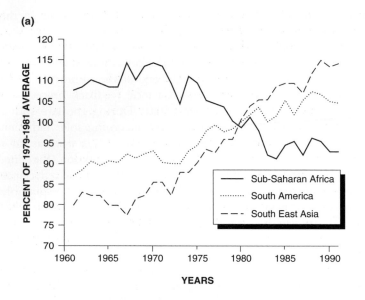

(b)

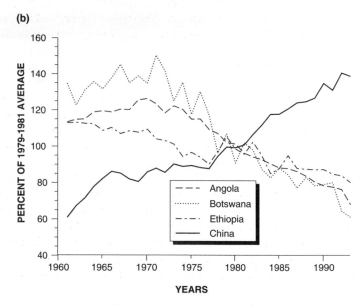

Figure 12.6 The continued downturn in *per capita* food production in sub-Saharan Africa: (a) shows the regional downturn, 1961–91, relative to equivalent indices for south-east Asia and South America; (b) shows the decline for some individual countries, 1961–93, compared to the trend for China. All values are expressed as a percentage of the 1979–81 average for each region.
Source: Compiled from data supplied by the Economic Research Service, US Department of Agriculture

The reliance on rain-fed agriculture throughout sub-Saharan Africa is also important. In the Sahel region rain-fed cropping covers about 95 per cent of the cultivated area and, during drought conditions, the management options are far fewer than in irrigated agricultural systems. Indeed, the only major options available in such a low-technology system are the selection of a particular crop type for sedentary cultivators and reduced stocking rates for pastoralists. Although agricultural science has made a successful contribution through the so-called 'Green Revolution' to the productivity of some tropical food supplies, much less progress has been made in the drier lands.

The traditional pattern of agricultural land use in the Sahel was well adapted to uncertain rainfall conditions. Generally speaking, the northern zone, with a mean annual rainfall of 100–350 mm was used for livestock, whilst the southern Sahel, with a rainfall of 350–800 mm, was used for rain-fed crops. This system permitted a degree of flexible interdependence. The pastoralists followed the rains by seasonal migration (transhumance) or the practice of full nomadism, whilst the cultivators grew a variety of drought-resistant subsistence crops, including sorghum and millet, to reduce the risk of failure. Long fallow periods were used to rest the land for perhaps as much as five years after cropping in order to maintain the fertility of the soil. In the absence of a cash economy, a barter system operated between pastoralists and sedentary farmers, leading to the exchange of meat and cereals.

This system has collapsed for a variety of reasons. Population growth, with the need for more food supplies, has led to increased pressure on the land. One consequence of this has been soil erosion as cultivation has spread into the drier areas formerly used for livestock. In turn, the rangelands have been overgrazed, with rapid degradation of the resource base. The need of national governments for export earnings and foreign exchange has produced a trend towards cash crops, which have competed for land with basic grains and reduced the fallowing system. Subsistence crops have been discouraged to the extent that farm produce prices have consistently declined in real value over many years. At the same time, the build-up of food reserves has been seriously neglected under pressure from international banks wanting loan repayments. In addition, a lack of government investment to improve the productivity of rain-fed agriculture and a failure to organise credit facilities for poor farmers have also tended to undermine the stability of the rural base.

Some socio-economic and political changes have increased the vulnerability of the nomadic herdsmen of Africa to drought. National governments have progressively legislated against nomadism whilst other bodies have also attempted to settle the herdsmen. In northern Kenya, for example, the Catholic Church has been influential in settling pastoralists in mission towns (Fratkin, 1992), although livestock remains the most important food production system in this dry area and these people still depend on their animals for subsistence and trade. But their mobility has been restricted and in many instances foreign aid has been earmarked for sedentary agriculture rather than pastoralism. The traditional system of animal accumulation was not understood, together with the emergency system of gifts and loans, so that governments have taxed animals in the belief that the herdsmen should be forced to sell. Increasingly strict game preservation laws have been introduced which restrict the possibility of hunting for meat during drought. Traditional forms of employment, such as in caravan trading, have declined as a result of the enforcement of international boundaries and customs duties, together with competition from lorries. Thus, African agriculture is facing many problems of which lack of rain is just one.

Poverty is also a key factor. Sub-Saharan Africa contains over two-thirds of the world's poorest countries. It has been said that Africa is locked into an economic system which obliges it to produce goods it does not consume and to consume goods it does not produce. There is a widespread view, particularly amongst those who support the structural paradigm of environmental hazards, that colonialism and the international trading system have reduced the innate ability of Africans to cope with fluctuations in their physical and societal environments. As in any disaster, the impact differs greatly; some prosper, some migrate to refugee camps and others die. Worst hit are the landless and jobless, especially the women and children in the rural areas, who lack the means to ensure their own food security.

Factors such as declining terms of trade for primary agricultural products, market protection by the industrialised countries, extreme commodity price fluctuations on international markets and the need to service enormous overseas debts have all restricted the ability of African governments to address their internal problems. All the African countries afflicted by drought are heavily in debt due to a loss of purchasing power for primary products (Brown, 1985). Even when clear national economic policies have been pursued, they have not favoured the rural sector and have not led to increased agricultural production. When investment has been made in agriculture it has often been to encourage cash crops, either for urban elites or for the export market, and there has been little incentive to increase local food crops.

LOSS-SHARING ADJUSTMENTS

Disaster aid

Loss acceptance and loss sharing are important strategies for dealing with drought. For many LDCs, disaster aid has become almost synonymous with drought relief and, in 1985, two out of every five Africans south of the Sahara were living on foreign aid. The 1991–2 El Niño-related drought in southern and eastern Africa threatened 30 million people and generated a major international aid effort. From April 1992 to June 1993 roughly five times more food and relief goods were successfully shipped into southern

Africa than were shipped to the Horn of Africa during the 1984–5 famine (IFRCRCS, 1994). In Zambia, the government informed the major donors about the food security situation in March 1992 and, by August, it was believed that sufficient food was arriving in the country to prevent famine, although there were internal problems with distribution due to congested railheads and poor road transport. A 'Programme to Prevent Malnutrition' (PPM) was established to coordinate activities between the agencies representing each of the geographic areas targeted for food assistance (IFRCRCS, 1994). This structure gave more than fifty NGOs access to nearly 250,000 tonnes of maize for distribution to about 2 million people, together with the flexibility to carry out activities consistent with their own priorities.

Although such programmes have been judged successful, food aid has been described as a 'blunt instrument' by de Waal (1989). This is because it is predicated on the conventional western view of famine as a mass starvation event. Given this interpretation of famine, the large-scale distribution of food appears to be a sensible strategy. But, if famine-related deaths are highly age-specific and are heavily dependent on other factors such as the presence of endemic disease, the indiscriminate distribution of food may not always help those at greatest risk. On the basis of their experience in northern Sudan in 1991, Kelly and Buchanan-Smith (1994) have argued that, in the apparent absence of many excess deaths due to starvation (that is, a

Plate 10   Food aid is often an essential component of emergency relief, especially for victims of drought. In the past, it has been a recurrent feature of hazard mitigation in sub-Saharan Africa, partly as a result of the progressive downturn in *per capita* food production. (Photo: International Federation)

substantial 'body count'), donors were unwilling to accept the real needs and to contribute fully to the relief operation. This situation arose partly because of an under-reporting of deaths and despite the fact that child mortality rates rose by at least 20 per cent on average, often as a direct result of an acute shortage of food. Clearly, it is desirable to strive for a better understanding of famine drought impacts and to deliver food aid, as well as other relief, to those most vulnerable. But it is difficult, in practice, to provide selective assistance at the individual household level (Kelly, 1992). For example, targeting food according to standard anthropometric criteria, such as weight for height indices, is not a method which is equally applicable everywhere and, in some cases, has led to the deliberate underfeeding of children to ensure that the household qualifies for rations.

Apart from inefficiency in emergency distribution and the danger of creating an 'aid culture', longer-term drought relief has not always been invested wisely. Comparatively little has been spent directly on agriculture and forestry or on field action at a local level. In the short term, the better deployment of disaster relief can be achieved only when those in most need have been identified and transportation methods have been improved. In the longer run, aid should be redirected to small farmers so that the rural sector can be stabilised again. Only lip service has really been paid to this. Part of the difficulty is that students selected for overseas training come from the urban elites. After their return, the great temptation is to remain in the cities. So the transfer of agricultural technology is from city to city, rather than into the rural areas where the food must be grown.

More attention must be paid to sustainable development in the rural areas. In the short term, this might well mean the provision of food aid via work programmes. This is the main method of distributing free maize in Zambia to those without food or cash resources. Recipients are required to participate in self-help projects such as repairing feeder roads, digging pit latrines to improve sanitation, drilling boreholes and wells and constructing dip tanks for cattle. A similar cash-for-work scheme operates in north-east Brazil. Here the programme is meant to guarantee a small salary and, in 1993, an estimated 2 million people, with another 4–6 million dependants, were employed in this way (IFRCRCS, 1994). In the longer term, more investment in research on basic staple grains and better dryland farming techniques, such as terracing, strip cropping and soil erosion control, suitable to rain-fed agriculture is needed. It may even be more productive to support nomadic pastoralism rather than irrigation schemes. This requires new attitudes because it goes against the normal funding priorities and does not show a conventional return on investment. Improving the physical infrastructure in areas at risk by the provision of better roads will not only reduce short-term vulnerability by helping the distribution of emergency food aid but will also allow the optimum location of new facilities, such as well-equipped health clinics, which will lead to more long-term resilience in the face of drought. Above all, there is a need to release local initiatives in order to encourage more self-reliance in the people so as to release them from dependence on famine aid.

Even in the MDCs, emergency drought relief has traditionally been viewed as a priority for governments. In a comparative analysis of drought policy in the USA and Australia, Wilhite (1986) showed that government actions have typically had a loss-sharing character, dependent on loans and grants, and that most drought mitigation has occurred in a crisis-management framework similar to that for emergency aid. In severe droughts, it is only governments which have the necessary resources to intervene at the scale required and the costs can be high. For example, the total cost of federal drought relief programmes in the form of loans and grants during the 1974–7 drought in the USA has been estimated at US$7–8 billion. During 1976–/ no fewer than forty programmes were available to provide assistance to the private sector, administered by sixteen different federal agencies. Not surprisingly, the exercise was not efficient.

More recently, the rising costs of drought relief, combined with the difficulties of coordinating the activities, have led to radical reviews of national drought policy by several of the industrialised countries. In turn, this has produced a general trend away from the use of emergency subsidies provided by the taxpayer towards greater long-term self-reliance by rural communities. For example, in 1989 the Australian government removed drought from the terms of the Natural Disaster Relief Arrangements because of increasing concerns about the cost to the federal purse (O'Meagher *et al.*, 2000). In a rather belated recognition that drought is an endemic feature of the Australian climate, the new National Drought Policy aims to have drought recognised as an integral part of all agricultural decisions rather than be treated as a random factor requiring an emergency response. According to Haylock and Ericksen (2000), central government responsibility for drought assistance in New Zealand has also been declining since the 1970s, with the definition of a drought eligible for support being progressively tightened. In 1996 this definition was restricted to a 1:50-year event (2 per cent annual probability of exceedance). The present policy is to devolve drought response as far as possible to local government, rural communities and individuals within an overall strategy of a more sustainable management of natural resources. At present, it is not clear how these changes in national policy will affect future drought management in either Australia or New Zealand.

EVENT MODIFICATION ADJUSTMENTS

Environmental control

The most fundamental option is the artificial stimulation of rainfall by cloud-seeding during the event. In order to have any chance of success, this technique must be applied to clouds with precipitation potential. In practice, this means either well-developed cumulus or deep orographic clouds. These clouds are unlikely to be present in large numbers during drought conditions and there is, therefore, little scope for the physical modification of atmospheric processes.

Indeed, the supply of water is not necessarily a solution. Where water has been made available, often through aid schemes, it has sometimes compounded the problems. Although every year about 5,000 ha of new land comes under irrigation in the Sahel, this is balanced by about the same area going out of use through waterlogging or soil salinity. The drilling of new boreholes in dry areas is an example of how aid and technology, without proper local management, can actually increase disaster. Along the southern edge of the Sahara desert new tube wells were constructed to provide water points so that the last reserves of rangeland could be opened up. Without the imposition of effective controls, the borehole sites provided an attractive focus for many cattle and humans. The water encouraged the growth of herds beyond the available feed until the new areas were stripped and the cattle died. Other inappropriate uses of irrigation water exist. Some of the new supplies have been used to irrigate export crops, such as pineapples, and rice grown for the urban elites. Both these crops consume large quantities of water and have done nothing to alleviate food shortages in the rural areas.

Hazard-resistant design

The standard defence against hydrological drought has been the construction of dams and pipelines for the artificial storage and transfer of water supplies. The emphasis on 'tech-fix' engineering solutions is symbolised by the global spread of large dams and the associated increase in regulated rivers, especially during the period 1945–70 (Beaumont, 1978). Regulated rivers smooth out the seasonal variations in riverflow and, in particular, provide artificially enhanced dry-weather flows which provide benefits for either the freshwater ecology or for water abstraction purposes. Figure 12.7 shows the half-monthly regulated flows for the river Blithe, England, during the 1976 drought at a point downstream from a regulating reservoir. When these actual flows are compared with the modelled 'naturalised' flows that would have occurred at this point in the absence of a reservoir, it can be seen that the reservoir was able to contain river discharges within a narrow band right through the year. During the winter months, reservoir storage retained the flood peaks that would otherwise have gone down the river and reduced average flows. Between May and September the regulated river discharge was enhanced above the natural flow regime. The river never fell below the designated minimum acceptable flow of $0.263 \text{ m}^3 \text{ s}^{-1}$, despite the severity of the drought conditions which would have reduced natural flows below this level for three months.

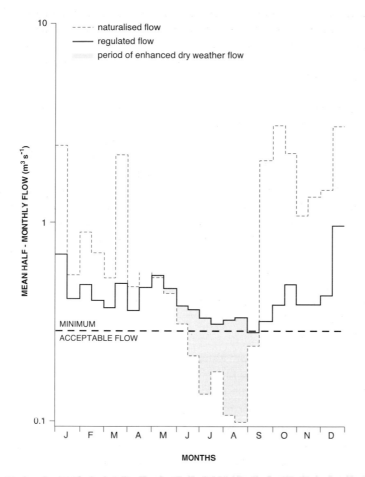

Figure 12.7 The effect of reservoir storage and flow regulation on the river Blithe, England, during the hydrological drought of 1976 in relation to the prescribed minimum acceptable flow. Regulated flow is the actual mean half-monthly discharge; naturalised flow is the estimated discharge in the absence of a reservoir.
Source: After David J. Gilvear (personal communication, 1990)

Water supply reservoirs have been used extensively to maintain urban water supplies. The greatest buffering against drought exists for those areas with a large margin between the daily supply capacity of the system and the maximum daily use. Many reservoir-based urban water systems are designed to provide a predetermined minimum supply during roughly 98 per cent of the time (2 per cent probability of failure), although relatively minor shortages may be accepted more frequently. With an element of overdesign and careful crisis management, it is possible for these systems to perform well during droughts of a magnitude beyond the 1:100 event.

Figure 12.8 shows how the enforcement of domestic water intake restrictions on the river Tone, Japan, during the summer 1994 drought helped to maintain supplies (Omachi, 1997). Without these drought responses, the content of the reservoirs would have declined more quickly and the water supply storage would have been exhausted by 12 August 1994.

VULNERABILITY MODIFICATION ADJUSTMENTS

Community preparedness

As with all hazards which offer limited scope for prevention, this is an important strategy, although there are large differences between rural areas in the LDCs and urban areas in the MDCs in the way that communities prepare for and react to drought.

Over time, many traditional societies have evolved 'coping' strategies which anticipate food insecurity and provide for survival in dry rural areas. However, given the drastic actions required for famine drought, the term 'coping' is perhaps unduly optimistic. For example, to cope with a 'normal' drought, nomadic people in the Sahel have adopted the practice of herd diversification, involving camels, cattle, sheep and goats, all with different grazing habits, water requirements and breeding cycles, which helped to spread any risk of pasture failure. During years with abundant rainfall, the tribes would increase their herds for food storage and as an insurance against

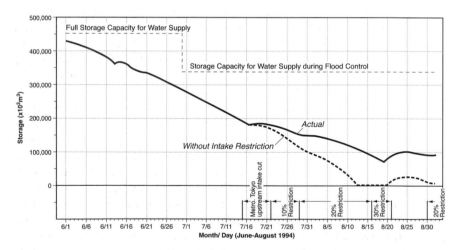

Figure 12.8   Changes in water storage in the reservoirs along the upper Tone river, Japan, in relation to severe restrictions in water use during the summer drought of 1994. In the absence of these restrictions, water supplies would have been exhausted by mid-August.
Source: After Omachi (1997)

drought. When drought did occur such people regularly migrated to find good pasture and, in the most severe episodes, could either eat or sell off the surplus livestock. Informal systems of communal loss-sharing allowed the transfer of gifts or loans of any spare animals available for those in greatest distress, and various fall-back activities, such as gazelle hunting or caravan trading, were intensified as temporary measures to help survive the drought. In a similar way, villagers in rural Mali have adjusted to decreasing harvests by diversifying their income sources from non-agricultural activities (Cekan, 1992).

Under severe conditions, all rural people have to do more. They often start by simply eating less, in an attempt to conserve existing food stocks. For farmers, agricultural adjustments include crop replacement (drought-resistant crops preferred at the normal planting time), gap filling in fields (where germination of an earlier crop has been poor) and resowing or irrigating crops. When food stocks have been exhausted, they turn to a wide variety of wild 'famine' foods which are not normally part of the diet because of their low nutritional value. In Zambia, for example, this includes eating honey mixed with soil, wild fruits and wild roots, some of which are poisonous unless boiled for several hours before eating. The selling of any livestock is usually well underway by this stage. Such action exacerbates the fall in the prices of livestock; this occurs at the time when the cost of grain is rising and produces a change in the relative terms of trade which is particularly disadvantageous for nomadic people. Poor pastoralists have to sell a larger proportion of their animals in order buy to food than do the wealthy. During a severe drought, therefore, many of the poor are squeezed out of the pastoral economy and forced to settle in towns to live on famine relief or from wages paid to herders or labourers. Without food and other resources, drought-stricken rural dwellers routinely turn to local wage labour for support, rather than work on their own unproductive land. Typically, supply overwhelms demand and leads to the rapid collapse of an already precarious labour market (Cutler, 1984).

For some households, outstanding debts and other favours can be called in and cash or food entitlements may be borrowed from more prosperous relatives, neighbours or other support groups in a non-agricultural response strategy. Without access to external help, there is little option but to resort to the trading of valuables, such as jewellery, or other capital assets, such as radios, bicycles or firearms, which can be sold to buy grain. As incomes decline, health conditions also deteriorate. This deterioration is exacerbated by poor nutrition and growing competition for declining, and increasingly polluted, water supplies. Wherever possible, villagers also poach wild game in order to survive. Table 12.1 shows that during the severe drought of 1994–5 in Bangladesh, which affected over 10 per cent of the country, households adopted a variety of these non-agricultural adjustments. Over half of those questioned sold livestock and over 70 per cent of the respondents either sold or mortgaged land (Paul, 1995).

Table 12.1   Adoption of non-agricultural adjustments to drought by households in Bangladesh

Adjustment	Number of households	Percentage
Sold livestock	166	55
Sold land	112	37
Mortgaged land	106	35
Mortgaged livestock	2	1
Sold belongings	26	9
Family members migrated	1	0

Source: Paul (1995)

Note
N = 265 households. Multiple responses are possible.

Ultimately, the family will start to break up. Some children may be sent to distant relatives out of the famine zone and male members may seek work in the towns. This can lead to large-scale migration which may prove to be permanent if families lose their land rights due to moving. In a study of rural households which had migrated from famine-affected communities in northern Darfur, Sudan, it was found that asset wealth did not enhance famine resistance as some of the earliest migrations were undertaken by 'wealthy' families (Pyle, 1992). As with other hazards, prior experience of coping with drought enhances the chances of survival but some traditional responses are now less available than in the past. For example, livestock raiding by dryland pastoralists has often been a means of rebuilding herds destroyed by drought but, as shown by Hendrickson *et al.* (1998), this subsistence-based activity has recently been disrupted in parts of Kenya by external raiders seeking cattle either to feed private armies or to sell for profit.

Famine drought is unknown in urban areas, even within the LDCs. Short-term adjustments used by water authorities during hydrological droughts are aimed mainly at the domestic consumer. They include both supply management and demand management practices. Supply management methods tend to concentrate on the more flexible use of available supplies and storage, as shown above. This is achieved by switching water abstraction between surface and ground sources and transfers between different water supply authority areas to ease the greatest shortages. Temporary engineering, such as the laying of emergency pipelines, is often necessary to import water from more distant sources. Other technical measures available to the water supply industry include reducing the water pressure in the main supply pipes and repairing all possible leaks in the distribution system. When all else has failed, it may be necessary to ration water in the worst-hit areas by rota cuts, which interrupt supplies for part of the day, or by the erection of standpipes in the street so that people can only draw small amounts of water under relatively inconvenient conditions.

Attempts to manage (that is, reduce) consumer demand during drought episodes normally include a mix of legal measures and public appeals to conserve water. At an early stage, local ordinances may be used to ban non-essential domestic uses of water, such as the washing of cars or the watering of gardens. As the drought continues, special legislation may be rapidly introduced, such as the Drought Act rushed through the British Parliament in August 1976 to prohibit the non-essential uses of water. Combined with 'save water' publicity campaigns, these management techniques can cut the residential demand by up to one-third for short periods.

But crisis management is no substitute for a real state of preparedness and longer-term planning for water conservation in urban areas. Where hydrological drought is a more common feature, such as in Adelaide, South Australia, the continued management of water demand is a central plank of policy. During the summer months, when rainfall is almost entirely absent, as much as 80 per cent of the water consumed within the metropolitan area is used to irrigate domestic gardens. As part of an overall conservation strategy, this proportion can be reduced through a combination of financial measures (seasonal peak pricing), technical measures (curbs on inefficient water-using appliances and advice on suitable watering methods) and social measures (persuading people to grow native plants in their gardens rather than more water-demanding European varieties).

Forecasting and warning

In order to be effective, drought forecasts need to be available many months ahead in order to aid seasonal decisions on crop planting and water management. The best hopes probably lie with the use of meteorological models which couple the atmosphere and the oceans, and much effort has been put into the refinement of ENSO-based methods. According to Ropelewski and Folland (2000), such methods already show some skill in seasonal rainfall prediction but they are not effective in all years and only provide rainfall results for broad regions averaged over several months. This information lacks precision for individual decision makers at specific sites.

In the case of famine drought, better early-warning systems are needed for crop failure. The best-known of these is the Famine Early Warning System (FEWS) operated by the UN Food and Agriculture Organization which issues monthly bulletins on rainfall, food production and famine vulnerability in threatened areas. These systems rely increasingly on sophisticated methods such as remote sensing to indicate where vegetative growth is reduced. The importance of regional factors in African drought encouraged the World Meteorological Organization to establish the AGRHYMET programme with its central base in Niamey, Niger. As the name implies, this initiative uses external expertise in agronomy, hydrology and meteorology to train local staff to improve agricultural production, water availability and drought response. But positive intervention will only be achieved through knowledgeable local field workers who can alert others of the need to mobilise resources.

Many countries in sub-Saharan Africa have installed famine early warning systems following the major drought of the mid-1980s. For example, in 1986 Chad and Mali set up comprehensive food and nutrition surveillance systems (Autier *et al.*, 1989). The premise is that the prompt detection of failing food supplies will ensure an equally prompt reaction from donors which, in turn, will preclude the emergence of a famine. On the other hand, as already suggested by Kelly and Buchanan-Smith (1994), 'donor fatigue' might require more visible evidence of excess mortality before relief efforts are fully deployed. These multi-disciplinary programmes are intended to detect food shortages as early as possible and are based on the concept of 'rising-risk monitoring'. Detection of the downward spiral into famine is heavily dependent on the outcome of local nutritional field surveys, as indicated by height for age, weight for age and weight for height measurements, to identify those, especially pre-school children, with the greatest needs. Other reasonably reliable famine precursors are rising grain prices, combined with falling livestock prices and wages, as the economic balance shifts from assets and services, such as jewellery and labour, to food, which rises in value both absolutely and relatively. In the longer term, investment in such early warning systems should probably be redirected to more comprehensive monitoring of overall rural development. This may help to explain why the longer-term resistance to famine is not increasing in households living in vulnerable areas.

Land use planning

Drought increases the pressure on land resources. Overgrazing, poor cropping methods, deforestation and improper soil conservation techniques may not create drought but they frequently amplify drought-related disaster. There is a need, therefore, for better agricultural land use practices. Sustained dryland farming is dependent on soil conservation measures against water and wind erosion. A grass or legume cover is an effective control against water erosion, as are strip cropping and contour cultivation which retard the flow of water down the slope. Wind erosion can be greatly reduced by maintaining a trash cover at the soil surface and the use of crop rotations and of shelter belts to lower the wind velocity.

Rural areas rarely have the massive water storage facilities and the options for reducing consumer demand that are available to cities in the MDCs. Therefore, the most prudent long-term drought strategies are those which relate agricultural production and management practices to withstand unexpected shortfalls of precipitation. This involves the adoption of appropriate stocking rates, so that the pasture is not easily exhausted, the build up of a reserve of fodder and the improvement of on-farm water supplies. The installation of an irrigation system may offer some security against drought but the reliability of supplies may not be high enough to provide complete drought-proofing. Pigram (1986) cited the heavy losses sustained by irrigators in the inland valleys of New South Wales, Australia, during the latter stages of the 1979–83 drought when water allocations were suspended in the middle of the irrigation season. Growers of rice and cotton crops were heavily penalised in this situation. Flexible decision-taking is necessary to make the most of predicted water shortages and drought resilience will be strengthened by a greater diversity of cropping patterns and income sources in drought-prone areas. For example, there is still much scope for the development of more drought-resistant crops and of crops with varying production cycles.

13 Technological hazards

NATURE OF TECHNOLOGICAL HAZARDS

Technological hazards are major 'man-made' accidents. To the extent that the initiating event arises from a human agency, they have links with some 'social' hazards, such as crime or terrorism. Beyond this common feature, the term 'technological hazard' has been widely interpreted. Thus, the 'technology' itself may range from a single toxic chemical to an entire industry such as nuclear power. The degree of human involvement also varies. Sometimes occupational and life-style risks are included. Some workers, for example Cutter (1993), include health issues involving long-term exposure to chemical pollutants or low-level hazardous waste as well as the safety issues which threaten group deaths by concentrated releases of energy or materials. Other workers have drawn attention to the 'na-tech' or 'hybrid' disasters which occur when natural hazards, such as earthquakes or floods, act as the release mechanism for dangerous spills of oil, chemicals or radiological materials (Showalter and Myers, 1994).

All technological innovation creates risks as well as benefits. The construction of a dam across a river may bring benefits, such as water supply and hydro-power, but it also carries the risk of a flood disaster from structural failure. The true balance between the risks and the benefits is not always apparent. When the internal combustion engine was introduced, no one could have foreseen either our current degree of dependence on the automobile or that vehicle accidents on the world's roads would now account for more than 250,000 deaths every year. The risk of a road accident versus the benefit of personal mobility has parallels with the balance that has to be struck for natural hazards between the certain, long-term advantages of a riverside location and the uncertain, short-term disadvantages of suffering an occasional flood.

Hohenemser et al. (1983) identified a number of characteristics of technological hazards, including the fact that they were defined by a causal sequence of events and that a basic distinction could be made between energy and materials releases. The overall taxonomy identified seven major classes of technological hazard ordered on a three-fold scale of severity

(Table 13.1). Few of these classes follow the definition of environmental hazard in Chapter 1 of this book which stresses extreme events posing an involuntary threat to human life on a group or community scale. Most of the listed threats involve some voluntary features (usually occupation or life-style) and represent diffuse rather than concentrated risks.

The 'rare catastrophes' category suggests that most rapid-onset technological accidents stem from failures in three areas of human activity where public safety, rather than public health, is important. These are:

- *large-scale structures*: public buildings, bridges, dams. In this case, risk is usually defined as the probability of failure during the life-time of the structure;
- *transport*: road, air, sea, rail. In this case, risk is usually defined as the probability of death or injury per kilometre travelled;
- *industry*: manufacturing, power production, storage and transport of hazardous materials. In this case, risk is usually defined as the probability of death or injury per person per number of hours exposed.

All these areas can provide potentially life-threatening releases of energy and materials. The energy may be released in either mechanical impact form (dam burst, waste tip slippage, vehicle deceleration) or chemical impact form (explosion, fire). It is generally accepted that the most hazardous materials are high-level radioactive materials, explosives and a limited number of gases and liquids that are poisonous when inhaled or ingested. Many chemicals are a hazard because they are flammable, explosive, corrosive or toxic in low concentrations. In order to constitute a community-scale risk, such substances must be present in large quantities and must be stored or transported in a less than secure manner. Glickman *et al.* (1992) focused on an important sub-set of these when they analysed major industrial

Table 13.1 A seven-class taxonomy of technological hazard

Class	Examples
Multiple extreme hazards	Nuclear war (radiation), recombinant DNA, pesticides
Extreme hazards	
Intentional biocides	Chain saws, antibiotics, vaccines
Persistent teratogens	Uranium mining, rubber manufacture
Rare catastrophes	LNG explosions, commercial aviation crashes
Common killers	Auto crashes, coal mining (black lung)
Diffuse global threats	Fossil fuel (CO_2 release), SST (ozone depletion)
Hazards	Saccharin, aspirin, appliances, skateboards, bicycles

Source: After Hohenemser *et al.* (1983). Reprinted with permission from *Science* 220: 378–84. © 1983 by AAAS

accidents in which a hazardous material caught fire, exploded or was released as a toxic cloud. Most of these events occurred at refineries and manufacturing plants or during transportation and tended to be linked with the nature and scale of industrial activity.

Hazardous materials are most likely to cause a disaster if transferred to the affected population by severe air or water pollution. This transfer reinforces the links between technological hazards and the natural environment. An important feature of severe pollution episodes is that the adverse effects, both on the human body and on the environment, can considerably outlast the impacts associated with natural disasters. If 'na-tech' events are excluded, the 'trigger' for a technological disaster is likely to arise for one of the following reasons:

- defective design;
- inadequate management;
- sabotage or terrorism.

Of these, only failures of design and management fall neatly within the framework of accidental cause. Thus, for the purposes of this book, technological hazards are best defined as: 'accidental failures of design or management affecting large-scale structures, transport systems or industrial activities which present life-threatening risks on a community scale'.

TECHNOLOGICAL HAZARDS IN THE TWENTIETH CENTURY

Technological failure is not a new phenomenon. As illustrated by Nash (1976), certain technological systems, such as river dams and other public structures, have been built – and have failed – since antiquity. To amplify this point, Table 13.2 contains an arbitrary list of disaster events which occurred before the end of World War I. On the other hand, important changes have taken place through time in the nature and scale of risk (Lagadec, 1982). For example, in the case of buildings destroyed by fire, because of improved fire regulations and more efficient fire-fighting services, whole areas of cities no longer burn down but individual buildings remain vulnerable. The advent of high-rise apartments and hotels, sometimes built of flammable materials that give off toxic fumes, has created a special type of fire hazard. Lack of adequate fire regulations and lax enforcement is still a problem in the LDCs and led, for example, to the death of 300 people in a cinema blaze in China in December 1994. It seems likely that, whenever people congregate in towns and occupy large buildings, there will always be some risk of death and damage from fire.

During the twentieth century, improvements in engineering design and a growing awareness of health and safety issues, reinforced by increasingly

Table 13.2 Some early examples of technological accidents

Structures (fire)
1666 Fire of London, England – 13,200 houses burned down
1772 Zaragoza theatre, Spain – 27 dead
1863 Santiago church, Chile – 2,000 dead
1871 Chicago fire, USA – 250–300 dead, 18,000 houses burned
1881 Vienna theatre, Austria – 850 dead

Structures (collapse)
Dam:
1802 Puentes, Spain – 608 dead
1864 Dale Dyke, England – 250 dead
1889 South Fork, USA – >2,000 dead

Building:
1885 Palais de Justice, Thiers, France – 30 dead

Bridge:
1879 Tay bridge, Scotland – 75 dead

Public transport
Air:
1785 Hot air balloon, France – 2 dead
1913 German airship LZ-18 – 28 dead

Sea:
1912 *Titanic* liner, Atlantic ocean – 1,500 dead

Rail:
1842 Versailles to Paris, France >60 dead
1903 Paris Metro, France – 84 dead
1915 Quintinshill junction, Scotland – 227 dead

Industry
1769 San Nazzarro, Italy (gunpowder explosion) – 3,000 dead
1858 London docks, England (boiler explosion) 3,000 dead
1906 Courrières, France (coal-mine explosion) – 1,099 dead
1907 Pittsburgh steelworks, USA (explosion) – 259 dead
1917 Halifax harbour, Canada (cargo explosion) – >1,200 dead

---

strict government legislation, have made engineered structures much safer than in the past. Table 13.3 shows the record for dam safety within the first 20 years of service, indicating how design and construction improvements progressively contributed to fewer failures. At the present time, about one-third of all dam failures are caused by overtopping due to inadequate spillway design, about one-third are caused by seepage through the dam and the remaining third are due to foundation problems and other effects, such as the liquefaction of earth dams as a result of earthquakes or land-slide-generated waves within the reservoir.

In the case of public transport, individual cars, ships, trains and aircraft are all safer than a few decades ago. Whilst transport-related deaths continue

Table 13.3 Failure of dams constructed during specific periods

Period	Dam failures (%)
1850–99	4.0 (out of 600 built)
1900–09	3.5 (out of 400 built)
1910–19	2.6 (out of 600 built)
1920–29	1.9 (out of 1,000 built)
1930–49	0.7 (out of 1,900 built)

Source: After Lagadec (1987)

to increase (Yagar, 1984), this is mainly a function of the rise in the distance travelled. The huge increase in business travel, together with the greater amounts of leisure time and disposable wealth in the MDCs, has led to more mobility. Car ownership is widespread. Air travel is as commonplace today as rail travel was for the previous generation. Therefore, the total exposure to transport-related risks has grown. Also many passenger vehicles are bigger and carry more passengers, so when an accident occurs it tends to create more victims. This feature was illustrated in late 1994 when the sinking of the ferry ship *Estonia* claimed 800 lives in Europe.

Within industry, considerable advances have been made and most MDCs have enacted comprehensive safety laws. Thus, in the UK the 1974 Health and Safety at Work Act gave considerable autonomy to the newly formed Health and Safety Executive to improve industrial safety under the ministerial authority of the Department of Employment. Section 2 placed a duty on 'every employer to ensure, so far as is reasonably practicable, the health, safety and welfare at work of all his [*sic*] employees'. Section 3 imposed a similar duty to 'ensure, so far as is reasonably practicable, that persons not in his employment who may be affected thereby are not thereby exposed to risks to their health or safety'. The Health and Safety Executive was given powers to set up a number of technical and consultative bodies to provide advice about major risks including the Major Hazard Branch and the Advisory Committee on Major Hazards.

Developments in occupational safety have to be set against certain trends which have increased industrial risks. The emergence of the modern chemical and petrochemical industry has created a suite of entirely new technologies. This industry has tended to group on large sites near significant concentrations of population. For example, Canvey Island is a major chemical and oil-refining complex on the north shore of the river Thames about 40 km downstream from London, England. Over 20 years ago, a study revealed that the quantities of flammable and toxic materials either in process, store or transport created a severe public safety hazard (Health and Safety Executive, 1978). The hazards ranged from fire, explosion and missiles to the spread of toxic gases. The most significant conclusion was that the existing industrial installations possessed a quantifiable risk of killing up to 18,000 people.

The rise of the nuclear industry also created new risks. Large nuclear power stations have the capability to cause many deaths and extreme social disruption. Because of this, the nuclear industry has been highly regulated and is rarely sited in close proximity to urban areas. The demand for other forms of energy has also resulted in industrial hazards. Oil and gas explosions have been the cause of many technological disasters. Whilst coal mining has become progressively safer, although still a dangerous occupation, the exhaustion of easily won fossil fuel sources has pushed the exploitation of hydrocarbon deposits into increasingly hostile physical environments.

Oil and gas have been developed offshore in areas such as Alaska and the North Sea as a result of innovations, such as large drilling platforms, which have proved vulnerable to human safety. In 1988 the Piper Alpha platform disaster in the North Sea claimed 167 lives. According to Paté-Cornell (1993), this was a largely self-inflicted disaster resulting from accumulated design and management errors, which ranged from insufficient protection of the structure against intense fires to a lack of communication about a critical piece of equipment that had been turned off for repair. Such failures in the MDCs raise even greater implications for the LDCs. The global expansion of multinational corporations has meant the spread of advanced industrial production techniques into countries which lack the safeguards necessary to handle the associated risks, as illustrated by the incident at Bhopal, India, in 1984.

Another factor contributing to the twentieth-century growth of technological hazard has been the increased transportation of hazardous materials, including radioactive waste. Table 13.4, from Cutter and Ji (1997), lists the number of reported transportation incidents in the USA between 1971 and 1991. Most of the accidents and most of the deaths occurred as a result of road transportation but water carrier incidents harmed people the most. The regional incidence of risk across the USA varied according to the extent of dependence on the chemical industry, the number of hazardous waste facilities and the length of rail track in each state. In other countries, major

Table 13.4 Accidents involving the transportation of hazardous materials in the USA, 1971–91

Carrier	Incidents (number)	Injuries (number)	Injury rate per incident (%)	Damages (US$ million)[a]	Deaths (number)
Air	2,961	276	9.3	1.79	1
Water	244	94	38.5	1.01	1
Highway	162,265	6,736	4.1	143.32	331
Rail	18,903	2,897	15.3	59.74	42
Total	184,373	10,003	5.4	205.86	375

Source: Cutter and Ji (1997)

Note
a Damages estimates adjusted for inflation to 1987 prices.

disasters have resulted from transportation accidents. For example, in 1978 more than 200 people were killed and 120 injured when a road tanker containing liquid propane gas exploded near a camp site in Spain. In November 1979, a rail freight train carrying a mix of hazardous materials including propane and chlorine was derailed in Mississauga, near Toronto, Canada. Although no lives were lost, the incident created a week-long emergency during which almost 250,000 people had to be evacuated from local homes and hospitals.

Early attention to the safety of hazardous materials transport was achieved through engineering techniques designed to improve the security of road and rail vehicles and their loads, but Glickman (1988) pointed out the need for other management strategies. The main problem is that there is often a conflict between selecting the routes which minimise the risk of an accident compared to those which offer the lowest operating costs. Along with emergency planning and preparedness, the current emphasis lies in placing various restrictions on the routing of such materials. These restrictions can include: prohibiting the use of specific roads, tunnels or bridges for the transport of certain materials; ordinances which require advance warning of hazardous shipments; special speed limits on permitted routes; and curfews which control the hours during which specified routes and facilities can be used for the transfer of hazardous materials. In a discussion of the problems arising from the transportation of high-level nuclear wastes in the USA, Fitzsimmons and Kirby (1987) highlighted the additional conflict which arises from the imposition by the federal government of specific routes and regulations on state and local governments without the provision of any support which would enable the directives to be met.

The year 1984 marked a clear watershed in the emergence of technological hazards (Lagadec, 1987). Figure 13.1 shows all industrial accidents up to 1984 causing more than fifty deaths (to workers and third parties), excluding accidents in the former USSR and those involving explosives, mining and gas distribution. There was an increasing frequency of such accidents but it was not until 1948 that more than one such accident occurred in any one year. Furthermore, it was not until 1957 that the first incident occurred outside the developed world, defined as Europe, the USA, the Soviet bloc and Japan. In 1984 three incidents caused a total of around 3,500 deaths:

*   *Cubatao, Brazil, 25 February*: petroleum spillage and fire in a shanty town built illegally on the industrial company's land – 500 deaths.
*   *Mexico City, Mexico, 19 November*: multiple explosions of liquefied petroleum gas in an industrial site in a heavily populated poor area – at least 452 deaths, 31,000 homeless, 300,000 evacuated.
*   *Bhopal, India, 2–3 December*: release of toxic gas from an urban factory – well over 2,000 immediate deaths, 34,000 eye defects, 200,000 people voluntarily migrated.

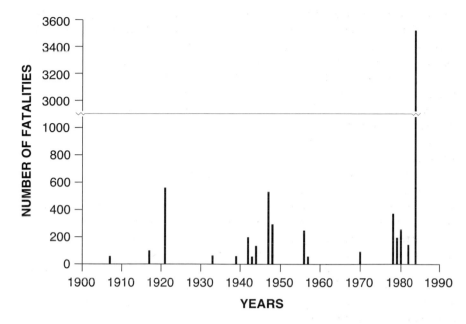

Figure 13.1   The annual total of deaths from industrial accidents worldwide which
caused more than 50 fatalities in the period 1900–84.
Source: Compiled from data presented in Lagadec (1987)

These events not only produced more fatalities in one year than all the
other technological disasters combined since World War II, but they also
confirmed a number of trends. First, they showed that technological hazard
was no longer confined to the MDCs. Second, they confirmed that, as with
natural disasters, poor people suffer most. Third, the Mexico City incident
marked the conclusive arrival of the 'domino' disaster which occurs when
a loss-of-containment accident, such as a small leak of flammable gas, inter-
feres with nearby systems and storages, causing a further loss of containment
as a result of ignition and explosion. Through such disaster chains, it is
possible for hazard impacts to be transmitted over very large distances.

THE SIGNIFICANCE OF TECHNOLOGICAL HAZARDS

Most technology, and most of the concern about technology, is concentrated in the MDCs. Fritzsche (1992) has shown that, in the two most
developed continents (Europe and North America), the fatality rate is about
the same for both natural and man-made disasters (Table 13.5). This
contrasts with the situation in the LDCs where natural hazards are more
prominent and the average fatality rate from all disasters is perhaps twenty
times higher than in the industrialised countries. Even so, compared with

Table 13.5 Annual death toll due to natural (N) and man-made (M) disasters (averages 1970–85)

Cause	World		North America		Europe	
	N	M	N	M	N	M
Fatalities per year	88,900	5,500	220	310	450	540
Population (millions)		4,264		245		477
Fatality rate per 100,000 per year	2.1	0.13	0.09	0.13	0.09	0.11

Source: After Fritzsche (1992)

the overall annual mortality rate of about 900 fatalities per 100,000 of the population in North America, the number due to major technological accidents is small. This pattern is repeated worldwide. Table 13.6 shows that although the average annual loss of life from technological accidents – along with other disasters of a 'non-natural' origin – increased across the globe during the 1973–97 period, the numbers were insignificant relative to the toll exerted by geophysical forces. Indeed, over the same 25–year period, technological accidents claimed only 0.75 per cent of the lives lost in natural disasters.

In terms of perception, people often view technological risks as tolerable when compared with the expected benefits (see Chapter 3). This attitude is particularly apparent in the medical field when patients, in order to secure treatment, willingly take chemical substances in prescribed measures and voluntarily expose themselves to low doses of ionising radiation through medical X-rays, in the knowledge that larger amounts may be extremely harmful. Such public acceptance is critical for technological hazards. Compared to natural hazards, there is usually less statistical evidence on which to base an objective probability assessment of technological risks and the public perception of the balance between the advantages and disadvantages of a particular technology may be very different from that of

Table 13.6 The annual average number of people reported killed by disasters with a non-natural cause, 1973–97

Period	Accident	Technological accident	Urban fire	Total
1973–1977	868	152	271	1,291
1978–1982	1,713	387	305	2,405
1983–1987	5,310	1,006	419	6,735
1988–1992	7,268	1,223	472	8,963
1993–1997	6,978	419	766	8,163

Source: CRED and IFRCRCS (1999)

scientists and technicians. Using a large sample of the United States population, Gardner and Gould (1989) confirmed that lay people take a more complex view of such matters, with an emphasis on the so-called 'dread' risks. This means that the degree to which risks are understood – and agreed upon – within the scientific community, the potential for catastrophe, as well as the possible benefits, all exert an influence on the lay perception of technological risk.

Some technological hazards create special problems, either because they exert a higher toll on society than is generally perceived or, conversely, because the hazards are perceived as posing a threat far in excess of experience. In these cases, society does not appear to have struck an entirely rational or comfortable balance between the benefits and the risks. This feature can be demonstrated by considering in more detail what is a relatively large *actual* risk (transport) and what is a relatively large *perceived* risk (nuclear power).

Transport

As already indicated, most forms of transport are getting safer. Table 13.7 shows that, with the exception of rail travel, which reflects the importance of two major accidents in the period concerned, the risk of death per passenger distance travelled in the UK fell during the late twentieth century (Cox *et al.*, 1992). Air travel is particularly safe. According to Lewis (1990), despite the media attention paid to air crashes, the average risk in the USA is one fatality per billion passenger miles and this seems to be expressed in a trust in commercial airlines. For example, Barnett *et al.* (1992) investigated the public response in the USA to the Sioux City disaster of 1989, which was the third DC 10 crash caused by the loss of hydraulic power and killed 112 out of 282 passengers on board. Despite adverse publicity and reduced booking rates in the first weeks after the disaster, within two months bookings recovered to about 90 per cent of what could have been

Table 13.7 Deaths per 10^9 kilometres travelled in the UK

	1967–71	1972–6	1986–90
Railway passengers	0.65	0.45	1.1
Passengers on scheduled UK airlines	2.3	1.4	0.23
Bus or coach drivers and passengers	1.2	1.2	0.45
Car or taxi drivers and passengers	9.0	7.5	4.4
Two-wheeled motor vehicle passengers	375.0	359.0	104.0
Pedal cyclists	88.0	85.0	50.0
Pedestrians[a]	110.0	105.0	70.0

Source: After Cox *et al.* (1992)

Note

a Assuming travel at 8.7 km per person per week.

expected in the absence of the incident. This appears to be an effective demonstration of the 'willingness-to-pay' principle in safety management which lets the market place adjudicate on what is an acceptable risk (McDaniels *et al.*, 1992).

Road travel is much more risky. In fact, if technological disaster is expressed through premature deaths alone, it is the familiar motor vehicle which has most to answer for. Traffic accidents claimed over 30 million lives worldwide during the twentieth century (IFRCRCS, 1998) but the personal convenience of travelling by road is widely perceived to outweigh the risk of death or injury. Indeed, the spread of car ownership is so associated with human progress that over 70 per cent of all road deaths now occur in the LDCs, where the annual cost of traffic accidents (over US\$53 billion) now rivals the amount of international aid received by these countries.

In the USA, road collisions account for about half of all accidental deaths whilst in Japan traffic accidents account for 0.01 per cent of all deaths, compared with a death rate of only 0.00025 per cent for natural disasters (Mizutani and Nakano, 1989). In the UK the average driver faces a risk of about 8:100,000 per year of being killed in a car accident. This compares with a 100:100,000 risk of being seriously injured, and a 2:100,000 risk of severe disablement in a traffic accident. The threats to other persons are even greater. The average risk of killing someone else in a road accident is 13:100,000 and of seriously injuring another person as high as 151:100,000. Such risks are strongly age-dependent. For example, driving accidents account for about three-quarters of all accidents in the 16–19 age group, and drivers age 21 years or under are responsible for about one-quarter of all road deaths. Collectively, these are the highest risks faced by the public from modern technology.

But the public perception does not fit the facts. Although road deaths in private vehicles dwarf public transport deaths in most MDCs, the greatest public concern is with the latter. This mis-emphasis probably stems from the larger group deaths associated with public transport accidents and also from the extra opportunity for blaming large corporations in an age when litigation is growing. Because of the high risks, investment in highway safety is a 'good buy'. The reduced risk in recent years has been achieved at relatively low cost through improvements in car design, more travel on motorways, and through legislation such as the compulsory wearing of seat belts and the stricter enforcement of drink-driving laws. This is not to say that other highway risks are not emerging. In parts of the MDCs, the number of vehicle miles travelled by large trucks is increasing at a faster rate than that for other vehicles. Joshua and Garber (1992) have used fault-tree analysis to show that increasing competition between such trucks and other vehicles for road space is likely to create more multi-vehicle collisions involving large trucks. In the UK the government already spends more on rail than on road safety. It has been claimed that the cost of introducing a fail-safe facility on trains – lobbied for after the Paddington rail disaster in 1999 – would cost as much as £14 million for every life saved.

Nuclear power

Practical problems are caused by the extreme public perception of nuclear risks and toxic waste sites. For example, the risk from hazardous-waste sites has been indicated as the most worrying environmental problem in polls of public opinion conducted in the USA (Dunlap and Scarce, 1991). According to Lewis (1990), the risk from a properly constructed nuclear waste repository is 'as negligible as it is possible to imagine . . . [and] a non-risk'. But there is widespread opposition to this technology under-pinned by public concern that the risks are very great (Slovic *et al.*, 1991). For example, in a wide-ranging study of technological and other hazards in Japan and the United States, Hinman *et al.* (1993) found that people in both countries had the greatest dread of nuclear waste and nuclear acci-dents at a level of perception that exceeded their fear of crime or AIDS.

Nuclear power poses a risk of an additional dose of ionising radiation beyond the background levels emitted via the earth and its atmosphere. High-level nuclear wastes are products from nuclear reactors, including spent or used fuel, with a radioactive half-life (the period taken for half the atoms to disintegrate) of more than 1,000 years (Cutter, 1993). Intermediate-level waste has a shorter half-life but exists in larger quantities. The general solution to the disposal of nuclear waste has been to store it for several years in pools of water near to the power plant, so that the temperature can cool and some of the radioactivity decay. Then the plan has usually been to transport it to a permanent storage site. This means that public highways are increasingly used for the transport of radioactive waste, which is a highly contentious issue. For example, in a study of radioactive waste transport through Oregon, MacGregor *et al.* (1994) found no reduction in public concern with distance away from the transport corridor.

However, it is the permanent storage risks which arouse most anxiety. By the year 2000, the USA expected to have some 40,000 tonnes of spent nuclear fuel stored at about seventy sites awaiting disposal. Many workers, including Flynn *et al.* (1993a), have found wide differences in attitude between the public and the technical community with respect to the long-term storage of high-level nuclear waste. Such differences have been exacerbated by the decision of the US Congress to designate Yucca Mountain, Nevada, as the sole potential site for the nation's first high-level nuclear waste repository. Flynn *et al.* (1993b) described how the people of Nevada opposed such an underground store and reported on the failure of an adver-tising campaign, largely through lack of trust, designed to change public perception. At the present time, the US Department of Energy's programme has been more or less halted by opposition to this strategy. Such opposi-tion is likely to be strengthened by the conclusions of Keeney and Winterfeldt (1994), who argued that the storage of nuclear waste above ground for the next 100 years, then burying it in a permanent repository, is likely to be US$10–50,000 million cheaper than building a permanent repository at Yucca Mountain now.

THE DISASTERS AT BHOPAL AND CHERNOBYL

The Bhopal disaster

Methyl isocyanate (MIC) is a fairly common industrial chemical used in the production of pesticides; but it has important qualities which make it hazardous (Lewis, 1990). First, it is extremely volatile and vaporises easily. Since MIC can boil at a temperature as low as 38°C, it is important for it to be kept cool. Second, MIC is active chemically and reacts violently with water. Third, it is highly toxic, perhaps 100 times more lethal than cyanide gas and more dangerous than phosgene, a poison gas used in World War I. Fourth, MIC is heavier than air and, when released, stays near ground level.

During the early morning of 3 December 1984, some 45 t of methyl isocyanate gas leaked from a pesticide factory in the industrial town of Bhopal, India, and created the world's worst industrial disaster in a town of over 1 million people (Hazarika, 1988). The MIC was stored in an underground tank, which became contaminated with water. This contamination produced a chemical reaction, followed by a rise in gas pressure and a subsequent leak. An investigative report indicated that the safety devices failed through a combination of faulty engineering and inadequate maintenance, although the company claimed that the cause was sabotage. A contributory factor was that the air-conditioning system, normally in use to keep the MIC cool, was shut down at the time of the accident. It is likely that the real trigger of this disaster will never be known, but safety measures were inadequate: for example, the Bhopal plant lacked the computerised warning and fail-safe system used in the company's factory in the USA.

In Bhopal, the factory was built within 5 km of the city centre by Union Carbide, a multinational company based in the USA. The dense cloud of gas drifted over an area with a radius of some 7 km. It is now believed that up to 6,400 people may have been killed by cyanide-related poisoning with a further 200,000 injured. In fact, a total of 600,000 injury claims and 15,000 death claims were ultimately filed with the Indian government. The greatest number of casualties occurred in the low-lying parts of the old city which comprised poor neighbourhoods, including a shanty town of some 12,000 people that had grown up near the gates of the factory. Most of the victims were the very young and the very old, although pregnant women suffered badly too. The disaster was severe because of the large numbers of people at immediate risk and the lack of any preparation for such an emergency. There was no local knowledge of the nature of the chemicals in the factory, no adequate warning or evacuation plans and no understanding of the specialised medical treatment which was needed by the victims. For example, oxygen for the urgent treatment of respiratory problems was in short supply.

Immediate steps were taken. The plant was unprofitable at the time of the accident and, because cut-backs had been made in maintenance, blame was attached to the local Indian management. Over the following two years, the parent company slimmed down, partly by distributing assets to shareholders and creditors, who were mainly banks. This strategy was claimed to be necessary in order to fend off a hostile takeover bid but it also served to off-load assets, which were then unavailable for compensation claims. At the same time, the US legal system overturned precedent by opposing compensation claims for such an overseas liability, on the grounds that it would unfairly tax the US courts, and the responsibility was passed back to the Indian government.

The Indian government made itself the sole representative of the victims and filed compensation claims against the company both in the USA and in India. Union Carbide made a final out-of-court compensation payment in 1989 of US$470 million, which compares, for example, with the US$5 billion awarded in the USA after the *Exxon Valdez* oil spill. Special courts were set up to hear compensation claims, which were typically settled at £500 for injury and £2,000 for death. In the meantime, the Indian government distributed relief at about £4 per month for each family affected. But the government failed to organise efficient legal or medical aid for the victims. As a result, victims found it difficult to have their cases brought to court without resorting to bribes, commonly £10 to a middleman, or paying private lawyers. Medicines which should be supplied free to victims still suffering from lung disorders, anaemia and gynaecological disorders were sold by doctors and chemists on the black market. Families sometimes needed to spend double their monthly government allowance on medicines. Ten years after the event it was estimated that less than one-quarter of the total claims had been settled and that less than 10 per cent of the damages paid by Union Carbide reached the victims. A combination of corruption, administrative failings and political apathy prevailed and, according to C. Thomas (1994), the total sum paid up until then was less than a quarter of the interest earned on the original compensation payment.

There are still few controls on chemical industries within India. On the other hand, there were significant safety improvements within the chemical industry in the United States. The main Union Carbide plant in West Virginia was forced to close but about US$5 million was spent on improvements at another Union Carbide plant in West Virginia. Based on a concern that a similar accident could occur in the United States, the Emergency Planning and Community Right-to-Know Act was passed in 1986 but has proved difficult to implement (Cutter, 1993). Given the widespread concern raised by Bhopal, it seems more than a coincidence that, over the past decade, the American chemical industry has turned to other LDCs, notably Mexico, in which to open new plants to deal with toxic substances.

The Chernobyl disaster

During the night of 25–26 April 1986, the world's worst nuclear accident to date occurred at Chernobyl, about 130 km north of the city of Kiev, in the Republic of Belarus (formerly the Ukrainian SSR). It was an example of a major transcontinental pollution incident stemming largely from human error. The immediate cause was that workers at the nuclear plant were conducting an unauthorised experiment on one of the reactors to determine the length of time that mechanical inertia would keep a steam turbine free-wheeling and the amount of electricity it would produce before the diesel generators needed to be switched on. During the experiment, the routine supply of steam from the reactor was turned off and the power level was allowed to drop below 20 per cent, well within the unstable zone for this type of water-cooled, graphite-moderated reactor design.

During the experiment, the reactor was not shut down and a number of the built-in safety devices were deliberately overridden. In this situation vast quantities of steam and chemical reactions built up sufficient pressure to create an initial explosion which blew the 1,000 t protective slab off the top of the reactor vessel. Lumps of radioactive material were ejected from the reactor and deposited within 1 km of the plant, where they started other fires. The main plume of radioactive dust and gas was sent into the atmosphere. This plume was rich in fission products and contained iodine-131 and caesium-137, both of which can be readily absorbed by living tissue.

Immediate efforts were made to control the release of radioactive material but a major limitation was that water could not be used on the burning graphite reactor core as this would have created further clouds of radioactive steam. Instead, the fire had to be starved of oxygen by dumping from helicopters many tonnes of material, including lead, boron, dolomite, clay and sand. In this early emergency period, thirty-one people died trying to contain the accident and a further 200 people sustained serious injuries through exposure to over 2,000 times the normal annual dose from background levels of radiation. Eventually some 135,000 people were evacuated within a 30 km radius of the plant.

In the two weeks following the accident, the radioactive plume circulated over much of north-western Europe. Away from Chernobyl itself, the greatest depositions of radioactive material occurred in areas affected by rain, which flushed much of the particulate material out of the atmosphere. These areas included Scandinavia, Austria, Germany, Poland, the UK and Ireland. Some of the heaviest fall-out was experienced in the Lappland province of Sweden, where it affected the grazing land of reindeer, contaminated the meat and dealt the Lapp culture a great blow. More widely, the immediate consequence was a general contamination of the food chain, and restrictions on the sale of vegetables, milk and meat were imposed. Some countries also issued a ban on grazing cattle out of doors and warnings to avoid contact with rainwater. It has so far proved difficult to estimate the long-term health risks, notably the increase in fatal cancers, which can be attributed to the Chernobyl accident.

Near Chernobyl itself, the town of Pripyat has been abandoned. In the surrounding area, the evidence of thyroid cancers and a catastrophic drop in the human birth rate are constant reminders of the disaster. As a result of political change, there is now a greater spirit of openness. It is accepted that there is a need to research and treat thyroid cancer in children and to create an economic development area in the affected zone. Since 1993, despite a shortage of funds, the United Nations has appointed a Co-ordinator of International Cooperation to act as a catalyst in the regular exchange of information and collaboration between organisations and member states in addressing such issues. The main task is to find funds and to bring in other agencies such as the EU, the Organization for Economic Cooperation and Development (OECD) and the World Bank. The United Nations has taken some steps: the World Health Organization has investigated the health effects, including the plight of the 'liquidators', that is, the 600–800,000 people who took part in the immediate clean-up efforts and who were never properly registered; UNESCO has created nine socio-psychological rehabilitation centres; and the FAO has developed projects aimed at restoring contaminated land to safe agricultural use. But poor safety conditions still plague nuclear power plants in the Third World.

LOSS-SHARING ADJUSTMENTS

Loss-sharing arrangements differ between natural and technological hazards. For example, the explicit allocation of blame is much more likely after tech nological disasters. This is because they are more obviously a product of human decisions and actions. The need to attach blame is also reflected in the increasing public pressure for corporate manslaughter charges to be brought against organisational failure. Such a case was brought in Britain after the *Herald of Free Enterprise* disaster, when 193 people died in a North Sea ferry disaster, but the trial collapsed halfway through. It was not until 1994 that 400 years of legal history were finally swept away when the first successful case of manslaughter was brought in the UK against a company running an outdoor activity centre that was held responsible for the deaths of four school children in a canoeing accident. Since then, other cases have been proved.

Because of the perceived importance of corporate failings in technological accidents, whether by commercial companies or governments, the victims of such incidents tend to attract less public sympathy than those suffering from natural disasters. Therefore, the role of international disaster aid is reduced and even the LDCs have to rely more on their own resources. Thus, after the Bhopal disaster, it was the Indian government which eventually set up a relief fund. The question of corporate responsibility raises the importance of legal compensation to the point where it tends to replace the loss-sharing function provided by aid.

Compensation

Compensation is a much less spontaneous form of loss-sharing than disaster aid. Indeed, it is often resisted by the donor and has to be legally enforced. Where litigation is involved, the final compensation settlement may include a punitive element which goes beyond the recovery of costs. Many persons in the MDCs now seek compensation for actual and perceived harm, including emotional distress, caused by industrial emissions (Baram, 1987). In the USA, the legal system allows people, either injured or at risk, to bring 'toxic tort' actions against industry and secure high monetary damages. Governments are also suing industry for large sums in order to clean up hazardous waste sites. Some of the firms involved are multinationals, so the repercussions of such actions on profits and jobs may be felt worldwide.

Although legal compensation can provide high monetary returns, it is not a very effective mechanism for loss-sharing. Quite apart from the major economic losses faced by industrial plants losing liability suits, it is not always to the advantage of the disaster victim. For example, litigation can delay settlements for years. In some cases, the plaintiff may die or the firm may go out of business before compensation can be paid. In India the lack of formal documentation held by many Bhopal victims, combined with inertia and inefficiency, delayed the settlement of many claims more than 10 years after the accident. Clearly, it would be better to have compensation schemes capable of discharging sums quickly to help such victims whilst also safeguarding the financial future of companies responsible for producing or using hazardous substances.

Such compensation schemes are unlikely to exist without government intervention. Theoretically, a government could create a compensation fund which spreads the risk in a variety of ways. For example, the government could itself establish a national technological disaster fund or it could ensure that the industry sets one up based on a levy imposed on the product. Neither of these arrangements is wholly satisfactory since they devolve the cost on to innocent third parties, respectively the taxpayer or the safe industrial plant. There are some instances where a contribution from the taxpayer is appropriate. Thus, it might be deemed equitable to compensate from general taxation a community that assumes local risks, perhaps from a nuclear power plant, on behalf of the nation as a whole. But, ideally, any direct government intervention should be motivated by the principle of making the hazard-maker pay. In practice, this suggests that government involvement might be best directed to ensuring that individual plants carry full insurance cover against civil liability for death, injury and environmental damage arising from industrial activities and emissions.

Insurance

Insurance can be used to spread the financial risks associated with technological hazards. Within the MDCs, many people are likely to have insurance because most personal life and accident policies cover all risks. This means that cover is normally available for exposure to hazardous substances, although such policies invariably exclude exposure to radiation. Property insurance similarly tends to cover all risks including hazardous materials, unless they are specifically excluded. On the other hand, personal insurance has some practical disadvantages. For example, it may be difficult to establish a link between liability and damage, particularly in more delayed-effect cases involving toxins such as carcinogens, and especially if the damage results from long-term, low-level leakage rather than from a rapid-onset accident. Personal insurance has little part to play in helping victims from the LDCs.

Insurance against technological risks can also be taken out by industry and this has considerable potential for risk spreading. With increasing public demands for safety, there is a growing economic need for industrial plants to have full insurance cover against civil liability for human injury or environmental pollution. A proactive partnership between industry and insurers could well result in insurers forcing a more responsible attitude towards hazard reduction on to industry. But, as with other types of hazard insurance, the difficulty lies in setting realistic premiums for industry to pay. Unless gross premiums are fully economic, an industry will not take technological hazards seriously and insurance companies will either fail commercially or withdraw from this market. Equally, unless premiums are weighted according to the actual risks involved at the level of individual plants, insurance will amount to little more than an unfair tax on the safe, well-managed sites within a particular industry.

EVENT MODIFICATION ADJUSTMENTS

Technological hazards offer more potential for modifying the events at source than exists for many geophysical hazards. Most technological disasters have their roots in a combination of faulty engineering and human weakness. Since the latter includes basic human flaws such as greed and carelessness, against which there are few reliable defences, it is the engineering route which offers the better chance of success.

There is no possibility of totally risk-free design and construction because it would be too expensive to build against any possibility of failure. But industrial plant and transport design has all too often been changed after, rather than before, an event. The risks at Chernobyl would have been less had the reactor been surrounded by a protective shield. The risks at Bhopal would have been reduced, although not eliminated, if the factory had been equipped with an effective gas exhaust facility, including a very high chimney stack which would have pierced through the nocturnal inversion layer and dispersed the toxic material through a much larger volume of air. There is a special need to ensure that, when multinational corporations operate within the LDCs, the safety standards in the subsidiary plants at least match those at the parent site.

The mitigation of frost hazards on highways is an example of event modification. Icing on road surfaces causes an increase in deaths and vehicle damage from skidding accidents during periods of low temperature. Salt is an effective de-icing agent down to temperatures of −21.0°C. In recent decades, many countries have increased their use of salting and, during the mid- to late 1980s, snow and ice control on roads cost about US$1.5 billion during an average winter in North America and perhaps as much as US$3 billion globally. Increased salt spreading has also led to road damage and pollution. Since as much as 90 per cent of the salt will be washed or blown off the road surface, there will be damage to vegetation and contamination of water sources. Therefore, it is important that salt is spread as sparingly as possible.

The most efficient use of road salt occurs when it is spread as an *anti-icing* agent at low application rates of about 10 g m^2, which is sufficient to prevent the formation of a thin film of ice. When ice has already formed, salt has to be used as a *de-icing* agent at application rates which are some five times higher per unit area. It is clear that salt should be applied in advance of ice formation, and recent advances in technology have enabled road engineers to monitor and forecast localised road temperature gradients (Perry and Symons, 1991). The use of ice detection sensors and site-specific road temperature forecasts has led to economies in winter salt usage amounting to 20 per cent or more in Europe and elsewhere.

VULNERABILITY MODIFICATION ADJUSTMENTS

Community preparedness

A pre-planned, preventative approach to all technological hazards is desirable. In the UK, legal responsibilities designed to enhance safety are normally required to be carried out so far as is reasonably practicable (Soby *et al.*, 1993). Within industry, increased attention is being given to the harmonisation of control measures designed to eliminate or control major accidents. For example, within the European Community, the Seveso Directive of 1982, and subsequently amended, requires that premises storing or using more than certain specified quantities of very hazardous substances are designated 'major hazard sites'. The operators of such sites must prepare emergency plans and communicate the dangers and appropriate responses to the public nearby. The EU's Major Accident Reporting System (MARS) keeps a database on accidents occurring at such sites (Drogaris, 1991). In the USA the Chemical Emergency Preparedness Program (CEPP) has been developed by the Environmental Protection Agency to increase the understanding of and to lower the threat from chemical risks. One focus for community planning has been the preparation of a list of acutely toxic chemicals that might endanger public health in the event of an accidental air-borne release.

Although in-plant preparedness may be high, the level of preparation for chemical emergencies at the community level is much lower. Several initiatives are required. For example, Glickman and Sontag (1995) devised a methodology for evaluating the cost–risk trade-offs involved in re-routing the road transport of hazardous materials. Another priority for improved preparedness must be a better, more scientific, training for all the local emergency responders – such as the police, fire and medical services – who may have to deal with the consequences of the accidental release of toxic substances. In addition, effective public response depends on more 'freedom of information' with regard to industrial hazards. Despite legislative strides in this direction by certain countries, such as the USA, there are still important restrictions on information when commercial competitiveness might be involved or when terrorist activity might be facilitated.

As far as off-site safety is concerned, most progress in developing emergency response plans has been achieved by the nuclear power industry. Following the accident at the Three Mile Island, Pennsylvania, nuclear power plant in 1979, all reactors in the USA are now required to produce emergency response plans which meet criteria laid down by the Federal Emergency Management Agency and the Nuclear Regulatory Commission (NRC). Formal approval of these plans is a condition of granting and maintaining operating licences for commercial nuclear power plants. They normally include the three protective measures used in any radiological emergency: indoor shelter (to protect against the short-term release of radionuclides); medical treatment (use of potassium iodide as a thyroid blocking agent); and evacuation (to remove the population from longer-term exposure to the pollution plume).

These response measures are to be taken within the context of two standardised Emergency Planning Zones (EPZs). The first EPZ (the plume exposure pathway) extends over an approximate 10 mile (16 km) radius of the plant downwind from the plant (Figure 13.2). It represents the area within which whole-body exposure and particle inhalation might be expected to occur. The second EPZ (the ingestion exposure pathway) extends to approximately 50 miles (80 km) from the plant, where the hazard would be largely due to contamination of water supplies and crops. These standard EPZs have been based on broad assumptions about meteorological conditions and types of potential material and releases. It is expected that they will be modified in individual plans to meet local circumstances. However, federal guidelines for warning effectiveness in a nuclear emergency have been specified as:

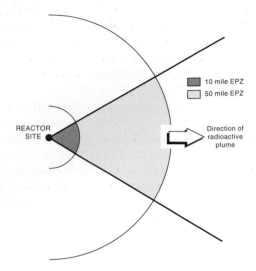

Figure 13.2   Diagrammatic representation of the emergency planning zones (EPZs) applied to US nuclear reactors. The 10-mile EPZ indicates the area within which direct human exposure to radiation might be expected to occur.

1   the capability to disseminate messages to the population inside the 10-mile zone within 15 minutes;
2   assurance of direct coverage of 100 per cent of the population within a 5 mile (8 km) zone around the plant;
3   arrangements to ensure 100 per cent coverage within 45 minutes of all who live within the 10-mile radius and who may not have heard or received the initial warning.

Despite their relative sophistication, these plans have come under criticism, especially with respect to the expectations and provisions for evacuation. Cutter (1984) claimed that there is little evidence that the public will follow the prescribed evacuation procedures, including the time-frame specified and the identified routes, as laid down by the authorities. Several reactors are sited in areas of high population density and there seems very little chance of safe evacuation from these areas. In some cases, EPZs cross state lines but no regional response plans have been formulated to permit inter-state cooperation in emergency planning. In general, there seems to have been a failure to learn from the findings of social science with regard to evacuation behaviour after natural disasters, although it could be equally misleading to transfer that experience directly to technological hazards. The perceived fear of radiation (the 'dread factor') is so great that 196,000 people evacuated in response to the Three Mile Island incident, although no formal and comprehensive evacuation order was issued (Cutter and Barnes, 1982).

As a proportion of the local population, this is a far greater response than could be expected when people are ordered to evacuate after natural disasters.

A detailed survey of the local population affected by the emergency response plan for the Diablo Canyon nuclear power plant located on the southern Californian coast has revealed other weaknesses of nuclear preparedness (Belletto de Pujo, 1985). Only one-third of the households had any familiarity with the plan and less than 6 per cent claimed they had information telling them what action to take in an emergency. In terms of response, only about half of the households questioned felt they would follow emergency instructions from the authorities, despite the fact that 40 per cent perceived the risk of a major accident at Diablo Canyon to be either high or very high. These findings show that public involvement and education with respect to emergency planning have not been effective. Although the plan was ready for action, the people clearly were not.

Forecasting and warning

Some technological accidents, such as industrial explosions, provide little scope for forecasting and warning. For certain types of structural failure and the release of toxic materials, a warning to the local population may be given by sirens or other audible means, but the limited timescale between the initiating event and the advent of the hazard often precludes preventative action. Where longer lead times are available, warnings are likely to be more beneficial. D.I. Smith (1989) quoted work on the effectiveness of warning systems for major dam failures following flash floods in the United States. In the cases where more than 90 minutes warning had been possible, the average loss of life averaged as few as two people per 10,000 residents in the inundation zone. On the other hand, when the local community received either less than 90 minutes warning or no warning at all, the average number of lives lost rose to the equivalent of 250 per 10,000 residents.

Land use planning

The purpose of land use planning is to resolve the conflicts and reduce the risks associated with the siting of dangerous facilities. Ideally, land use zoning should aim to separate densely populated residential areas from hazardous industrial activities and their associated transport routes. This implies the creation of buffer zones around hazardous facilities, and Cutter (1992) has described the concept of toxic parks, similar to the idea of national parks, as applied to industrial plants in the USA.

In practice, the vested interests in the urban land market often preclude such straightforward responses. Hazardous facilities will almost certainly be unwanted by most of the local population. The least acceptable developments tend to be nuclear waste and toxic chemical disposal sites, plus nuclear power plants. There is already an enormous legacy of problems arising from largely unregulated waste treatment and disposal sites. Landfill methods have been used for waste disposal for decades and there may be as many as 10,000 sites in USA alone which require cleaning up.

Even non-toxic waste can be hazardous. In 1966, 144 people – including 116 children at school – were suffocated at Aberfan, south Wales, when a large dump of coal mining waste slid down a hillside to overwhelm the village. The waste had been dumped over a 30-year period on steep slopes above the settlement across a natural spring-line, and water saturation at the base of the material finally produced the rapid, liquefied slide. At the time, such dumps were seen as creating aesthetic planning problems rather than a public safety risk, although the tip had suffered significant movement on at least two previous occasions and warnings had been given. No regulations were in force for either geophysical testing or stabilisation of the material through drainage. The resulting Public Inquiry revealed that the tip had not even been routinely inspected for four years prior to the disaster. Subsequent investigations by the National Coal Board indicated a further 100–200 unsafe coal tips within the south Wales area alone.

There is a growing need for better planning of all hazardous installations, which includes a requirement for improved public information and acceptance. In the UK the Flixborough disaster of 1974 marked a watershed. Before this explosion at a chemical plant, which killed twenty-nine people and injured more than 100 others, there had been no real concern for the siting of hazardous facilities. Immediately after the event, the Health and Safety Executive began to examine major industrial risks and to consider how best to intersperse hazardous sites with other land uses. But an essentially pragmatic approach is still adopted. Whilst it is self-evident that chemical plants should not be located near schools, hospitals or densely populated areas, there is no universally valid rule which determines exactly how far away development should be permitted. Because of this, it is necessary to develop emergency response plans in parallel with the siting of hazardous facilities.

Like all risk assessment, land use control must rely on a balanced appraisal of the probability of large escapes of toxic material or explosions from a site, the local consequences of a major accident and the necessity for accepting particular types of risk in the regional or national interest. Economies of scale mean that large-scale sites often provide cheaper manufacturing and transport costs but bigger operations also tend to create larger risks. In the USA the Environmental Protection Agency has used a GIS to map the releases of toxic chemicals in eight south-eastern states (Stockwell *et al.*, 1993). This work has shown that the largest releases have taken place near densely populated areas.

Some of the general issues concerning the conflict between state and community interests in land planning may be illustrated by the November 1984 disaster at a liquid petroleum gas plant operated by the national oil corporation (PEMEX) in Mexico City. In this event a series of explosions resulted in around 500 deaths and some 2,500 injuries, together with the destruction of a seven-block area of a nearby working-class district (Johnson, 1985). Within a few days of the disaster the government decided not to rebuild this devastated area and to create a 14 ha 'commemorative park'. However, this change in land use failed to reflect the residents' wishes for a resiting of the PEMEX facility, involved the demolition of the remaining dwellings in the damaged zone and the resettlement of almost 200 families in other parts of the city. Such centralised decisions, made with virtually no input from the residents of the area and while the community was still recovering from the immediate aftermath of the disaster, can be seen as both arbitrary and insensitive from a local perspective.

The perceived risks of nuclear power have generally led to the location of these facilities in relatively remote or rural areas, but even greenfield industrial sites tend to attract later development and create planning tensions. In the absence of strong planning control, it is likely that unwanted hazardous facilities will continue to be placed by developers and governments where local resistance is less than at other candidate sites. Therefore, small rural communities with low income and high unemployment, which may also be remote from political influence, are most likely to have industrial hazards imposed upon them in the future.

14 Conclusion

THE CHALLENGES AND OPPORTUNITIES OF HAZARD

At the present time, the challenges and the opportunities for reducing the loss from environmental hazards have never been greater. Theoretically, the challenge is easily defined: to eliminate all disasters which cause death and injury or damage to property or the environment. In practice, this goal is impossible to achieve. Although many risks are potentially avoidable, global environmental change and uncertainty about future hazardous events, together with the central role played by human failings in all disasters, make the total elimination of hazard an unrealistic task. The question then is: How safe is safe enough? As Charlton (1990) suggests, this question is a little like asking an athlete: 'How fast is fast enough?' All he or she can answer is: 'As fast as I can run today and then faster still tomorrow.'

The existing databases are inadequate for the precise determination of spatial and temporal patterns of disaster worldwide but the available evidence suggests that the overall losses remain high. Despite increased investment in hazard mitigation measures, the deaths and material damages show few signs of a sustained decline. The continued growth of population and the encroachment of humans into hazardous zones plus the rise in exposed wealth resulting from economic growth are the main causes. The greatest absolute financial losses continue to occur in the MDCs whilst the most severe impacts, in terms of deaths and relative economic loss, are experienced by Third World countries. In some cases these impacts are sufficiently severe to jeopardise economic development efforts. Social and political systems are undergoing rapid transition everywhere and many traditional hazard responses, ranging from indigenous cultural attitudes in the LDCs to national civil defence organisations in the MDCs, are now seen as outdated. In addition, new threats are emerging for the twenty-first century.

On the other hand, opportunities do exist for disaster reduction. There is currently a widespread awareness of risk in the environment and a growing recognition that the toll exerted by disasters, especially in the LDCs, is unacceptable. It is defeatist not to bring hazard awareness into development planning because continuing disaster losses simply reinforce poverty and vulnerability. Disaster impact in the Third World needs to be reduced to the point where stable investment can take place and the indigenous skills and energy, on which sustainable long-term development depends, can be released. It seems likely that, following the International Decade for Natural Disaster Reduction (IDNDR) of 1990–2000, environmental hazards will remain high on the public policy and political agenda for some time to come.

THE CHALLENGES OF HAZARD

The future holds the prospect of environmental change across the earth which will be unprecedented in historical times. This change is likely to arise from a complex combination of physical and human processes and, in many cases, may lead to increased exposure to hazard. An awareness of climate change, including global warming and stratospheric ozone depletion, has been followed by concern about growing stresses on the world's oceans and land areas leading to environmental degradation and a loss of biological diversity. Although many of these processes operate on longer time-scales than are considered in this book, global environmental change highlights the fact that the so-called 'elusive' hazards often provide the context in which short-term disasters can flourish. For example, global warming and rising sea levels may well create conditions under which more hurricanes, ENSO events and coastal flooding emerge in the coming decades.

The two great drivers of these adverse changes are population growth and the associated increase – both absolute and *per capita* – in the human use of natural resources. If current trends are projected forward, they raise fundamental questions about the carrying capacity of the earth and its ability to guarantee sustainable development in the future. In particular, any further environmental degradation, accompanied by more inequality in human wealth and welfare, will produce a downward spiral creating greater hazard vulnerability. This is because poverty forces people to adopt unsustainable land use practices which, in turn, lead to more fragile natural environments and human livelihoods.

Population change and urbanisation

Over the next few decades, the world's population is likely to increase by 50 per cent, with a doubling predicted for the year 2100. This growth will be accompanied by significant ageing and urbanising trends. For example, during the 1990s, 80 per cent of the world's population growth occurred in urban areas as a result of an excess of births over deaths and inward migration. Most of the inward migration arose as rural people were forced out of a countryside that offered a degraded resource base with decreasing opportunities for survival. At the present time, the Pacific Rim, exposed to earthquakes, volcanoes and floods, is the most rapidly urbanising region on earth (IFRCRCS, 1999).

We have little detailed understanding of the hazard-related consequences of these demographic changes. But the warning signs are clear. Not only will more people be exposed to environmental hazards but there is also likely to be increased hazard vulnerability amongst the poor, the illiterate and the unemployed. Already some 25,000 children die every day from malnutrition or diseases which are relatively easy to cure. According to data reported in IFRCRCS (1999), by the year 2025 80 per cent of the world's population will be in the LDCs and up to 60 per cent of these people will be vulnerable to disaster.

At the start of the twentieth century, there were only eleven metropolitan areas with populations in excess of 1 million and almost all of these were in the MDCs. At the start of the twenty-first century, there were between 300 and 400 cities each housing more than one million people, totalling approximately 20 per cent of the world population. Most of these were in the LDCs. Large cities are now defined as those with 3–8 million people. Beyond this level, there are presently some 28 mega-cities each with at least 8 million inhabitants. As shown in Figure 14.1, these major agglomerations emerged during the second half of the twentieth century

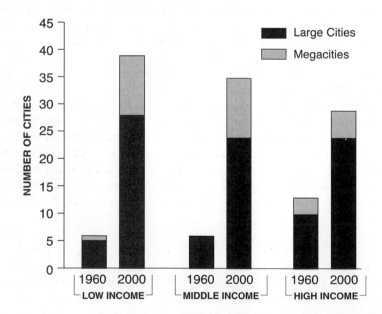

Figure 14.1   The growth in the number of large cities (>3 million inhabitants) and mega-cities (>8 million inhabitants) in low-, medium- and high-income countries between 1960 and 2000. About two-thirds of all mega-cities are now in the Third World.
Source: After Institution of Civil Engineers (1995)

in low- and middle-income countries to the extent that about two-thirds of the mega-cities are located in the LDCs. Asia now accounts for nearly 45 per cent of the world's urban population which lives in cities of at least 1 million people (IFRCRCS, 1998). Many of these cities have high annual growth rates. UN sources indicate that between 1990 and 2020 the urban population of the developing world will grow by 2,200 million people, a trend which will raise the LDC's proportion of the world's urban population from 34 per cent to 57 per cent of a much larger total (Institution of Civil Engineers, 1995). Overall, it is estimated that by 2020 about 30 per cent of the global population will live in large cities.

As explained by Mitchell (1999), the concentration of people and infrastructure on this scale not only creates an impressive symbol of the apparent human domination of nature but it also provides a daunting future prospect in terms of hazard exposure. On the positive side, well-planned cities with strong buildings, efficient public facilities and good emergency services provide a physical shield against natural disasters. Their growth is often a reflection of economic success and the rising *per capita* income of the residents while the resulting economies of scale offer opportunities for a better and safer quality of life for many people. On the negative side, most of the world's major cities have long outgrown what may originally have been a safe site and now rely – for day-to-day support – on the importation of basic commodities, such as water and power, along vulnerable lifelines. In some cases, the urban expansion has been accompanied by deforestation, slope modification and interference with natural drainage channels on a scale sufficient to increase, rather than reduce, the threat from natural processes. Where such cities have grown in a very rapid, uncontrolled fashion and contain large 'informal settlements' of poor and disadvantaged people, the risks from external and internal threats increase greatly. According to IDNDR (1996), the proportion of the total population housed in 'shanty towns' in several large cities throughout the LDCs ranges from 30–60 per cent. These low-income communities frequently live in fragile homes, often built of flammable materials, usually without piped water or reliable facilities for the disposal of sewage and other wastes. They have little security of tenure and the congested streets are difficult to access by emergency services.

Many of these large urban concentrations are located in known hazard zones. By year 2000, half of the population of the world's fifty largest cities lived within 200 km of geological faults capable of producing an earthquake of M = 7 or greater (Tucker *et al.*, 1994) and many others were in low-lying coastal cities (Timmerman and White, 1997). Mega-cities are a recent phenomenon but, since 1939, more than 100 natural disasters have affected large cities (Parker and Mitchell, 1995). The high densities of population, the rapid spread to unsafe locations and the influx of disadvantaged groups combine to ensure that many mega-cities lack the financial resources and the management expertise necessary to plan for safe growth.

This implies not only a greater exposure to natural hazards but also a risk from internally generated 'hybrid' disasters such as air and water pollution, fire, disease and traffic accidents. A high dependency on lifeline supplies means that sudden power failures, or any other interruptions to the assumed safe delivery of goods and services, can quickly dislocate urban activities. In addition, the difficulty of fulfilling the aspirations of floods of rural refugees creates additional social pressures likely to lead to urban crime and even terrorism. It is these concerns which create doubts about the long-term sustainability of large cities.

Environmental change and sea-level rise

Every year almost 0.5 per cent of the world's arable land is lost from soil erosion or other forms of land degradation. This statistic illustrates the pressure on natural resources in a world where the average citizen of the wealthiest countries consumes 200 times more fossil fuel, and perhaps 100 times more fresh water, than the typical citizen of the poorest countries. These resources are not infinite. It has been estimated that by 2025 about 5 billion people, out of a world total of some 8 billion, will be living in countries experiencing 'water stress', that is, using at least 20 per cent of the available resources (Arnell, 1999). If global environmental change reinforces these patterns and creates more inequality, it seems inevitable that the greatest disaster impacts in the future will be felt in those parts of the world which already bear a disproportionate burden of environmental stress.

The tropics appear especially vulnerable to future changes, for example in the availability and the quality of fresh water supplies. The arid and semi-arid regions of the tropics, already prone to seasonal or longer periods of drought, are home to about 350 million people in some of the most poverty-stricken countries on earth. Other LDCs are found in the water surplus areas of the tropics, which suffer from recurrent floods. Any changes in hydroclimate which increase the risk of droughts or floods will be most evident in these countries, notably in the drainage basins of large, unregulated rivers. In the poorest countries, hydroclimatic change has the potential to impinge adversely on the quality of life and future development prospects. For example, effective water management is critical for health, and any greater failures of water services could create more outbreaks of diseases such as cholera. As already indicated, many vector-borne diseases, such as malaria, are prevalent mainly in tropical areas but could spread to higher latitudes with shifts in air temperature and rainfall patterns.

The most certain outcome of climate change is a further rise in sea-level. As shown in Chapter 11, over the last 100 years the global sea-level has risen by 0.10–0.20 m. Current projections suggest additional increases above the 1990 level of about 0.21 m by the 2050s and about 0.38 m by the 2080s. This will create an extra risk of marine flooding for the one-fifth of the world's population living within 30 km of the sea (Nicholls, 1998). Much of this threatened population is contained within the two-thirds of the world's mega-cities which lie on the coast (Timmerman and White, 1997). The calculation of future risk contains numerous uncertainties but, without an adaptive human response – such as improved sea defences or a managed retreat from the shoreline – current estimates indicate that, by the 2080s, sea-level rise could have increased the number of people flooded by storm surge in a typical year by five times over the 1990 total (Nicholls *et al.*, 1999).

Such estimates depend not only on the amount of sea-level rise itself but also on related assumptions for the coastal zone, such as the expected frequency of storm surges and the future standard of coastal protection achieved by any adaptive responses. If storm climatology is assumed to remain constant and the coastal population at risk is assumed to grow at twice the national rate (roughly in line with recent experience), the main uncertainty lies in the assumptions made about sea defences. Nicholls (1998) has presented results for two cases:

- *constant protection*: where no changes are assumed from 1990 levels)
- *evolving protection*: where sea defences are upgraded in line with projected increases in economic growth measured by GDP. This latter case mimics historical development but makes no extra allowance for future sea-level rise.

Given these assumptions, Figure 14.2 shows that, with *constant protection*, the annual number of people at risk of flooding increases from 10 million in 1990 to 78 million in the 2050s. With *evolving protection* the number of people at risk in the 2050s is limited to 50 million. By the 2080s the numbers at risk will have increased to about 220 and 100 million respectively.

Table 14.1 The world regions most vulnerable to coastal flooding due to future sea-level rise

Region	Average annual number of people flooded (million)		
	1990	2080s	
		Constant protection	Evolving protection
S. Mediterranean	0.2	13	6
West Africa	0.4	36	3
East Africa	0.6	33	5
South Asia	4.3	98	55
Southeast Asia	1.7	43	21

Source: Nicholls *et al.* (1999)

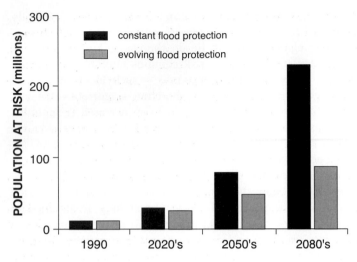

Figure 14.2   The estimated growth in the annual number of people at risk from coastal flooding as a result of sea-level rise from 1990 to the 2020s, 2050s and 2080s. The left-hand bar indicates the scenario assuming constant flood protection and the right-hand bar indicates the scenario assuming evolving flood protection.
Source: After Nicholls (1998)

These impacts will not occur uniformly around the globe. The greatest problems exist in the low-lying coastal zones with high concentrations of population, such as the delta areas of Egypt and Bangladesh. These zones are expensive to protect because of the relatively long shoreline and the need for on-going management of the fresh and saline waters held behind the coastal barriers. Some islands in the south Pacific and Indian oceans, such as the Maldives, with villages only a few metres above the sea, are also at direct risk from any further rise in ocean level. These communities typically depend on a narrow range of primary products, the agricultural land is often of poor quality, fresh water supplies are under stress and investment is limited. Table 14.1 summarises the parts of the world where most people are likely to be flooded by the 2080s. These five regions contain more than 90 per cent of the average annual number of potential flood victims, irrespective of which flood protection scenario is considered.

Future hazards and sustainability

Most attempts to anticipate hazards in the twenty-first century have recognised the increasing potential impact of known problems, especially those arising from climate-related natural disasters and the threat from relatively new sources, particularly technological-type hazards. For example, Rubin (1998) anticipated a rise in natural hazard impacts but gave most attention to relatively new sources such as na-tech hazards, emerging biolog-ical and chemical hazards and the risk of technological failures which threaten the flow of key services such as power, communications and information. There appears to be a general tendency to broaden the existing definitions of technological hazard beyond innovation and new products to include many of the 'slower-onset hazards' associated with modern urban life, such as air and water pollution, fire and congestion, together with some 'social hazards', such as crime, unemployment and terrorism. It cannot be denied that these are real threats, especially in the mega-city context, but it is important to retain a perspective. The impact of such hazards would have to grow by many times before they claimed the worldwide deaths and damages due to so-called 'natural disasters'.

The significance of current trends in both natural and 'technological-type' hazards, and their potential future combination, lies in the fear of unsustainability. At present there is little understanding of the likely future balance between traditional 'rapid-onset' natural disasters, rapid-onset 'surprises' resulting from new technology and the build up of 'slow-onset' threats posed by unsustainable development, but the combined effects could be serious. For sustainable development, three types of stability are required:

1   *Economic stability* This means financial effectiveness, the provision of jobs and other local opportunities for betterment. It is threatened by past failures to control disaster losses, which can be taken as a sign that expenditure on hazard mitigation has been ineffective. This is a special problem for the LDCs where repeated disaster strikes undermine the development process. Here the challenge is to create development policies that rely on indigenous efforts and reduce hazard vulnerability.

2   *Ecological stability* This means the retention of natural resources and biodiversity. Land degradation, resource exhaustion and pollution create greater vulnerability to hazard impacts. Beatley (1995) emphasised the 'green' issues which should underpin sustainable urban planning and growth. Some of these 'green' perspectives are of most immediate rele-vance for communities in the MDCs, although Berke (1995) has explored how aid can be linked to hazard reduction and more sustainable devel-opment in poorer countries.

3   *Social stability* This means the promotion of general human well-being, such as social inclusion and continuing improvements in the quality of life. It has relevance throughout the world, especially in the largest cities which are likely to suffer most from hazards associated with ongoing functions such as waste disposal and the control of pollution. The size of the informal sector is of particular concern in the poorest cities of the LDCS. McGranahan and Songsore (1994) have claimed that the normal concept of the sustainable city does not capture the needs of the urban poor who cannot afford the luxury of a concern for conven-tional 'green' issues. Instead, they are burdened by intractable 'brown'

issues at the household level, such as the provision of safe water supplies, better sanitation and an improvement of the indoor environment.

THE OPPORTUNITIES OF HAZARD

The IDNDR and its lessons

The IDNDR was established by the United Nations to run for a period of 10 years from 1 January 1990. The basic aim of the Decade, summarised by Housner (1989) and Lechat (1990), was to shift natural disaster management from a reactive strategy of post-disaster improvisation, which relied heavily on relief aid, to a more proactive strategy of pre-disaster planning and preparedness. The five main goals of the Decade were:

1   To improve the capacity of each country to mitigate the effects of natural disasters expeditiously and effectively, paying special attention to assisting developing countries in the assessment of disaster damage potential and in the establishment of early-warning systems and disaster-resistant structures when and where needed.
2   To devise appropriate guidelines and strategies for applying existing scientific and technical knowledge, taking into account the cultural and economic diversity among nations.
3   To foster scientific and engineering endeavours aimed at closing critical gaps in knowledge in order to reduce the loss of life and property.
4   To disseminate existing and new technical information related to measures for the assessment, prediction and mitigation of natural disasters.
5   To develop measures for the assessment, prediction, prevention and mitigation of natural disasters through programmes of technical assistance and technology transfer, demonstration projects, and education and training, tailored to specific disasters and locations, and to evaluate the effectiveness of those programmes.

The IDNDR was critically dependent on financial and other support provided by the member governments. It was also the responsibility of individual states to formulate their own national disaster-mitigation programmes and to establish committees to oversee these activities. The Decade had a slow start but national committees were eventually set up in over 130 countries. These committees differed in their effectiveness and probably only one-quarter were fully active. Demonstration projects were brought forward for support but, again, only about one-quarter of the total were successful, often on a co-funded basis.

As the IDNDR progressed, several criticisms of its key objectives emerged:

1   The Decade's emphasis on the exploitation of scientific solutions and the transfer of hazard-mitigating technologies to the LDCs attracted adverse comment because of its reliance on capital-intensive measures

which are inappropriate when resources are scarce. At worst, such projects can reflect undesirable motives when support is tied to a specified reliance on foreign specialists and products.

2   A related criticism emphasised the neglect of the social, economic and political dimensions of hazards (Bates *et al.*, 1991). It is no longer sufficient to rely on structural and technical solutions for hazard reduction, and mitigation strategies have to be understood by, and accessible to, all who need them. Local participation is increasingly seen as important in redevelopment planning and recovery after disaster (Berke *et al.*, 1993).

3   The omission of technological hazards from the IDNDR brief was also interpreted as a defect. For example, despite adopting a so-called integrated hazards approach, which recognises a common framework for all natural hazards, the IDNDR failed to extend the concept to its logical conclusion and embrace all environmental hazards.

4   Another criticism was the lack of any integration of hazard reduction with the basic human needs of food, fuel, shelter and health provision in order to achieve sustainable development (Oaks and Bender, 1990). When dealing with the LDCs, the need for people to live more harmoniously with nature might well have been given more prominence because the best hope for disaster reduction in such countries often lies in the protection of the natural resource base.

These and other criticisms were recognised during a mid-Decade review of progress (Bruce, 1994). In the second half of the Decade, disaster reduction was promoted at a higher economic and political level embracing legislators and key environmentalists, as well as relief professionals and development officials. At the end of the Decade, there was an acceptance that 10 years is a rather short period for real progress to be attained. The IDNDR was only a signpost near the start of a very long journey. Moreover, the route indicated is unlikely to be either the best or only road, bearing in mind that hazard policies will have to evolve and, hopefully, improve through time. These requirements were reflected by the announcement during 2000 of a successor programme, called the International Strategy for Disaster Reduction (ISDR), to be implemented by an Inter-Agency Task Force chaired by the UN's Under-Secretary-General for Humanitarian Affairs. The main policy objective of the ISDR is likely to be increased community resilience in the face of disaster, as opposed to an IDNDR focus on more direct physical confrontations of hazard, and will be based on reducing socio-economic losses through an improved public awareness of risk.

A new consensus on old problems?

Despite the fact that academic and professional protectionism continues to provide some limited, discipline-bound approaches to hazard mitigation, there are encouraging signs of a growing consensus on key policy issues. The last 50 years have witnessed a progressive widening of attitudes to environmental risk and its management. The most enlightened local governments now produce hazard-management programmes which are more flexible and comprehensive than ever before and which attempt to consider macro issues, such as global change and sustainability. The choice is no longer between either risk reduction or safe development. In the future, both of these goals have to be achieved in the context of sustainability and self-sufficiency. A basis for modest optimism exists in the following areas.

Science and technology

Science is at last taking a more appropriate place in hazard reduction where the main task is to understand and predict future environmental change. Amongst other things, this means disentangling the relationships between natural processes and human activities with a view to providing policy-makers with firm evidence on which to base human adaptations to change. More basic research is also required, for example leading to a better understanding of the dynamics of the land–ocean interface in order to predict the effects that sea-level rise will have on many vulnerable communities. The improved forecasting of key elements of tropical climates (such as monsoon rainfall and ENSO events) is also a priority.

Science will continue to improve the knowledge base for policy making but, more generally, it can be argued that the effective transfer of knowledge is not solely from the MDCs to the LDCs. There is a need to disseminate knowledge among developing disaster-prone countries where numerous traditional practices of risk reduction are being lost in the face of imported technologies. Much of the knowledge already exists to reduce the damaging impacts from most environmental hazards. What is needed is a more effective implementation of hazard-reducing capability; in other words, either a shift from research to action or a wider recognition that good hazards research increasingly implies the better 'application' of existing knowledge as well as the 'generation' of new knowledge. For example, scientific knowledge, especially that which relates to uncertainty, must be expressed in ways that are more useful to local decision makers.

A culture of prevention

Another emerging consensus surrounds the need to create a culture of disaster 'prevention'. It is now believed that each country and community should be assisted to build its own capacities and develop the self-reliance necessary to undertake proactive hazard reduction measures. For this to happen, a better targeting of effort is required on the part of the MDCs and the aid agencies to ensure that disaster mitigation measures are seen as central to the achievement of sustainable development.

One way forward is to recognise that the most vulnerable people are not only the victims of hazard but are also the main players in disaster prevention. For all such people, effective mobilisation prior to a disaster strike is essential. Thus, in the Philippines, a Citizen's Disaster Response Network has been established to stimulate a variety of non-governmental organisations to formulate preparedness plans and create disaster volunteer teams (Delica, 1993). Beyond the national level, countries with similar hazardous environments and cultural attributes need to involve local community leaders and other decision makers in a shared experience through regional workshops, demonstration projects and improved networking to ensure the optimum blend of technical and traditional responses. Even in the wealthy countries of the world, education and public information is important to ensure that the user communities become better skilled at defining their own needs.

A much greater prior commitment to reducing risk is highly relevant. Important mitigation measures, such as better building design or land use restrictions, are already well proven but all too often they are not applied because of cost and lack of strong enforcement. To ensure a safer world, disaster reduction will have to be given some priority over more traditional commercial and social goals. Better training of disaster personnel plus improved organisation and communication are all necessary for achieving a significant decrease in the losses from hazards in the future. But without a considerable increase in safety motivation by managers and victims alike, backed by a strong political will to achieve effective results, little progress is likely to be made.

The opportunity exists to put integrated disaster management at the heart of modern societies. This can only be done if hazard workers set aside the fragmented 'professional culture' approach of the past and work together on complex issues such as the anticipation of 'mega' disasters in urban areas and the halting of the degradation of valuable ecosystems in order to stabilise the population of rural areas. At the international level, there is the need for greater recognition of a shared responsibility for the safety of nuclear technologies. The Chernobyl incident is a reminder that such accidents occur, and will continue to do so, and that all nations need to engage in managing these global risks.

Helping the most vulnerable

The most urgent requirement is to improve the situation of the most vulnerable people at risk. At the start of the twenty-first century, 1.3 billion people, almost one-quarter of the world's population, were living on less than US$1 per day. Every year millions of displaced persons flee violence, drought and environmental degradation and there are several million refugees in Africa alone, many living in large 'temporary' camps. Such camps impose an insupportable burden on the local areas, being reliant on imported food and medical support to maintain life and control disease, whilst competing with the local people for scarce natural resources, such as fuelwood and water. In turn, such activities create further deforestation and soil erosion. Women make up the largest proportion of these vulnerable people. When human resources are being strengthened, this means not only better training for staff and aid volunteers but also measures designed to give an increased role to all refugees, including women and young people, to make better use of their often neglected local skills and knowledge with a view to integrating them with the local community.

Despite the proven success of preparedness programmes, emergency aid will continue to be required. The various agencies involved in relief work need to become more proactive so that the allocation and distribution of disaster relief can be made more effective. Some advances have been made. For example, the southern African drought emergency of 1992, when food production was halved and 18 million people in ten countries were directly affected, prompted an impressive importation and distribution of emergency food – five times the amount normally handled in the region – and a major famine was averted. However, support and protection for the most disadvantaged groups in society, such as the very young and the elderly, is needed at all stages of the disaster mitigation process (Eldar, 1992), and Uitto (1998) described a pilot study of disaster vulnerability amongst the homeless of Tokyo as a pilot method for helping the more marginalised groups in large cities.

The most intractable problems – and the greatest opportunities – exist in the countries which are too poor to help themselves. In many of these areas economic development and industrialisation threaten to sweep away the traditional life-styles and the indigenous responses to disaster. Without this accumulated knowledge, people will lose their ability to absorb and recover from hazard impact. Disasters will never be eliminated but progress will be achieved if the weakest groups in society can be empowered, through information and a better access to resources of all kinds. By this means, they will better understand, and participate more actively in, the hazard reduction process. If disaster reduction can be coupled to policies which serve wider development goals, then the most disadvantaged people will have an opportunity to live with environmental hazards in a more harmonious and sustainable way in the future.

Appendix

Abridged Modified Mercalli
Intensity Scale

Average peak velocity (cm/s ^{1})	Intensity value and description	Average peak acceleration^a
	I. Not felt except by a very few under especially favourable circumstances.	
	II. Felt only by a few persons at rest, especially on upper floors of buildings. Delicately suspended objects may swing.	
	III. Felt quite noticeably indoors, especially on upper floors of buildings, but many people do not recognise it as an earthquake. Standing automobiles may rock slightly. Vibration like passing truck.	
1–2	IV. During day felt indoors by many, outdoors by few. At night some awakened. Dishes, windows doors disturbed; walls make creaking sound; sensation like heavy truck striking building. Standing automobiles rock noticeably.	0.015 g–0.02 g
2–5	V. Felt by nearly everyone, many awakened. Some dishes, windows and so on broken; cracked plaster in a few places; unstable objects overturned. Disturbance of trees, poles and other tall objects sometimes noticed. Pendulum clocks may stop.	0.03 g–0.04 g
5–8	VI. Felt by all, many frightened and run outdoors. Some heavy furniture moved; a few instances of fallen plaster and damaged chimneys. Damage slight.	0.06 g–0.07 g
8–12	VII. Everybody runs outdoors. Damage negligible in buildings of good design and construction; slight to moderate in well-built ordinary structures; considerable in poorly built or badly designed structures; some chimneys broken. Noticed by persons driving cars.	0.10 g–0.15 g

20–30	VIII.	Damage slight in specially designed stuctures; considerable in ordinary substantial buildings with partial collapse; great in poorly built structures. Panel walls thrown out of frame structures. Fall of chimneys, factory stacks, columns, walls, monuments. Heavy furniture overturned. Sand and mud ejected in small amounts. Changes in well water. Persons driving cars disturbed.	0.25 *g*–0.30 *g*
45–55	IX.	Damage considerable in specially designed structures; well-designed frame structures thrown out of plumb; great in substantial buildings, with partial collapse. Buildings shifted off foundations. Ground cracked conspicuously. Underground pipes broken.	0.50 *g*–0.55 *g*
>60	X.	Some well-built wooden structures destroyed; most masonry and frame structures destroyed with foundations; ground badly cracked. Rails bent. Landslides considerable from river banks and steep slopes. Shifted sand and mud. Water splashed, slopped over banks.	>0.06 *g*
	XI.	Few, if any, (masonry) structures remain standing. Bridges destroyed. Broad fissures in ground. Underground pipelines completely out of service. Earth slumps and land slips in soft ground. Rails bend greatly.	
	XII.	Damage total. Waves seen on ground surface. Lines of sight and level distorted. Objects thrown into the air.	

Note: *g* is gravity = 9.8 m/s^2

References

Abersten, L. 1984 Diversion of a lava flow from its natural bed to an artificial channel with the aid of explosives: Etna, 1983. *Bulletin of Volcanology* 47: 1165–74.

Adams, W.C. 1986 Whose lives count? TV coverage of natural disasters. *Journal of Communication* 36: 113–22.

Adams, W.M. 1993 Indigenous use of wetlands and sustainable development in West Africa. *Geographical Journal* 159: 209–18.

Ahearn, F.L. and Cohen. R.E. (eds) 1984 *Disasters and Mental Health: An Annotated Bibliography*. Rockville, MD: National Institute of Mental Health, US Department of Health and Human Sciences. Publication no. (ADM) 84–1311.

Alexander, D. 1987a Land of disasters. *Geographical Magazine* (May): 226–31.

Alexander, D. 1987b *The 1982 Urban Landslide Disaster at Ancona, Italy*. Working Paper 57, Boulder, CO: Institute of Behavioral Science, University of Colorado.

Alexander, D. 1989 Urban landslides. *Progress in Physical Geography* 13: 157–91.

Alexander, D. 1993 *Natural Disasters*. London: UCL Press Limited.

Alexander, D. 1997 The study of natural disasters 1977–1997: some reflections on a changing field of knowledge. *Disasters* 21: 284–304.

Allen, W.H. 1993 The Great Flood of 1993. *Bioscience* 43: 732–7.

All-Industry Research Advisory Council (AIRAC) 1986 *Catastrophic Losses: How the Insurance System would Handle Two $7 Billion Hurricanes*. Oak Brook, IL: AIRAC.

Ambrose, J. and Vergun, D. 1985 *Seismic Design of Buildings*. New York: John Wiley.

Anderson, M.B. 1991 Which costs more: prevention or recovery? In Kreimer, A. and Munasinghe, M. (eds), *Managing Natural Disasters and the Environment*, Washington, DC: Environmental Department, World Bank, 17–27.

Anderson, M.B. 1995 Vulnerability to disaster and sustainable development: a general framework for assessing vulnerability. In Munasinghe, M. and Clarke, C. (eds), *Disaster Prevention for Sustainable Development*, Washington, DC: IDNDR and World Bank, 41–59.

Anderson, M.B. 2000 Vulnerability to disaster and sustainable development. In Pielke, R.A. Jr and Pielke, R.A. Sr (eds), *Storms*, London and New York: Routledge, vol. 1, 11–25.

Anonymous, 1995 Disasters and property insurance: coping with the aftershocks. *Coastal Heritage* 9: 3–12.

Armstrong, B.R. 1984 Avalanche accident victims in the USA. *Ekistics* 51 (309): 543–6.

Arnell, N.W. 1984 Flood hazard management in the United States and the National Flood Insurance Program. *Geoforum* 15: 525–42.

Arnell, N.W. 1999 Climate change and global water resources. *Global Environmental Change* 9: 531–49.

Arnell N.W., Clark, M.J. and Gurnell A.M. 1984 Flood insurance and extreme events: the role of crisis in prompting changes in British institutional response to flood hazard. *Applied Geography* 4: 167–81.

Au, S.W.C. 1998 Rain-induced slope instability in Hong Kong. *Engineering Geology* 51: 1–36.

Autier, P., D'Altilia, J.-P., Delamalle, J.-P. and Vercruysse, V. 1989 The food and nutrition surveillance systems of Chad and Mali: the 'SAP' after two years. *Disasters* 13: 9–32.

Autier, P., Ferir, M.-C., Hairapetien, A., Alexanian, A., Agoudian, V., Schmets, G., Dallemagne, G., Leva, M.-N. and Pinel, J. 1990 Drugs supply in the aftermath of the 1988 Armenian earthquake. *Lancet* (9 June): 1388–90.

Ayscue, J.K. 1996 Hurricane damage to residential structures: risk and mitigation. Working Paper no. 94, Boulder, CO: Natural Hazards Research and Applications Information Center.

Baker, E.J. 2000 Hurricane evacuation in the United States. In Pielke, R.A. Jr and Pielke, R.A. Sr (eds), *Storms*, London and New York: Routledge, vol. 1, 306–19.

Bakun, W.H. 1988 Geophysical instrumentation near Parkfield. *Earthquakes and Volcanoes* 20: 60–71.

Bakun, W.H. and Lindh, A.G. 1985 The Parkfield, California, earthquake prediction experiment. *Science* 229: 619–29.

Baram, M.S. 1987 Chemical industry hazards: liability, insurance and the role of risk analysis. In Kleindorfer, P.R. and Kunreuther, H.C. (eds), *Insuring and Managing Hazardous Risks*, New York: Springer-Verlag, 415–42.

Barberi, F. and Carapezza, M.L. 1996 The problem of volcanic unrest: the Campi Flegrei case history. In Scarpa, R. and Tilling, R.I. (eds), *Monitoring and Mitigation of Volcano Hazards*, Berlin, Springer-Verlag, 771–86.

Bardsley, K.L., Fraser, A.S. and Heathcote, R.L. 1983 The second Ash Wednesday: 16 February 1983. *Australian Geographical Studies* 21: 129–41.

Barnett, A., Menighetti, J. and Prete, M. 1992 The market response to the Sioux City DC-10 crash. *Risk Analysis* 12: 45–52.

Barnett, B.J. 1999 US government natural disaster assistance: historical analysis and a proposal for the future. *Disasters* 23: 139–55.

Barrows, H.H. 1923 Geography as human ecology. *Annals of the Association of American Geographers* 13: 1–14.

Barry, R.G. and Chorley, R.J. 1987 *Atmosphere, Weather and Climate*. London: Methuen.

Bates, F.L., Dynes, R.R. and Quarantelli, E.L. 1991 The importance of the social sciences to the International Decade for Natural Disaster Reduction. *Disasters* 15: 288–9.

Bayulke, N. 1984 Earthquake emergency preparedness, rescue and relief: the Turkish experience. *Proceedings of the Seminar on Earthquake Preparedness*. Geneva: United Nations, 98–116.

BBC 1999 *Horizon* documentary programme on the Galtür avalanche. Broadcast 25 November.

Beatley, T. 1988 Ethical dilemmas in hazard management. *Natural Hazards Observer* 12: 1–3.

Beatley, T. 1995 Planning and sustainability: the elements of a new (improved?) paradigm. *Journal of Planning Literature* 9: 382–95.

Beaumont, P. 1978 Man's impact on river systems: a world view. *Area* 10: 38–41.

Beechley, R.W., van Bruggen, J. and Truppi, L.E. 1972 Heat island = death island? *Environmental Research* 5: 85–92.

Beinin, L. 1985 *Medical Consequences of Natural Disasters*. New York: Springer-Verlag.

Belletto de Pujo, J. 1985 *Emergency Planning: the Case of Diablo Canyon Nuclear Power Plant*. Natural Hazard Research Working Paper no. 51, Boulder, CO: Institute of Behavioral Sciences, University of Colorado.

Belt, C.B. Jr 1975 The 1973 flood and man's constriction of the Mississippi river. *Science* 189: 681–4.

Berke, P.R. 1995 Natural hazard reduction and sustainable development: a global assessment. *Journal of Planning Literature* 9: 370–82.

Berke, P.R., Kartez, J. and Wenger, D. 1993 Recovery after disaster: achieving sustainable development, mitigation and equity. *Disasters* 17: 93–109.

Bernard, E.N., Behn, R.R., Hebenstreit, G.T., Gonzales, F.I., Krumpe, P., Lander, J.F., Lorca E., McManamon, P.M. and Milburn H.B. 1988 On mitigating rapid onset natural disasters: project THRUST. *EOS Transactions of the American Geophysical Union* 69 (24): 649–61.

Bernknopf, R.L, Campbell, R.H., Brookshire, D.S. and Shapiro, C.D. 1988 A probabilistic approach to landslide hazard mapping in Cincinnati, Ohio, with applications for economic evaluation. *Bulletin of the Association of Engineering Geologists* 25: 39–56.

Berz, G. 1990 Natural disasters and insurance/reinsurance. *UNDRO News* Jan–Feb, 18–19.

Berz, G. 1999 Disasters and climate change: the insurance industry looks ahead. In IDNDR Secretariat, *1999 World Disaster Reduction Campaign 'Prevention Pays'*, Geneva, United Nations, 15–8.

Bhowmik, N.G. (ed.) 1994 *The 1993 Flood on the Mississippi River in Illinois*. Miscellaneous Publication 151, Champaign-Urbana, IL: Illinois State Water Survey.

Bigg, G.R. 1990 El Niño and the Southern Oscillation. *Weather* 45: 2–8.

Binder, D. 1998 The duty to disclose geologic hazards in real estate transactions. *Chapman Law Review* 1: 13–56.

Blaikie, P., Cannon, T., Davis I. and Wisner, B. 1994 *At Risk: Natural Hazards, People's Vulnerability and Disasters*. London and New York: Routledge.

Blong, R.J. 1984 *Volcanic Hazards: A Sourcebook on the Effects of Eruptions*. London: Academic Press.

Bolin, R. and Stanford, L. 1998 *The Northridge Earthquake: Vulnerability and Disaster on the Margins of Los Angeles*. London and New York: Routledge.

Bolt, B.A. 1993 *Earthquakes*. New York: W.H. Freeman

Bolt, B.A., Horn, W.L., Macdonald, G.A. and Scott, R.F. 1975 *Geological Hazards*. Berlin: Springer-Verlag.

Brabb, E.E. 1991 The world landslide problem. *Episodes* 14: 52–61.

Bradley, J.T. 1972 Hurricane Agnes: the most costly storm. *Weatherwise* 25: 174–84.

Bridger, C.A. and Helfand, L.A. 1968 Mortality from heat during July 1966 in Illinois. *International Journal of Biometeorology* 12: 51–70.

Brinkmann, W.A.R. 1975 *Hurricane Hazard in the United States: A Research Assessment.* Monograph no. NSF-RA-E-75–007, Boulder, CO: Institute of Behavioral Science, University of Colorado.

Britton, N.R. 1984 Australia's organised response to natural disasters: the constrained organisation on two wildfire settings. *Disasters* 8: 214–25.

Britton, N.R. 1987 Disaster in the South Pacific: impact of tropical cyclone 'Namu' on the Solomon Islands, May 1986. *Disasters* 11: 120–33.

Broecker, W.S. 1991 The Great Ocean Conveyor. *Oceanography* 4: 79–89.

Brooks, R. 1971 Human response to recurrent drought in north-eastern Brazil. *Professional Geographer* 23: 40–4.

Brotak, E.A. 1980 Comparison of the meteorological conditions associated with a major wildland fire in the United States and a major bushfire in Australia. *Journal of Applied Meteorology* 19: 474–6.

Brown, J. and Muhsin, M. 1991 Case study: Sudan emergency flood reconstruction program. In Kreimer, A. and Munasinghe, M. (eds), *Managing Natural Disasters and the Environment*, Washington, DC: Environment Department, World Bank, 157–62.

Brown, M.M. 1985 *Famine: a Man-made Disaster?* A report for the Independent Commission on International Humanitarian Issues. London: Pan Books.

Bruce, J.P. 1994 Challenges from Yokohama. *STOP Disasters* 19–20: 3.

Brugge, R. 1994 The blizzard of 12–15 March 1993 in the USA and Canada. *Weather* 49: 82–9.

Bryant, E.A. 1991 *Natural Hazards.* Cambridge and New York: Cambridge University Press.

Building Seismic Safety Council 1987 *Abatement of Seismic Hazards to Lifelines: Action Plan.* Earthquake Hazards Reduction Series 32, Washington DC: Federal Emergency Management Agency.

Buller, P.S.J. 1986 *Gale Damage to Buildings in the UK: An Illustrated Review.* Watford, Herts.: Building Research Establishment.

Burby, R.J. and Dalton, L.C. 1994 Plans can matter! The role of land use plans and state planning mandates in limiting the development of hazardous areas. *Public Administration Review* 54: 229–37.

Burby, R.J. with Cigler, B.A., French, S.P., Kaiser, E.J., Kartez, J., Roenigk, D., Weist, D. and Whittington, D. 1991 *Sharing Environmental Risks: How to Control Governments' Losses in Natural Disasters.* Boulder, CO: Westview Press.

Burton, I. and Kates, R.W. 1964a The perception of natural hazards in resource management. *Natural Resources Journal* 3: 412–41.

Burton, I. and Kates, R.W. 1964b The floodplain and the seashore: a comparative analysis of hazard-zone occupance. *Geographical Review* 54: 366–85.

Burton, I., Kates, R.W. and White, G.F. 1993 *The Environment as Hazard*, 2nd edn. New York and London: Guildford Press; 1st edn, 1978.

Buser, O., Fohn, P., Good, W., Gubler, H. and Salm, B. 1985 Different methods for the assessment of avalanche danger. *Cold Regions Science and Technology* 10: 199–218.

Bush, D.M., Webb, C.A., Young, R.S., Johnson, B.D. and Bates, G.M. 1996 Impact of hurricane 'Opal' on the Florida–Alabama coast. *Quick Response Report* no. 84, Boulder, CO: Natural Hazards Research and Applications Information Center.

Butler, D.R. 1987 Snow-avalanche hazards, Southern Glacier National Park, Montana: the nature of local knowledge and individual responses. *Disasters* II: 214–20.

Butler, D.R. 1997 A major snow-avalanche episode in north-west Montana, February, 1996. *Quick Response Report* no. 100, Boulder, CO: Natural Hazards Research and Applications Information Center.

Butler, J.E. 1976 *Natural Disasters*. Melbourne: Heinemann Educational.

Cairncross, S., Hardoy, J.E. and Satterthwaite, D. (eds) 1990 *The Poor Die Young: Housing and Health in Third World Cities*. London: Earthscan.

Cekan, J. 1992 Seasonal coping strategies in cental Mali: five villages during the 'Soudiere'. *Disasters* 16: 66–73.

Chaine, P.M. 1973 Glaze and its misery: the ice storm of 22–23 March 1972 north of Montreal. *Weatherwise* 26: 124–7.

Chang, S. 1984 Do disaster areas benefit from disasters? *Growth and Change* (Oct): 24–31.

Changnon, S.A. 1987 Climate fluctuations and record-high levels of Lake Michigan. *Bulletin of the American Meteorological Society* 68: 1394–402.

Changnon, S.A. 2000 Impacts of hail in the United States. In Pielke, R.A. Jr and Pielke, R.A. Sr (eds), *Storms*, London and New York: Routledge, vol. 2, 163–91.

Changnon, S.A. and Changnon, J.M. 1992 Temporal fluctuations in weather disasters: 1950–1989. *Climatic Change* 22: 191–208.

Changnon S.A. and Semonin, R.G. 1966 A great tornado disaster in retrospect. *Weatherwise* 19: 56–65.

Chappelow, B.F. 1989 Repair and restoration of supplies in Jamaica in the wake of hurricane Gilbert. *Distribution Developments* (June): 10–14.

Charlton, R.M. 1990 *How Safe is Safe Enough?* Selected Papers Group Public Affairs, London: Shell International Petroleum.

Charoenphong, S. 1991 Environmental calamity in southern Thailand headwaters. *Land Use Policy* 18: 185–8.

Cheney, N.P. 1979 Bushfire disasters in Australia, 1945–1975. In Heathcote, R.L. and Thom, B.G. (eds), *Natural Hazards in Australia*, Canberra: Australian Academy of Science, 72–93.

Chester, D. 1993 *Volcanoes and Society*. London: E. Arnold.

Clapperton, C.M. 1986 Fire and water in the Andes. *Geographical Magazine* 58: 74–9.

Clark, R.A. 1994 Evolution of the national flood forecasting system in the USA. In Rossi, G., Harmancioglu, N. and Yevjevich, V. (eds), *Coping with Floods*, NATO Advanced Study Institute, vol. 257, Dordrecht: Kluwer, 437–44.

Clarke, C.L. and Munasinghe, M. 1995 Economic aspects of disasters and sustainable development: an introduction. In Munasinghe, M. and Clarke, C. (eds) *Disaster Prevention for Sustainable Development*, Washington, DC: IDNDR and World Bank, 1–9.

Coburn, A.W., Hughes, R.E., Illi, D., Nash, D.F.T. and Spence, R.J.S. 1984 The construction and vulnerability to earthquakes of some building types in northern areas of Pakistan. In Miller, K.J. (ed.), *The International Karakorum Project*, vol. 2, Cambridge: Cambridge University Press, 228–37.

Comerio, M.C., Landis, J.D. and Firpo, C.J. 1996 *Residential Earthquake Recovery*. CPS Brief, California Policy Seminar, Berkeley, California.

Comfort, L.K. 1986 International disaster assistance in the Mexico City earthquake. *New World: A Journal of Latin American Studies* 1: 12–43.

Comfort, L.K. 1996 Self organisation in disaster response: the Great Hanshin, Japan, earthquake of January 17, 1995. *Quick Response Report* no. 78, Boulder, CO: Natural Hazards Research and Applications Information Center.

Cook, R.J., Barron, J.C., Papendick. R.I. and Williams, G.J. 1981 Impact on agriculture of the Mount St. Helens eruptions. *Science* 211: 16–22.

Cooke, R.U. 1984 *Geomorphological Hazards in Los Angeles.* London: Allen and Unwin.

Cotton, W.R. and Cochrane, D.A. 1982 Love Creek landslide disaster, January 5, 1982, Santa Cruz County. *California Geology* 35: 153–7.

Covello, V.T. and Mumpower, J. 1985 Risk analysis and risk management: an historical perspective. *Risk Analysis* 5: 103–120.

Cox, D., Crossland, B., Darby, S.C., Forman, D., Fox, A.J., Gore, S.M., Hambly, E.C., Kletz, T.A. and Neill, N.V. 1992 Estimation of risk from observation on humans. In *Risk*, London: Royal Society, 67–87.

Crandell, D.R., Mullineaux, D.R. and Miller, C.D. 1979 Volcanic-hazards studies in the Cascades Range of the western United States. In Sheets, P.D. and Grayson, D.K. (eds), *Volcanic Activity and Human Ecology*, London: Academic Press, 195–219.

Cross, J.A. 1985 *Residents' Acceptance of Hurricane Hazard Mitigation Measures.* Final Summary Report, NSF Grant no. CEE-8211441, University of Wisconsin, Oshkosh.

Cruden, D.M. and Krahn, J. 1978 Frank rockslide, Alberta, Canada. In Voight, B. (ed.), *Rockslides and Avalanches*, vol. 1: *Natural Phenomena*, Amsterdam: Elsevier, 97–112.

Cunningham, C.J. 1984 Recurring natural fire hazards: a case study of the Blue Mountains, New South Wales, Australia. *Applied Geography* 4: 5–57.

Cuny, F.C. 1991 Living with floods: alternatives for riverine flood mitigation. In Kreimer, A. and Munasinghe, M. (eds), *Managing Natural Disasters and the Environment*, Washington, DC: Environment Department, World Bank, 62–73.

Cutler, P. 1984 Famine forecasting: prices and peasant behaviour in northern Ethiopia. *Disasters* 8: 48–56.

Cutler, S.L. 1984 Emergency preparedness and planning for nuclear power plant accidents. *Applied Geography* 4: 235–45.

Cutter, S.L. 1992 Toxic monuments and the making of technological hazards. In Janelle, D. *et al.* (eds) *Geographical Snapshots of North America*, New York: Guilford Press, 117–21.

Cutter, S.L. 1993 *Living with Risk: The Geography of Technological Hazards.* London and New York: E. Arnold.

Cutter, S.L. and Barnes, K. 1982 Evacuation behaviour and Three Mile Island. *Disasters* 6: 116–24.

Cutter, S.L. and Ji, M. 1997 Trends in US hazardous materials transportation spills. *Professional Geographer* 49: 318–31.

Czerwinski, S.J. 1999 *Information on the Cost-effectiveness of Hazard Mitigation Projects.* GAO/T-RCED-99–106, U.S. Washington, DC: General Accounting Office.

Dando, W.A. 1980 *The Geography of Famine.* London: Edward Arnold.

David, E. and Mayer, J. 1984 Comparing costs of alternative flood hazard mitigation plans. *Journal of the American Planning Association* 50: 22–35.

Davis, I. 1978 *Shelter after Disaster.* Oxford: Oxford Polytechnic Press.

De Bruycker, M., Greco, D. and Lechat, M.F. 1985 The 1980 earthquake in southern Italy: mortality and morbidity. *International Journal of Epidemiology* 14: 113–17.

Decker, R.W. 1986 Forecasting volcanic eruptions. *Annals and Review of Earth and Planetary Science* 14: 267–91.

Degg, M. 1992 Natural disasters: recent trends and future prospects. *Geography* 77: 198–209.

Delica, Z.G. 1993 Citizenry-based disaster preparedness in the Philippines. *Disasters* 17: 239–47.

Department of Commerce 1994 *The Great Flood of 1993*. Natural Disaster Survey Report, Silver Spring, MD: Department of Commerce.

Department of Humanitarian Affairs 1994 *Strategy and Action Plan for Mitigating Water Disasters in Vietnam*. New York and Geneva: United Nations Development Programme.

De Scally, F.A. and Gardner, J.S. 1994 Characteristics and mitigation of the snow avalanche hazard in Kaghan valley, Pakistan Himalaya. *Natural Hazards* 9: 197–213.

De Vries, J. 1985 Analysis of historical climate–society interaction. In Kates, R.W., Ausubel, J.H. and Berberian, M. (eds), *Climate Impact Assessment*, New York: John Wiley, 273–91.

De Waal, A. 1989 *Famine that Kills: Dorfur, Sudan, 1984–85*. Oxford: Clarendon Press.

Dilley, M. and Heyman, B.N. 1995 ENSO and disaster: droughts, floods and El Niño–Southern Oscillation warm events. *Disasters* 19: 181–93.

Dohler, G.C. 1988 A general outline of the ITSU master plan for the tsunami warning system in the Pacific. *Natural Hazards* 1: 295–302.

Dolan, J.F., Sieh, K., Rockwell, T.K., Yeats, R.S., Shaw, J., Suppe, J., Huftile, G.J. and Gath, E.M. 1995 Prospects for larger or more frequent earthquakes in the Los Angeles Metropolitan Region. *Science* 267: 199–205.

Donald, J.R. 1988 Drought effects on crop production and the US economy. *The Drought of 1988 and Beyond*. Proceedings of a Strategic Planning Seminar, 18 October, 1988, Rockville, MD: National Climate Program Office, 143–62.

Dow, K. and Cutter, S.L. 1997 Repeat response to hurricane evacuation orders. *Quick Response Report* no. 101, Boulder, CO: Natural Hazards Research and Applications Information Center.

Drabek, T.E. 1991 *Microcomputers in Emergency Management* Monograph no. 51, Boulder, CO: Institute of Behavioral Science, University of Colorado.

Drabek, T.E. 1995 Disaster responses within the tourist industry. *International Journal of Mass Emergencies and Disasters* 13: 7–23.

Drabek, T.E. and Boggs, K.S. 1968 Families in disaster: reactions and relatives. *Journal of Marriage and the Family* 30: 443–51.

Drogaris, G. 1991 *Major Accident Reporting System: Lessons Learned from Accidents Notified*. Community Documentation Centre on Industrial Risk.

D'Souza, F. 1984 Disaster research: ten years on. *Ekistics* 51 (309): 496–99.

Dunlap, R.E. and Scarce, R. 1991 The polls-poll trends: environmental problems and protection. *Public Opinion Quarterly* 55: 651–72.

Dymon, U.J. 1999 Effectiveness of geographic information systems (GIS) applications in flood management during and after hurricane 'Fran'. *Quick Response Report* no. 114, Boulder, CO: Natural Hazards Research and Applications Information Center.

Earthquake Engineering Research Institute (EERI), 1986 *Reducing Earthquake Hazards: Lessons Learned from Earthquakes*. Publication no. 86–02, El Cerrito, CA: Earthquake Engineering Research Institute.

Echevarria, J.A., Norton, K.A. and Norton, R.D. 1986 The socio-economic consequences of earthquake prediction: a case study in Peru. *Earthquake Prediction Research* 4: 175–93.

Economic Commission for Latin America and the Caribbean 1992a *Economic Impacts of the Eruption of the Cerro Negro Volcano in Nicaragua*. Paper LC/L. 686, Geneva: United Nations.

Economic Commission for Latin America and the Caribbean 1992b *The Tsunami of September 1992 in Nicaragua and its Effects on Development*. Paper LC/L.708, Geneva: United Nations.

Eldar, R. 1992 The needs of elderly persons in natural disasters: observations and recommendations. *Disasters* 16: 355–8.

Elo, O. 1994 Making disaster reduction a priority in public policy. *STOP Disasters* 17: 14–15.

El-Sabh, M.I. and Murty, T.S. (eds) 1988 *Natural and Man-made Hazards*. Dordrecht: D. Reidel.

Elsom, D.M. 1993 Deaths caused by lightning in England and Wales, 1852–1990. *Weather* 48: 83–90.

Elsom, D.M. and Webb, J.D.C. 1993 Destructive hailstorms in Essex on 26 May 1985. *Weather* 48: 166–73.

Emanuel K.A. 1987 The dependence of hurricane intensity on climate. *Nature* 326: 483–5.

Emel, J. and Peet, R. 1989 Resource management and natural hazards. In Peet, R. and Thrift, N. (eds), *New Models in Geography*, vol. 1, London: Unwin and Hyman, 49–76.

Emergency Preparedness Canada 1998 *A National Mitigation Policy: Executive Summary*, Ottawa: Emergency Preparedness Canada.

Emmi, P.C. and Horton, C.A. 1993 A GIS-based assessment of earthquake property damage and casualty risk: Salt Lake City, Utah. *Earthquake Spectra* 9: 11–33.

Ericksen, N.J. 1986 *Creating Flood Disasters? New Zealand's Need for a New Approach to Urban Flood Hazard*. Wellington: National Water and Soil Conservation Authority.

Fahmi, K.J. and Alabbasi, J.N. 1989 Seismic intensity zoning and earthquake risk mapping in Iraq. *Natural Hazards* 1: 331–40.

Falck, L.B. 1991 Disaster insurance in New Zealand. In Kreimer, A. and Munasinghe, M. (eds), *Managing Natural Disasters and Environment*, Washington, DC: Environment Department, World Bank, 120–5.

Federal Emergency Management Agency (FEMA) 1989 *Alluvial Fans: Hazards and Management*. Washington, DC: Federal Emergency Management Agency.

Fischer, G.W., Morgan, M.G., Fischhoff, B., Nair, I and Lave, L.B. 1991 What risks are people concerned about? *Risk Analysis* 11: 303–14.

Fischhoff, B., Lichtenstein, S., Slovic, P., Derby, S.L. and Keeney, R.L. 1981 *Acceptable Risk*. Cambridge: Cambridge University Press.

Fisher, H.W. 1996 What emergency management officials should know to enhance mitigation and effective disaster response. *Journal of Contingencies and Crisis Management* 4: 209–17.

Fitzsimmons, A. and Kirby, A. 1987 *Have Waste, Will Travel: An Examination of the Implications of High-level Nuclear Waste Transportation*. Working Paper no. 59, Boulder, CO: Institute of Behavioral Science, University of Colorado.

Flynn, J., Slovic, P. and Mertz, C.K. 1993a Decidedly different: expert and public views of risks from a radioactive waste repository. *Risk Analysis* 13: 643–8.

Flynn, J., Slovic, P. and Mertz, C.K. 1993b The Nevada initiative: a risk communication fiasco. *Risk Analysis* 13: 497–502.

Fordham, M.H. 1998 Making women visible in disasters: problematising the private domain. *Disasters* 22: 126–43.

Foster, H.D. 1975 The forgotten ingredient: paying nature's rent. *Habitat* 18: 18–21.

Fothergill, A. 1996 Gender, risk and disaster. *International Journal of Mass Emergencies and Disasters* 14: 33–56.

Fournier d'Albe, E.M. 1979 Objectives of volcanic monitoring and prediction. *Journal of the Geological Society of London* 136: 321–6.

Foxworthy, B.L. and Hill, M. 1982 *Volcanic Eruptions of 1980 at Mount St Helens: The First 100 Days*. Geological Survey Professional Paper 1249, Washington, DC: Government Printing Office.

Fratkin, E. 1992 Drought and development in Marsabit District, Kenya. *Disasters* 16: 119–30.

French, S.P. 1995 Damage to urban infrastructure and other public property from the 1989 Loma Prieta (California) earthquake. *Disasters* 19: 57–67.

Fritzsche, A.F. 1992 Severe accidents: can they occur only in the nuclear production of electricity? *Risk Analysis* 12: 327–9.

Frutiger, H. 1980 Swiss avalanche hazard maps. *Journal of Glaciology* 26 (94): 518–19.

Fujita, T.T. 1973 Tornadoes around the world. *Weatherwise* 26: 56–62, 79–83.

Fukuchi, T. and Mitsuhashi, K. 1983 Tsunami countermeasures in fishing villages along the Sanriku coast, Japan. In Iida, K. and Iwasaki, T. (eds), *Tsunamis*, Boston, MA: D. Reidel, 389–96.

Funaro-Curtis, R. 1982 *Natural Disasters and the Development Process: A Discussion of Ideas* OFDA, Washington, DC: Agency for International Development.

Garcia, R.V. and Escudero, J.C. 1982 *Drought and Man*, vol. 2: *The Constant Catastrophe: Malnutrition, Famines and Drought*. Oxford: Pergamon Press.

Gardner G.T. and Gould, L.C. 1989 Public perceptions of the risks and benefits of technology. *Risk Analysis* 9: 225–42.

Garner, A.C. and Huff, W.A.K. 1997 The wreck of Amtrak's Sunset Limited: news coverage of a mass transport disaster. *Disasters* 21: 4–19.

General Accounting Office 1992 *Federal Buildings: Many are Threatened by Earthquakes but Limited Action Has Been Taken*. Report to Congressional Committees GAO/GGD-92–62, Washington, DC: General Accounting Office.

Gentry, R.C. 1970 Hurricane Debbie modification experiments. *Science* 168: 473–5.

Gere, J.M. and Shah, H.C. 1984 *Terra Non Firma: Understanding and Preparing for Earthquakes*. New York: W.H. Freeman.

Gibbs, W.J. 1984 The great Australian drought: 1982–1983. *Disasters* 8: 89–104.

Gilvear, D.J. and Black, A.R. 1999 Flood-induced embankment failures on the river Tay: implications of climatically-induced hydrological change in Scotland. *Hydrological Sciences Journal* 44: 345–62.

Gilvear, D.J., Davies, J.R. and Winterbottom, S.J. 1994 Mechanisms of floodbank failure during large flood events on the rivers Tay and Earn, Scotland. *Quarterly Journal of Engineering Geology* 27: 319–32.

Glantz, M.H. 1982 Consequences and responsibilities in drought forecasting: the case of Yakima, 1977. *Water Resources Research* 18: 3–13.

Glickman. T.S. 1988 Hazardous materials routing: risk management or mismanagement? *Resources* 93: 11–13.

Glickman, T.S. and Sontag, M.A. 1995 The trade-offs associated with re-routing highway shipments of hazardous materials to minimise risk. *Risk Analysis* 15: 61–7.

Glickman, T.S., Golding, D. and Silverman, E.D. 1992 *Acts of God and Acts of Man: Recent Trends in Natural Disasters and Major Industrial Accidents*. Discussion Paper CRM 92–02, Washington, DC: Resources for the Future.

Glickman, T.S., Golding, D. and Terry K.S. 1993 *Fatal Hazardous Materials Accidents in Industry: Domestic and Foreign Experience from 1945–1991*. Washington, DC: Resources for the Future.

Goddard, J. 1976 The national increasing vulnerability to flood catastrophe. *Journal of Soil and Water Conservation* 31: 48–52.

Golden J.H. 2000 Tornadoes. In Pielke, R.A. Jr and Pielke, R.A. Sr (eds), *Storms*, London and New York: Routledge, vol. 2, 103–32.

Graham, N.E. 1977 Weather surrounding the Santa Barbara fire: 26 July 1977. *Weatherwise* 30 (4) 158–9.

Grainger, A. 1990 *The Threatening Desert: Controlling Desertification*. London: Earthscan.

Gray, W.M. 1990 Strong association between West African rainfall and US land-fall of intense hurricanes. *Science* 249: 1251–6.

Gray, W.M. and Landsea, C.W. 1992 African rainfall as a precursor of hurricane-related destruction on the US east coast. *Bulletin of the American Meteorological Society* 73: 1352–64.

Greenberg, M.R., Sachsman, D.B., Sandman, P.M. and Salomone, K.L. 1989 Network evening news coverage of environmental risk. *Risk Analysis* 9: 119–26.

Greene, J.P. 1994 Automated forest fire detection. *STOP Disasters* 18: 18–19.

Griggs, G.B. and Gilchrist, J.A. 1977 *The Earth and Land Use Planning*. North Scituate, MA: Duxbury Press.

Gross, E.M. 1991 The hurricane dilemma in the United States. *Episodes* 14: 36–45.

Gruber, U. and Haefner, H. 1995 Avalanche hazard mapping with satellite data and a digital elevation model. *Applied Geography* 15: 99–114.

Gruntfest, E. 1987 Warning dissemination and response with short lead times. In Handmer, J.W. (ed.), *Flood Hazard Management*, Norwich: Geo Books, 191–202.

Gupta, R.P. and Joshi, B.C. 1990 Landslide hazard zoning using the GIS approach: a case study from the Ramganga catchment, Himalayas. *Engineering Geology* 28: 119–32.

Haas, J.E., Kates, R.W. and Bowden, M.J. (eds) 1977 *Reconstruction Following Disaster*. Cambridge, MA: MIT Press.

Haines D.A. *et al.* 1983 Fire danger rating and wildfire occurrence in the north-eastern US. *Forest Science* 29: 679–96.

Hamilton, L.S. 1987 What are the impacts of Himalayan deforestation on the Ganges–Brahmaputra lowlands and delta? Assumptions and facts. *Mountain Research and Development* 7: 256–63.

Hammerton, J.L., George, G. and Pilgrim, R. 1984 Hurricanes and agric losses and remedial actions. *Disasters* 8: 279–86.

Handmer, J.W. 1987a The flood problem in perspective. In Handmer J.W. (ed.), *Flood Hazard Management*, Norwich: Geo Books, 9–32.

Handmer, J.W. 1987b Guidelines for floodplain acquisition. *Applied Geography* 7: 203–221.

Handmer, J.W. 1999 Natural and anthropogenic hazards in the Sydney sprawl: is the city sustainable? In Mitchell, J.K. (ed.), *Crucibles of Hazard*, Tokyo: United Nations University Press, 138–85.

Handmer, J.W. and Ord, K.D. 1986 Flood warning and response. In Smith, D.I. and Handmer, J.W. (eds), *Flood Warning in Australia*, Canberra: Centre for Resource and Environmental Studies, Australian National University, 235–57.

Hardy, D. 1988 A better adobe home for earthquake regions. *UNDRO News* July-August: 12–13.

Harlin, B.W. 1952 The great southern glaze storm of 1951. *Weatherwise* 5: 10–13.

Harmancioglu, N.B. 1994 Flood control by reservoirs. In Rossi, G., Harmancioglu, N and Yevjevich, V (eds), *Coping with Floods*, Dordrecht: Kluwer, 637–52.

Haskell, R.C. and Christiansen, J.R. 1985 Seismic bracing of equipment. *Journal of Environmental Sciences* 9: 67–70.

Hay, I. 1996 Neo-liberalism and criticisms of earthquake insurance arrangements in New Zealand. *Disasters* 20: 34–48.

Haylock, H.J.K. and Ericksen, N.J. 2000 From state dependency to self-reliance. In Wilhite, D.A. (ed.), *Drought*, London and New York: Routledge, vol. 2, 105–14.

Hazard Mitigation Team 1994 *Southern California Firestorms*. FEMA-1005–DR-CA Report. San Francisco, CA: FEMA.

Hazarika, S. 1988 *Bhopal: The Lessons of a Tragedy*. New Delhi: Penguin Books.

Healey, D.T., Jarrett, F.G. and McKay, J.M. 1985 *The Economics of Bushfires: The South Australian Experience*. Melbourne: Oxford University Press.

Health and Safety Executive 1978 *Canvey: An Investigation of Potential Hazards from Operations in the Canvey Island/Thurrock Area*. London: HMSO.

Hearn, G. and Jones, D.K.C. 1987 Geomorphology and mountain highway design: some lessons from the Dharan–Dhankuta highway, east Nepal. In V. Gardiner (ed.), *International Geomorphology 1986*, Proceedings of the First International Conference on Geomorphology, Chichester: John Wiley, 203–19.

Hendrickson, D., Armon, J. and Mearns, R. 1998 The changing nature of conflict and famine vulnerability: the case of livestock raiding in Turkana District, Kenya. *Disasters* 22: 185–99.

Hestnes, E. and Lied, K. 1980 Natural-hazard maps for land-use planning in Norway. *Journal of Glaciology* 26 (94): 331–43.

Hewitt, K. (ed.) 1983 *Interpretations of Calamity*. Boston, MA, and London: Allen and Unwin.

Hewitt, K. 1992 Mountain hazards. *Geojournal* 27: 47–60.

Hewitt, K. 1997 *Regions of Risk: A Geographical Introduction to Disasters*. London: Longman.

Hewitt, K. and Burton, I. 1971 *The Hazardousness of a Place: A Regional Ecology of Damaging Events*. Toronto: Department of Geography, University of Toronto.

Heyman, B.N., Davis, C. and Krumpke, P.F. 1991 An assessment of worldwide disaster vulnerability. *Disaster Management* 4: 3–14.

Hill, B.T. 1998 *Catastrophic Wildfires Threaten Resources and Communities.* GAO/ T-RCED-98–273, US Washington, DC: General Accounting Office.

Hinman, G.W., Rosa, E.A., Kleinhesselink, R.R. and Lowinger, T.C. 1993 Perceptions of nuclear and other risks in Japan and the United States. *Risk Analysis* 14: 449–55.

Hobbs, J.E. and Lawson, S. 1982 The tropical cyclone threat to the Queensland Gold Coast. *Applied Geography* 2: 207–19.

Hodge, D., Sharp, V. and Marts, M. 1979 Contemporary responses to volcanism: case studies from the Cascades and Hawaii. In Sheets, P.D. and Grayson D.K. (eds), *Volcanic Activity and Human Ecology*, London: Academic Press, 221–48.

Hohenemser, C., Kates, R.W. and Slovic, P. 1983 The nature of technological hazard. *Science* 220: 378–84.

Hollis, G.E. 1975 The effect of urbanisation on floods of different recurrence interval. *Water Resources Research* 11: 431–4.

Holway, J.M. and Burby, R.J. 1993 Reducing flood losses: local planning and land use controls. *Journal of the American Planning Association* 59: 205–16.

Holzer, T.L. 1994 Loma Prieta damage largely attributed to enhanced ground shaking. *EOS: Transactions of the American Geophysical Union* 75 (26): 299–301.

Hoque, B.A., Sack, R.B., Siddiqui, M., Jahangir, A.M., Hazera, N. and Nahid, A. 1993 Environmental health and the 1993 Bangladesh cyclone. *Disasters* 17: 144–52.

Horikawa, K. and Shuto, N. 1983 Tsunami disasters and protection measures in Japan. In Iida, K. and Iwasaki, T. (eds), *Tsunamis*, Boston, MA: D. Reidel, 9–22.

Housner, G.W. (chair) 1987 *Confronting Natural Disasters: An International Decade for Natural Hazard Reduction.* Washington, DC: National Academy Press.

Housner, G.W. 1989 An international decade for natural disaster reduction, 1990–2000. *Natural Hazards* 2: 45–75.

Hoyt, W.G. and Langbein, W.B. 1955 *Floods.* Princeton, NJ: Princeton University Press.

Hughes, P. 1979 The great Galveston hurricane. *Weatherwise* 32: 148–56.

Hupp, C.R., Osterkamp, W.R. and Thornton, J.L. 1987 *Dendrogeomorphic Evidence and Dating of Recent Debris Flows on Mount Shasta, Northern California.* US Geological Survey Professional Paper 1396–B, Washington, DC: US Geological Survey.

IDNDR 1996 *Cities at Risk: Making Cities Safer Before Disaster Strikes.* Supplement 28, Geneva: STOP Disasters.

Iida, K. 1983 Some remarks on the occurrence of tsunamigenic earthquakes around the Pacific. In Iida. K. and Iwasaki, T. (eds), *Tsunamis*, Boston, MA: D. Reidel, 61–76.

Institution of Civil Engineers 1995 *Megacities: Reducing Vulnerability to Natural Disasters.* London: Thomas Telford.

International Federation of Red Cross and Red Cresent Societies (IFRCRCS) 1993 *World Disasters Report 1993.* Dordrecht: Martinus Nijhoff.

IFRCRCS 1994 *World Disasters Report 1994.* Dordrecht: Martinus Nijhoff.

IFRCRCS 1998 *World Disasters Report 1998.* Dordrecht: Martinus Nijhoff.

IFRCRCS 1999 *World Disasters Report 1999*. Geneva: International Federation of Red Cross and Red Crescent Societies.

Ives, J.D. and Messerli, B. 1989 *The Himalayan Dilemma: Reconciling Development and Conservation*. London: Routledge.

Ives, J.D., Mears, A.I., Carrara, P.E. and Bovis, M.J. 1976 Natural hazards in mountain Colorado. *Annals of the Association of American Geographers* 66: 129–144

Jiminez Dias, V. 1992 Landslides and the squatter settlements of Caracas. *Environment and Urbanisation* 4: 80–9.

Johnson, K. 1985 *State and Community during the Aftermath of Mexico City's November 19, 1984, Gas Explosion*. Special Publication no. 13, Boulder, CO: Institute of Behavioral Science, University of Colorado.

Jones, D.K.C. 1992 Landslide hazard assessment in the context of development. In McCall, G.J.H., Laming, D.J.C. and Scott, S.C. (eds), *Geohazards*, London: Chapman and Hall, 117 41.

Jones, D.K.C. 1993 Environmental hazards in the 1990s: problems, paradigms and prospects. *Geography* 339: 161–5.

Jones, D.K.C. 1995 The relevance of landslide hazard to the International Decade for Natural Disaster Reduction. In *Landslides Hazard Mitigation with Particular Reference to Developing Countries*, Proceedings of a Conference. London: Royal Academy of Engineering, 19–33.

Jones, D.K.C., Lee, E.M., Hearn, G.J. and Genc, S. 1989 The Catak landslide disaster, Trabzon province, Turkey. *Terra Nova* 1: 84–90.

Jones, F.O. 1973 Landslides of Rio de Janeiro and the Serra das Araras escarpment. Professional Paper 697, US Geological Survey.

Jones-Lee, M.W., Hammerton, M. and Philips, P.R. 1985 The value of safety: the results of a national survey. *Economic Journal* 95: 49–72.

Joshua, S.C. and Garber, N.J. 1992 A causal analysis of large vehicle accidents through fault-free analysis. *Risk Analysis* 12: 173–87.

Jowett, A.J. 1989 China: the demographic disaster of 1958–1961. In Clarke, J.I., Curson, P., Kayastha, S.L. and Nag, P. (eds), *Population and Disaster*, Oxford: Basil Blackwell, 137–58.

Kajoba, G.M. 1992 Food security and the impact of the 1991–92 drought in Zambia. Unpublished text of lecture delivered at the University of Stirling, October.

Karanci, A.N. and Rustemli, A. 1995 Psychological consequences of the 1992 Erzincan (Turkey) earthquake. *Disasters* 19: 8–18.

Karter, M.J. 1992 *Fire Loss in the United States during 1991*. Quincy, MA: Fire Analysis and Research Division, National Fire Protection Association.

Kasperson, R.E., Renn, O., Slovic, P., Brown, H.S., Emel, J., Goble, R., Kasperson, J.X. and Ratick, S. 1988 The social amplification of risk: a conceptual framework. *Risk Analysis* 8: 177–87.

Kates, R.W. 1962 *Hazard and Choice Perception in Flood Plain Management*. Paper 78, Chicago, IL: Department of Geography, University of Chicago.

Kates, R.W. 1971 Natural hazard in human ecological perspective: hypotheses and models. *Economic Geography* 47: 438–51.

Kates, R.W. 1978 *Risk Assessment of Environmental Hazard*. SCOPE report 8, New York: John Wiley.

Kates, R.W. and Kasperson, J.X. 1983 Comparative risk analysis of technological hazards (a review). *Proceedings of National Academy of Science USA* 80: 7027–38.

Keaton, J.R. 1994 Risk-based probabilistic approach to site selection. *Bulletin of the Association of Engineering Geologists* 31: 217–29.

Keefer, D.K. 1984 Landslides caused by earthquakes. *Bulletin of the Geological Society of America* 95: 406–21.

Keefer, D.K., Wilson, R.C., Mark, R.K., Brabb. E.E., Brown, W.M., Ellen, S.D., Harp, E.L., Wieczorek, G.F., Alger, C.S. and Zatkin, R.S. 1987 Real-time landslide warning during heavy rainfall. *Science* 238: 921–5.

Keeney, R.L. 1995 Understanding life-threatening risks. *Risk Analysis* 15: 627–37.

Keeney, R.L. and von Winterfeldt, D. 1994 Managing nuclear waste from power plants. *Risk Analysis* 14: 107–8.

Kelly, M. 1992 Anthropometry as an indicator of access to food in populations prone to famine. *Food Policy* 17: 443–54.

Kelly, M. and Buchanan-Smith, M. 1994 Northern Sudan in 1991: food crisis and the international relief response. *Disasters* 18: 16–34.

Key, D. (ed.) 1995 *Structures to Withstand Disaster*. London: Institution of Civil Engineers, Thomas Telford.

Kobayashi, Y. 1981 Causes of fatalities in recent earthquakes in Japan. *Journal of Disaster Science* 3: 15–22.

Kockelman, W.J. 1986 Some techniques for reducing landslide hazards. *Bulletin of the Association of Engineering Geologists* 23: 29–52.

Krall, S. 1995 Desert locusts in Africa: a disaster? *Disasters* 19: 1–7.

Kreimer, A. and Munasinghe, M. (eds) 1991 *Managing Natural Disasters and the Environment*. Washington, DC: Environment Department, World Bank.

Krewski, D., Clayson, D. and McCullough, R.S. 1982 Identification and measurement of risk. In Burton I., Fowle, C.D. and McCullough, R.S. (eds), *Living with Risk*, Environmental Monograph 3, Toronto: Institute of Environmental Studies, University of Toronto, 7–23.

Kunreather, H. 1978 *Disaster Insurance Protection: Public Policy Lessons*. New York: John Wiley.

Kunreather, H. 1995 The role of insurance in reducing losses from natural hazards. In Munasinghe, M. and Clarke, C. (eds), *Disaster Prevention for Sustainable Development*, Washington, DC: IDNDR and World Bank, 87–102.

Kusler, J. and Larson, L. 1993 Beyond the ark: a new approach to US floodplain management. *Environment* 35: 7–34.

Lagadec, P. 1982 *Major Technological Risk. An Assessment of Industrial Disasters*. Oxford: Pergamon Press.

Lagadec P. 1987 From Seveso to Mexico and Bhopal: learning to cope with crises. In Kleindorfer, P.R. and Kunreuther, H.C. (eds), *Insuring and Managing Hazardous Risks*, New York: Springer-Verlag, 13–27.

Landsea, C.W. 2000 Climate variability of tropical cyclones. In Pielke, R.A. Jr and Pielke, R.A. Sr (eds), *Storms*, London and New York: Routledge, vol. 1, 220–41.

Landsea, C.W., Gray, W.M., Mielke, P.W. and Berry, K.J. 1994 Seasonal forecasting of Atlantic hurricane activity. *Weather* 49: 273–84.

Lechat, M.F. 1990 The International Decade for Natural Disaster Reduction: background and objectives. *Disasters* 14: 1–6.

Lecomte, E. 1989 Earthquakes and the insurance industry. *Natural Hazards Observer* 14: 1–2.

Le Guern, F., Tazieff, H. and Faivre Pierret, R. 1982 An example of health hazard: people killed by gas during the phreatic eruption, Dieng Plateau (Java, Indonesia). *Bulletin of Volcanology* 45 (2): 153–6.

Leimena, S.L. 1980 Traditional Balinese earthquake-proof housing structures. *Disasters* 4: 147–50.

Lewis, H.W. 1990 *Technological Risk*. New York: W.H. Norton.

Lockett, J.E. 1980 Catastrophes and catastrophe insurance. *Journal of the Institute of Actuaries Student's Society* 24: 91–134.

Lockridge, P. 1985 Tsunamis: the scourge of the Pacific. *UNDRO News* (Jan/Feb): 14–17.

Lockwood, J.P. and Torgerson, F.A. 1980 Diversion of lava flows by aerial bombing: lessons from Mauns Loa volcano, Hawaii. *Bulletin of Volcanology* 43: 727–41.

Lott, J.N. 1994 The US summer of 1993: a sharp contrast in weather extremes. *Weather* 49: 370–83.

Lucas-Smith, P. and McRae, R. 1993 Fire risk problems in Australia. *STOP Disasters* 11: 3–4.

Luke, R.H. and McArthur, A.G. 1978 *Bushfires in Australia*. Canberra: Australian Government Publishing Service.

Lumb, P. 1975 Slope failures in Hong Kong. *Quarterly Journal of Engineering Geology* 8: 31–65.

McDaniels, T.L., Kamlet, M.S. and Fischer, G.W. 1992 Risk perception and the value of safety. *Risk Analysis* 12: 495–503.

McGranahan, G. and Songsore, J. 1994 Wealth, health and the urban household: Weighing environmental burdens in Accra, Jakarta and Sao Paulo. *Environment* 36: 4–11, 40–5.

MacGregor, D., Slovic, P., Mason, R.G., Detweiler, J., Binney, S.E. and Dodd, B. 1994 Perceived risks of radioactive waste transport through Oregon: results of a statewide survey. *Risk Analysis* 14: 5–14.

McKay, J.M. 1983 Newspaper reporting of bushfire disaster in south-eastern Australia: Ash Wednesday 1983. *Disasters* 7 (4): 283–90.

McKee, C.O., Johnson, R.W., Lowenstein, P.L., Riley, S.J., Blong R.J., de Saint Ours, P. and Talai, B. 1985 Volcanic hazards, surveillance and eruption contingency planning. *Journal of Volcanology and Geothermal Research* 23: 195–237.

McNutt, S.R. 1996 Seismic monitoring and eruption forecasting of volcanoes: a review of the state of the art and case histories. In Scarpa, R. and Tilling, R.I. (eds), *Monitoring and Mitigation of Volcano Hazards*, Berlin: Springer-Verlag, 99–146.

Malingreau, J.P. and Kasawanda, X. 1986 Monitoring volcanic eruptions in Indonesia using weather satellite data: the Colo eruption of July 28 1983. *Journal of Volcanology and Geothermal Research* 27: 179–94.

Malmquist, D.L. and Michaels, A.F. 2000 Severe storms and the insurance industry. In Pielke, R.A. Jr and Pielke, R.A. Sr (eds), *Storms*, London and New York: Routledge, vol. 1, 54–69.

Marco, J.B. 1994 Flood risk mapping. In Rossi, G., Harmancoglin, N. and Yevjevich, V. (eds), *Coping with Floods*, Dordrecht: Kluwer, 353–73.

Marin, A. 1986 Evaluating the nation's risk assessors: nuclear power and the value of life. *Public Money* 6: 41–5.

Marsh, G.P. 1864 *Man and Nature*. New York: Charles Scribner.

Martens, P., McMichael, A., Kovats, S. and Livermore, M. 1998 Impacts of climate change on human health: malaria. In *Climate Change and Its Impacts*, Bracknell, Berks.: Meteorological Office and Department of Energy, Transport and the Regions, 10.

Maskrey, A. 1989 *Disaster Mitigation: A Community Based Approach*. Development Guidelines no. 3, Oxford: Oxfam.

Mason, J. and Cavalie, P. 1965 Malaria epidemic in Haiti following a hurricane. *American Journal of Tropical Medicine and Hygiene* 14: 533–9.

Mathur, K. and Jayal, N.G. 1992 Drought management in India: the long term perspective. *Disasters* 16: 60–5.

Mears, A.I. 1984 Municipal avalanche zoning. *Ekistics* 51 (309): 539–42.

Mejía-Navarro, M. and Garcia, L.A. 1996 Natural hazard and risk assessment using decision-support systems: Glenwood Springs, Colorado. *Environmental and Engineering Geoscience* 1: 291–98.

Meltsner, A.J. 1978 Public support for seismic safety: where is it in California? *Mass Emergencies* 3: 167–84.

Mercer, D. 1971 Scourge of an arid continent. *Geographical Magazine* 45: 563–7.

Mileti, D.S. and Darlington, J.D. 1995 Societal responses to revised earthquake probabilities in the San Francisco Bay area. *International Journal of Mass Emergencies and Disasters* 13: 119–45.

Mileti, D.S., Darlington, J.D., Passerini, E., Forrest, B.C. and Myers, M.F. 1995 Toward an integration of natural hazards and sustainability. *Environmental Professional* 17: 117–26.

Mileti, D.S. *et al.* 1999 *Disasters by Design: A Reassessment of Natural Hazards in the United States*. Washington, DC: Joseph Henry Press.

Miller, C.D., Mullineaux, D.R. and Crandell, D.R. 1981 Hazards assessments at Mount St Helens. In Lipman, R.W. and Mullineaux, D.R. (eds), *The 1980 Eruption of Mount St Helens, Washington*, US Geological Survey Professional Paper 1250, 789–802.

Ministry of Transportation and Highways 1981 *Snow Avalanche Atlas (Golden East-Toby Creek)*. Victoria, BC: Snow Avalanche Section, Province of British Columbia.

Mitchell J.K. 1989 *Where Might the International Decade for Natural Disaster Reduction Concentrate its Activities?* Paper no. IDNDR2, Boulder, CO: Natural Hazards Research and Applications Center.

Mitchell J.K. 1990 Human dimensions of environmental hazards. In Kirby, A. (ed.), *Nothing to Fear*, Tucson: University of Arizona Press, 131–75.

Mitchell, J.K. 1999 Natural disasters in the context of mega-cities. In Mitchell, J.K. (ed.), *Crucibles of Hazard*, Tokyo: United Nations University Press, 15–55.

Mittler, E. 1997 A case study of Florida's homeowners' insurance since hurricane 'Andrew'. *Working Paper* no. 96, Boulder, CO: Natural Hazards Research and Applications Information Center.

Mizutani, T. and Nakano, T. 1989 The impact of natural disasters on the population of Japan. In Clarke, J.I., Curson, P., Kayastha, S.L. and Nag, P. (eds), *Population and Disaster*, Oxford: Basil Blackwell, 24–33.

Mogil, H.M. *et al.* 1984 The Great Freeze of '83: analysing the causes and the effects. *Weatherwise* 7: 304–8.

Monteverdi, J.P. 1973 The Santa Ana weather type and extreme fire hazard in the Oakland-Berkeley Hills. *Weatherwise* 26: 118–21.

Montz, B.E. 1992 The effects of flooding on residential property values in three New Zealand communities. *Disasters* 16: 283–98.

Montz, B. and Grundfest, E.C. 1986 Changes in American floodplain occupancy since 1958: the experiences of nine cities. *Applied Geography* 6: 325–38.

Moore, P.G. 1983 *The Business of Risk*. Cambridge: Cambridge University Press.

Morimiya, Y. 1985 Covering natural disasters in the Japanese market. *Risk Management* 32: 18–26.

Morris, J., West, G., Holck, S., Blake, P., Echeverria, P. and Karaulnik, M. 1982 Cholera among refugees in Rangsit, Thailand. *Journal of Infectious Diseases* 1: 131–4.

Morrow, B.H. 1999 Identifying and mapping community vulnerability. *Disasters* 23: 1–18.

Morton, A. 1998 Hong Kong: managing slope safety in urban systems. *STOP Disasters* 33: 8–9.

Mothes, P.A. 1992 Lahars of Cotopaxi volcano, Ecuador: hazard and risk evaluation. In McCall, G.J.H., Laming, D.J.C. and Scott, S.C. (eds), *Geohazards*, London: Chapman and Hall, 53–63.

Mulady, J.J. 1994 Building codes: they're not just hot air. *Natural Hazards Observer* 18: 4–5.

Munich Re 1999 *Topics 2000*. Report of the Geoscience Research Group, Munich. Munich Reinsurance Company.

Nakano, T. and Matsuda, I. 1984 Earthquake damage, damage prediction and counter measures in Tokyo, Japan. *Ekistics* 51: 415–20.

Nash, J.R. 1976 *Darkest Hours: A Narrative Encyclopaedia of Worldwide Disasters from Ancient Times to the Present*. Chicago, IL: Nelson Hall.

National Association of Independent Insurers 1994 *Mitigating Catastrophic Property Insurance Losses*. Des Plaines, IL: National Association of Independent Insurers.

National Wildlife Federation 1998 *Higher Ground: A Report on Voluntary Property Buyouts in the Nation's Floodplains*. Vienna, VA: National Wildlife Federation.

Neal J. and Parker, D.J. 1988 *Floodplain Encroachment: a Case Study of Datchet, UK*. Geography and Planning Paper no. 22, Enfield, Middx: Middlesex Polytechnic.

Newhall, C.G. and Self, S. 1982 The Volcanic Explosivity Index (VEI): an estimate of explosive magnitude for historical volcanism. *Journal of Geophysical Research* 87: 1231–8.

Newman, C.J. 1976 Children of disaster: clinical observations at Buffalo Creek. *American Journal of Psychiatry* 133: 306–12.

Nicholls, R.J. 1998 Impacts on coastal communities. In *Climate Change and its Impacts*, Blackwell, Berks.: Meteorological Office and Department of Energy Transport and the Regions, 9.

Nicholls, R.J., Hoozemans, F.M.J. and Marchand, M. 1999 Increasing flood risk and wetland losses due to global sea-level rise: regional and global analyses. *Global Environmental Change* 9: S69–S87.

Oaks, S.D. and Bender, S.O. 1990 Hazard reduction and everyday life: opportunities for integration during the Decade for Natural Disaster Reduction. *Natural Hazards* 3: 87–9.

Oaks, S.D. and Dexter, L. 1987 Avalanche hazard zoning in Vail, Colorado: the use of scientific information in the implementation of hazard reduction strategies. *Mountain Research and Development* 7: 157–68.

Oechsli, F.W. and Buechly, R.W. 1970 Excess mortality associated with three Los Angeles September hot spells. *Environmental Research* 3: 277–84.

Office of US Foreign Disaster Assistance (OFDA) 1994 *Annual Report Financial Year 1993*. Washington, DC: Office of US Foreign Disaster Assistance, Agency for International Development.

Okrent, D. 1980 Comment on societal risk. *Science* 208: 372–5.

Ollier, C. 1988 *Volcanoes*. Oxford: Basil Blackwell.

Olshansky, R.B. and Rogers, J.D. 1987 Unstable ground: landslide policy in the United States. *Ecology Law Quarterly* 13: 939–1006.

Omachi, T. 1997 *Drought Conciliation and Water Rights: Japanese Experience*. Water Series no. 1, Tokyo: Infrastructure Development Institute.

O'Meagher, B., Stafford-Smith, M. and White, D.H. 2000 Approaches to integrated drought risk management. In Wilhite, D.A. (ed.), *Drought*, London and New York: Routledge, vol. 2, 115–28.

Otero, R.B. and Martí, R.Z. 1995 The impacts of natural disasters on developing economies: implications for the international development and disaster community. In Munasinghe, M. and Clarke, C. (eds), *Disaster Prevention for Sustainable Development*, Washington, DC: IDNDR and World Bank, 11–35.

Othman-Chande, M. 1987 The Cameroon volcanic gas disaster: an analysis of a makeshift response. *Disasters* 11: 96–101.

Owen, J.A. and Ward, M.N. 1989 Forecasting Sahel rainfall. *Weather* 44: 57–64.

Palm, R. 1981 Public response to earthquake hazard information. *Annals of the Association of American Geographers* 71: 389–99.

Palm, R.I. 1990 *Natural Hazards: An Integrative Framework for Research and Planning*. Baltimore, MD: Johns Hopkins University Press.

Palmer, W.C. 1965 *Meteorological Drought*. Research Paper 45. Washington, DC: US Weather Bureau, Department of Commerce.

Pappas, R.G. 1978 The 1977–1978 Southern Californian winter. *Mariners Weather Log* 22: 317–24.

Parasuraman, S. 1995 The impact of the 1993 Latur-Osmanabad (Maharashtra) earthquake on lives, livelihoods and property. *Disasters* 19: 156–69.

Parker, D. 1999 Criteria for evaluating the condition of a tropical cyclone warning system. *Disasters* 23: 193–216.

Parker, D. and Mitchell, J.K. 1995 Disaster vulnerability of mega-cities: an expanding problem that requires rethinking and innovative responses. *Geojournal* 37: 295–301.

Parker, D.E. and Horton, E.B. 1999 Global and regional climate in 1998. *Weather* 54: 173–84.

Parker, K.T., Lord, W.B.H., Read, N.J. and Parsons, J. 1986 *Social and Economic Responses to Climatic Variability in the UK*. London: Technical Change Centre.

Paté-Cornell, M.E. 1993 Learning from the Piper Alpha accident: a post-mortem analysis of technical and organizational factors. *Risk Analysis* 13: 215–32.

Patel, T. 1995 Satellite senses risk of forest fires. *New Scientist* (11 March): 12.

Paul, B.K. 1995 Farmers' and public responses to the 1994–95 drought in Bangladesh: A case study. *Quick Response Report* no. 76. Boulder, CO: Natural Hazards Research and Applications Information Center.

Paul, B.K. 1997 Survival mechanisms to cope with the 1996 tornado in Tangail, Bangladesh: a case study. *Quick Response Report* no. 92. Boulder, CO: Natural Hazards Research and Applications Information Center.

Peacock, W.G., Morrow, B.H. and Gladwin, H. (eds) 1997 *Hurricane Andrew: Ethnicity, Gender and the Sociology of Disasters* London and New York: Routledge.

Pearce, F. 1988 Cool oceans caused floods in Bangladesh and Sudan. *New Scientist* (8 September): 31.

Penn, S. 1984 Colour-enhanced infra-red photgraphy of landslips. *Quarterly Journal of Engineering Geology* 17: 2–5.

Penning-Rowsell, E.C. 1986 The development of integrated flood warning systems. In Smith, D.I. and Handmer, J.W. (eds), *Flood Warning in Australia*, Canberra: Centre for Resource and Environmental Studies, Australia National University, 15–36.

Penning-Rowsell, E.C. 1987 The power behind the flood scene. In Handmer, J.W. (ed.), *Flood Hazard Management*, Norwich: Geo Books, 61–73.

Penning-Rowsell, E.C., Chatterton, B.J. and Parker, D.J. 1978 *The Effect of Flood Warning on Flood Damage Reduction*. Report for the Central Water Planning Unit, Middlesex Polytechnic, London.

Perla, R.I. 1978 Failure of snow slopes. In Voight, B. (ed.), *Rockslides and Avalanches*, vol. 1: *Natural Phenomena*, Amsterdam: Elsevier, 23–32.

Perla, R.I. and Martinelli, M. Jr. 1976 *Avalanche Handbook*. Agriculture Handbook 489. Washington, DC: US Department of Agriculture (Forest Service).

Perrow, C. 1984 *Natural Accidents: Living with High-risk Technologies*. New York: Basic Books.

Perry, A.H. and Symons, L.J. (eds) 1991 *Highway Meteorology*. London: E. and F.N. Spon.

Perry, C.A. 1994 *Effects of Reservoirs on Flood Discharges in the Kansas and the Missouri River Basins, 1993*. Circular 1120E. Denver, CO: US Geological Survey.

Perry, R.W. and Lindell, M.K. 1990 Predicting long-term adjustment to volcano hazard. *International Journal of Mass Emergencies and Disasters* 8: 117–36.

Petak, W.J. 1984 Geologic hazard reduction: the professional's responsibility. *Bulletin of the Association of Engineering Geologists* 21: 449–58.

Peterson, D.W. 1996 Mitigation measures and preparedness plans for volcanic emergencies. In Scarpa, R. and Tilling, R.I. (eds), *Monitoring and Mitigation of Volcano Hazards*, Berlin: Springer-Verlag, 701–18.

Pielke, R.A. Jr 1997 Reframing the US hurricane problem. *Society and Natural Resources* 10: 485–99.

Pielke, R.A. Jr and Landsea, C.W. 1998 Normalised hurricane damages in the United States 1925–95. *Weather and Forecasting* 13: 621–31.

Pielke, R.A. Jr and Pielke, R.A. Sr 1997 *Hurricanes: Their Nature and Impacts on Society*. Chichester and New York: John Wiley.

Pielke, R.A. Jr and Pielke, R.A. Sr 2000 *Storms*, vols 1 and 2, London and New York: Routledge.

Pierson, T.C. 1989 Hazardous hydrologic consequences of volcanic eruptions and goals for mitigative action: an overview. In Starosolszky, O. and Melder, O.M. (eds), *Hydrology of Disasters*, London: James and James, 220–36.

Pigram, J.J. 1986 *Issues in the Management of Australia's Water Resources*. Melbourne: Longman Cheshire.

Plafker, G. and Eriksen, G.E. 1978 Nevados Huascaran avalanches, Peru. In Voight, B. (ed.), *Rockslides and Avalanches*, vol. 1: *Natural Disasters*, Amsterdam: Elsevier, 48–55.

Platt, R.H. 1999 Natural hazards of the San Francisco Bay mega-city: trial by earthquake, wind and fire. In Mitchell, J.K. (ed.), *Crucibles of Hazard*, Tokyo: United Nations University Press, 335–74.

Platt, R.H. and McMullen, G.M. 1980 *Post-flood Recovery and Hazard Mitigation: Lessons from the Massachusetts Coast, February 1978*. Publication no. 115. Amherst, MA: Water Resources Research Center, University of Massachusetts.

Pomonis, A., Coburn, A.W. and Spence, R.J.S. 1993 Seismic vulnerability, mitigation of human casualties and guidelines for low-cost earthquake resistant housing. *STOP Disasters* 12, March-April: 6–8.

Porfiriev, B.N. 1992 The environmental dimensions of national security: a test of systems analysis methods. *Environmental Management* 16: 735–42.

Preuss, J. 1983 Land management guidelines for tsunami hazard zones. In Iida, K. and Iwasaki, T. (eds), *Tsunamis*, Boston, MA: Reidel, 527–39.

Purtill. A. 1983 A study of the drought. *Quarterly Review of the Rural Economy* 5: 3–11.

Pyle, A.S. 1992 The resilience of households to famine in El Fasher, Sudan, 1982–89. *Disasters* 16: 19–27.

Quarantelli, E.L. 1984 Perceptions and reactions to emergency warnings of sudden hazards. *Ekistics* 51 (309): 511–15.

Quarantelli, E.L. (ed.) 1998 *What is a Disaster?* London and New York: Routledge.

Quayle, R. and Doehring, F. 1981 Heat stress: a comparison of indices. *Weatherwise* 34: 120–4.

Rahn, P.H. 1984 Floodplain management program in Rapid City, South Dakota. *Bulletin of the Geological Society of America* 95: 838–43.

Rahn, P.H. 1994 Flood Plains. *Bulletin of the Association of Engineering Geologists* 31: 171–181.

Ramachandran, R. and Thakur, S.C. 1974 India and the Ganga floodplains. In White, G.F. (ed.) *Natural Hazards*, New York: Oxford University Press, 36–43.

Ramsli, G. 1974 Avalanche problems in Norway. In White, G.F. (ed.), *Natural Hazards*, New York: Oxford University Press, 175–80.

Rapanos, D. *et al.* 1981 *Floodproofing New Residential Buildings in British Columbia*. Victoria, BC: Ministry of Environment, Province of British Columbia.

Rappaport, E.N. 1994 Hurricane 'Andrew'. *Weather* 49: 51–61.

Rasid, H. and Pramanik, M.A.H. 1990 Visual interpretation of satellite imagery for monitoring floods in Bangladesh. *Environmental Management* 14: 815–21.

Ratcliffe, R.A.S. 1978 Meteorological aspects of the 1975–76 drought. *Proceedings of the Royal Society of London*, Series A, 363: 3–20.

Ray, P.S. and Burgess, D.W. 1979 Doppler radar: research at the National Severe Storms Laboratory. *Weatherwise* 32: 68–75.

Reyes, P.J.D. 1992 Volunteer observers program: a tool for monitoring volcanic and seismic events in the Philippines. In McCall, G.J.H., Laming, D.J.C. and Scott, S.C. (eds), *Geohazards*, London: Chapman and Hall, 13–24.

Riebsame, W.E., Diaz, H.F., Moses, T. and Price, M. 1986 The social burden of weather and climate hazards. *Bulletin of the American Meteorological Society* 67: 1378–88.

Rodrigue, C.M. and Rovai, E. 1995 The 'Northridge' earthquake: differential geographies of damage, media attention and recovery. *National Social Science Perspectives Journal* 7: 98–111.

Rogers, P., Lydon, P. and Seckler, D. 1989 *Eastern Waters Study: Strategies to Manage Flood and Drought in the Ganges–Brahmaputra Basin*. Washington, DC: US Agency for International Development.

Rohrmann, B. 1994 Risk perception of different societal groups: Australian findings and cross-national comparisons. *Australian Journal of Psychology* 46: 150–63.

Romme, W.H. and Despain, D.G. 1989 The Yellowstone fires. *Scientific American* 261: 37–46.

Rooney, J.F. 1967 The urban snow hazard in the United States: an appraisal of disruption. *Geographical Review* 57: 538–59.

Ropelewski, C.F. and Folland, C.K. 2000 Prospects for the prediction of meteorological drought. In Wilhite, D.A. (ed.), *Drought*, London and New York: Routledge, vol. 1, 21–40.

Rose, G.A. (ed.) 1994 *Fire Protection in Rural America: A Challenge for the Future*. Rural Fire Protection in America Steering Committee, Report to Congress sponsored by the National Association of State Foresters, Washington, DC.

Rothery, D.A., Francis, P.W. and Wood, C.A. 1988 Volcano monitoring using short wavelength infrared data from satellites. *Journal of Geophysical Research* 93: 7993–8008.

Royal Society 1992 *Risk: Analysis, Perception and Management*. Report of a Royal Society Study Group. London: Royal Society.

Rubin, C.B. 1998 What hazards and disasters are likely in the twenty-first century – or sooner? *Working Paper* no. 99. Boulder, CO: Natural Hazards Research and Applications Information Center.

Rubin, C.B., Yezer, A. M., Hussain, Q. and Webb, A. 1986 *Summary of Major Natural Disaster Incidents in the US 1965–1985*. Special Publication no. 17. Boulder, CO. Institute of Behavioral Science, University of Colorado.

Rural Fire Protection in America 1994 *Fire Protection in Rural America: A Challenge for the Future*. Washington, DC: Rural Fire Protection in America.

Russell, L.A., Goltz, J.D. and Bourque, L.B. 1995 Preparedness and hazard mitigation actions before and after two earthquakes. *Environment and Behaviour* 27: 744–70.

Sagan, L.A. 1984 Problems in health measurements for the risk assessor. In Ricci, P.F., Sagan, L.A. and Whipple, C.G. (eds), *Technological Risk Assessment*, The Hague: Martinus Nijhoff, 1–9.

Sanders, J.F. and Gyakum, J.R. 1980 Synoptic-dynamic climatology of the bomb. *Monthly Weather Review* 108: 1598–606.

Sapir, D.G. and Misson, C. 1992 The development of a database on disasters. *Disasters* 16: 74–80.

Sassa, K. 1992 Landslide volume–apparent friction relationship in the case of rapid loading on alluvial deposits. *Landslide News* 6: 16–19.

Sauchyn, D.J. and Trench, N.R. 1978 LANDSAT applied to landslide mapping. *Photogrammetric Engineering and Remote Sensing* 44: 735–41.

Scarpa, R. and Gasparini, P. 1996a A review of volcano geophysics and volcano-monitoring methods. In Scarpa, R. and Tilling, R.I. (eds), *Monitoring and Mitigation of Volcano Hazards*, Berlin: Springer-Verlag, 3–22.

Schaerer, P.A. 1981 Avalanches. In Gray, D.M. and Male, D.H. (eds), *Handbook of Snow*, Toronto: Pergamon, 475–518.

Schmidlin, T., King, P.S., Hammer, B.O. and Ono, Y. 1998 Risk factors for death in the 22–23 February 1998 Florida tornadoes. *Quick Response Report* no. 106. Boulder, CO: Natural Hazards Research and Applications Information Center.

Schuster, R.L. and Fleming, R.W. 1986 Economic losses and fatalities due to landslides. *Bulletin of the Association of Engineering Geologists* 23: 11–28.

Seaman, J., Leivesley, S. and Hogg, C. 1984 *Epidemiology of Natural Disasters. Contributions to Epidemiology and Biostatistics*, vol. 5, Basel: S. Karger.

Seeley, M.W. and West, D.O. 1990 Approach to geologic hazard zoning for regional planning, Inyo National Forest, California and Nevada. *Bulletin of the Association of Engineering Geologists* 27: 23–35.

Sehmi, N.S. 1989 The hydrology of disastrous floods in Asia: an overview. In Starosolszky, O. and Melder, O.M. (eds), *Hydrology of Disasters*, London: James and James, 106–22.

Showalter, P.S. and Myers, M.F. 1994 Natural disasters in the United States as release agents of oil, chemicals or radiological materials between 1980–1989: analysis and recommendations. *Risk Analysis* 14: 169–82.

Siebert, L. 1992 Threats from debris avalanches. *Nature* 356: 658–59.

Sigurdsson, H. 1988 Gas bursts from Cameroon crater lakes: a new natural hazard. *Disasters* 12: 131–46.

Sigurdsson, H. and Carey, S. 1986 Volcanic disasters in Latin America and the 13 November eruption of Nevado del Ruiz volcano in Colombia. *Disasters* 10: 205–16.

Simon, H.A. 1956 Rational choice and the structure of the environment. *Psychological Review* 63: 129–38.

Slovic, P. 1986 Informing and educating the public about risk. *Risk Analysis* 6: 280–85.

Slovic, P., Flynn, J.H. and Layman, M. 1991 Perceived risk, trust, and the politics of nuclear waste. *Science* 254: 1603–7.

Smets, H. 1987 Compensation for exceptional environmental damage caused by industrial activities. In Kleindorfer, P.R. and Kunreuther, H.C. (eds), *Insuring and Managing Hazardous Risks*, New York: Springer-Verlag, 79–138.

Smith, B.J. and de Sanchez, B.A. 1992 Erosion hazards in a Brazilian suburb. *Geographical Review* 6: 37–41.

Smith, D.I. 1989 A dam disaster waiting to break. *New Scientist* 11 November: 42–6.

Smith, D.I., Den Exter, P., Dowling, M.A., Jelliffe, P.A., Munro, R.G. and Martin, W.C. 1979 *Flood Damage in the Richmond River Valley, New South Wales*. Canberra: Centre for Resource and Environmental Studies, Australian National University.

Smith, K. 1997 Climatic extremes as hazards to humans. In Thompson, R.D. and Perry, A. (eds) *Applied Climatology: Principles and Practice*, London and New York: Routledge, 304–16.

Smith, K. and Ward, R. 1998 *Floods: Physical Processes and Human Impacts*. Chichester and New York: John Wiley.

Smith, W.D. and Berryman, K.R. 1986 Earthquake hazard in New Zealand: inferences from seismology and geology. *Bulletin of the Royal Society of New Zealand* 24: 223–42.

Soby, B.A., Ball, D.J. and Ives, D.P. 1993 Safety investment and the value of life and injury. *Risk Analysis* 13: 356–70.

Sokolowska, J. and Tyszka, T. 1995 Perception and acceptance of technological and environmental risks: Why are poor countries less concerned? *Risk Analysis* 15: 733–43.

Solis, G.Y., Hightower, H.C., Sussex, J. and Kawaguchi, J. 1996 *Disaster Debris Management*. Ottawa: Emergency Preparedness Canada.

Somers, E. 1995 Perspectives on risk management. *Risk Analysis* 15: 677–84.

Sommer, A. and Mosely, W.H. 1972 East Bengal cyclone of November 1970: epidemiological approach to disaster assessment. *Lancet* 1: 1029–36.

Southern, R.L. 2000 Tropical cyclone warning-response strategies. In Pielke, R.A. Jr and Pielke, R.A. Sr (eds), *Storms*, London and New York: Routledge, vol. 1, 259–305.

Spangle, W. and Associates Inc. 1988 *California at Risk: Steps to Earthquake Safety for Local Government*. Sacramento, CA: California Seismic Safety Commission.

Stark, K.P. and Walker, G.R. 1979 Engineering for natural hazards with particular reference to tropical cyclones. In Heathcote, R.C. and Thom, B.G. (eds), *Natural Hazards in Australia*, Canberra: Australian Academy of Science, 189–203.

Starosolszky, O. 1994 Flood control by levees. In Rossi, G., Harmancogliu, N. and Yevjevich, V. (eds), *Coping with Floods*, Dordrecht: Kluwer, 617–35.

Starr, C. 1969 Social benefit versus technological risk. *Science* 165: 1232–8.

Starr, C. 1979 General philosophy of risk-benefit analysis. In *Perspectives on Benefit-Risk Decision Making*. US National Academy of Engineering Report and Stanford University; quoted in Moore, P.G., *The Business of Risk*, Cambridge: Cambridge University Press.

Starr, C. and Whipple, C. 1980 Risk of risk decisions. *Science* 208: 1114–9.

Stephenson, R. and Anderson, P.S. 1997 Disasters and the information technology revolution. *Disasters* 21: 305–34.

Stockwell. J.R., Sorenson, J.W., Eckert, J.W. Jr. and Carreras, E.M. 1993 The US EPA Geographic Information System for mapping environmental releases of toxic chemical release inventory (TRI) chemicals. *Risk Analysis* 13: 155–64.

Street-Perrott, F.A. and Perrott, R.A. 1990 Abrupt climate fluctuations in the tropics: the influence of the Atlantic circulation. *Nature* 343: 607–12.

Stretton, A.B. 1979 Ten lessons from the Darwin disaster. In Heathcote R.L. and Thom, B.G. (eds), *Natural Hazards in Australia*, Canberra: Australian Academy of Science, 503–7.

Suryo, I. and Clarke, M.C.G. 1985 The occurrence and mitigation of volcanic hazards in Indonesia as exemplified at the Mount Merapi, Mount Kelut and Mount Galunggung volcanoes. *Quarterly Journal of Engineering Geology* 18: 79–98.

Susman, P., O'Keefe, P. and Wisner, B. 1983 Global disasters: a radical interpretation. In Hewitt, K. (ed.), *Interpretations of Calamity*, Boston, MA, and London: Allen and Unwin, 263–83.

Sweeney J. 1985 The 1984 drought on the Canadian Prairies. *Weather* 4: 302–9.

Sylves, R. 1996 The politics and administration of presidential disaster declarations. *Quick Response Report* no. 86. Boulder, CO: Natural Hazards Research and Information Center.

Tayag, J.C. and Punongbayan, R.S. 1994 Volcanic disaster mitigation in the Philippines: experience from Mt Pinatubo. *Disasters* 18: 1–15.

Teng, W.L. 1990 AVHRR monitoring of US crops during the 1988 drought. *Photogrammetric Engineering and Remote Sensing* 56: 1143–6.

Thomas, C. 1994 Corruption makes mockery of claims by Bhopal victims. *The Times* 3 December: 14.

Thomas, M.F. 1994 *Geomorphology in the Tropics*. Chichester and New York: John Wiley.

Thompson, S.A. 1982 *Trends and Developments in Global Natural Disasters 1947 to 1981*. Working Paper no. 45. Boulder, CO: Institute of Behavioral Science, University of Colorado.

Thorarinsson, S. 1979 On the damage caused by volcanic eruptions with special reference to tephra and gases. In Sheets, P.D. and Grayson, D.K. (eds), *Volcanic Activity and Human Ecology*, London: Academic Press, 125–59.

Timmerman, P. 1981 *Vulnerability, Resilience and the Collapse of Society*. Environmental Monograph no. 1. Toronto: Institute for Environmental Studies, University of Toronto.

Timmerman, P. and White, R. 1997 Megahydropolis: coastal cities in the context of global environmental change. *Global Environmental Change* 7: 205–34.

Tinsley, J.C., Youd, T.L., Perkins, D.M. and Chen, A.T.F. 1985 Evaluating liquefaction potential. In Ziony, J.I. (ed.), *Evaluating Earthquake Hazards in the Los Angeles Region*, Washington, DC: Department of the Interior, 263–315.

Tobin, G.A. and Montz, B.E. 1997 The impacts of a second catastrophic flood on property values in Linda and Olivehurst, California. *Quick Response Report* no. 95. Boulder, CO: Natural Hazards Research and Applications Information Center.

Tomblin, J. 1988 UNDRO's role in responding to volcanic emergencies. *UNDRO News* March–April: 7–10.

Torry, W.I. 1979 Hazards, hazes and holes: a critique of *The Environment as Hazard* and general reflections on disaster research. *Canadian Geographer* 23: 368–83.

Torry, W.I. 1980 Urban earthquake hazard in developing countries: squatter settlements and the outlook for Turkey. *Urban Ecology* 4: 317–27.

Toulmin, L.M. 1987 Disaster preparedness and regional training on nine Caribbean islands: a long-term evaluation. *Disasters* 11: 221–34.

Trenberth, K. 2000 Short-term climate variations. In Pielke, R.A. Jr and Pielke, R.A. Sr (eds), *Storms*, London and New York: Routledge, vol. 1, 126–41.

Trenberth, K., Branstator, G.W. and Arkin, P.A. 1988 Origins of the 1988 North American drought. *Science* 242: 1640–5.

Tucker, B.E., Erdik, M. and Hwang, C.N. (eds) 1994 *Issues in Urban Earthquake Risk*, vol. 271, NATO ASI Series E: Applied Sciences. Berkeley, CA: Dordrecht, Kluwer.

Turner, R.H., Nigg, J.M., Paz, D.H. and Young, B.S. 1979 *Earthquake Threat: The Human Response in Southern California*. Los Angeles, CA: Institute for Social Science Research, University of California.

Turner, R.H., Nigg, J.M. and Paz, D.H. 1986 *Waiting for Disaster*. Berkeley, CA: University of California Press.

Twigg, J. (ed.) 1998 *Development at Risk? Natural Disasters and the Third World*. London: UK National Coordination Committee for the IDNDR.

Uitto, J.I. 1998 The geography of disaster vulnerability in mega-cities. *Applied Geography* 18: 7–16.

United Nations Disaster Relief Organization (UNDRO) 1985 *Volcanic Emergency Management*. New York: United Nations.

—— 1990 *Disaster Prevention and Preparedness Project for Ecuador and Neighbouring Countries*. Project Report. Geneva: Office of the Disaster Relief Co-ordinator.

United Nations Environment Programme (UNEP) 1987 *Environmental Data Report*. Oxford: Basil Blackwell.

US Department of Commerce 1994 *The Great Flood of 1993*. Natural Disaster Survey Report, Silver Spring, MD: Department of Commerce.

Valery, N. 1995 Earthquake engineering: a survey. *The Economist* 22 April.

Varnes, D.J. 1978 Slope movements and types and processes. In *Landslides: Analysis and Control*. Transportation Research Board, Special Report 176. Washington, DC: National Academy of Sciences, 11–13.

Vesely, W.E. 1984 Engineering risk analysis. In Ricci, P.F., Sagan, L.A. and Whipple, C.G. (eds), *Technological Risk Assessment*, The Hague: Martinus Nijhoff, 49–84.

Voight, B. 1996 The management of volcano emergencies: Nevado del Ruiz. In Scarpa, R. and Tilling, R.I. (eds), *Monitoring and Mitigation of Volcano Hazards*, Berlin: Springer-Verlag, 719–69.

Waddell, E. 1983 Coping with frosts, governments and disaster experts: some reflections based on a New Guinea experience and a perusal of the relevant literature. In Hewitt, K. (ed.), *Interpretations of Calamity*, Boston, MA, and London: Allen and Unwin, 33–43.

Wadge, G. 1993 Remote sensing for natural hazard assessment and mitigation. *STOP Disasters* 16 (Nov–Dec): 9–10.

Warrick, R.A., Anderson, J., Downing, T., Lyons, J., Ressler, J., Warrick, M. and Warrick, T. 1981 *Four Communities under Ash*. Monograph no. 34. Boulder, CO: Institute of Behavioral Science, University of Colorado.

Weaver, D.C. 1968 The hurricane as an economic catalyst. *Journal of Tropical Geography* 27: 66–71.

Weihe, W.H. and Mertens, R. 1991 Human well-being, diseases and climate. In Jager, J. and Ferguson, H.L. (eds), *Climate Change*, Cambridge: Cambridge University Press, 345–59.

Wells, M. 1984 We can improve relief efforts – if we try. *Ekistics* 51 (309): 501–6.

White, G.F. 1936 The limit of economic justification for flood protection. *Journal of Land and Public Utility Economics* 12: 133–48.

White, G.F. 1945 *Human Adjustment to Floods: A Geographical Approach to the Flood Problem in the United States*. Research Paper 29. Chicago, IL: Department of Geography, University of Chicago.

White, G.F. (ed.) 1974 *Natural Hazards: Local, National, Global*. New York: Oxford University Press.

White, G.F. and Haas, J.E. 1975 *Assessment of Research on Natural Hazards*. Cambridge, MA: MIT Press.

Whittow, J. 1980 *Disasters*. Harmondsworth, Middx: Penguin Books.

Whyte, A. V. and Burton, I. 1982 Perception of risk in Canada. In Burton, I., Fowle, C.D. and McCullough, R.S. (eds), *Living with Risk*, Toronto: Institute of Environmental Studies, University of Toronto, 39–69.

Wigley, T.M.L. 1983 The role of statistics in climate impact analysis. *Proceedings of the 2nd International Meeting on Statistical Climatology*, Lisbon, 8.1–1 to 8.1–10.

Wigley, T.M.L. 1985 Impact of extreme events. *Nature* 316: 106–7.

Wijkman, A. and Timberlake, L. 1984 *Natural Disasters: Acts of God or Acts of Man?* London: Earthscan.

Wilhite, D.A. 1986 Drought policy in the US and Australia: a comparative analysis. *Water Resources Bulletin* 22: 425–38.

Wilhite, D.A. and Glantz, M.H. 1985 Understanding the drought phenomenon: the role of definitions. *Water International* 10: 111–20.

Wilhite, D.A. and Vanyarkho, O. 2000 Drought: pervasive impacts of a creeping phenomenon. In Wilhite, D.A. (ed.), *Drought*, London and New York: Routledge, vol. 1, 245–55.

Williams, M. 1986 Emergency food aid to Africa: high risk of permanent dependence. *UNDRO News* Mar/Apr: 17–20.

Willoughby, H.E., Jorgensen, D.P., Black, R.A. and Rosenthal, S.L. 1985 Project STORMFURY: scientific chronicle 1962–1983. *Bulletin of the American Meteorological Society* 66: 505–14.

Witt, V.M. and Reiff, F.M. 1991 Environmental health conditions and cholera vulnerability in Latin America and the Caribbean. *Journal of Public Health Policy* 12: 450–64.

Wrathall. J.E. 1988 Natural hazard reporting in the UK press. *Disasters* 12: 177–82.

Wu, Q. 1989 The protection of China's ancient cities from flood damage. *Disasters* 13: 193–227.

Yagar, S. (ed.) 1984 *Transport Risk Assessment*. Waterloo, Ontario: University of Waterloo Press.

Zaman, M.Q. 1991 The displaced poor and resettlement policies in Bangladesh. *Disasters* 15: 117–25.

Zeckhauser, R. and Shepard, D.S. 1984 Principles for saving and valuing lives. In Ricci, P.F., Sagan, L.A. and Whipple, C.G. (eds), *Technological Risk Assessment*, The Hague: Martinus Nijhoff, 133–68.

Zimmerman, R. 1990 *Governmental Management of Chemical Risk*. London: Lewis; quoted in Royal Society, *Risk*, London: Royal Society, 1992.

Zupka, D. 1988 Economic impact of disasters. *UNDRO News* Jan–Feb: 19–22.

Subject index

Aa lava 161, 167–168
'Acts of God' syndrome 11, 71, 196
adobe structures 141, 261
adverse selection 94
AGRHYMET programme 313
agriculture, loss in disaster from 92, 126,
    160–162, 165, 213–215, 225, 242–244,
    254–255, 259, 274, 290, 294–296,
agricultural drought 294–296
aid following disaster 84–93: categories of
    88–91; after drought 304–307; after
    earthquakes 138–139; effectiveness of
    91–93; after floods 274, 275; purposes
    of 84–87; after tropical cyclones 228;
    after volcanic eruptions 166–167; after
    wildfires 255–256
AIDS 245
air transport hazards 325–326
ashfall 150, 158, 159, 160–161, 165,
    170–171, 179
avalanches snow 191–195, 199–202; causes
    of 181, 191–195; deaths from 182–183;
    environmental control of 199–202;
    forecasting and warning for 204; hazard-
    resistant design for 201–202; land use
    planning for 204–208; preparedness for
    202–203; remote sensing of 203, 205;
    types of 193–194; zoning systems
    205–207

band of tolerance 12–14, 66
bounded rationality 67–68
brushfire 248; see also wildfires
building codes 97, 102–106, 146, 233,
    238, 256
bushfire(s) 248; see also wildfires

caesium-137 330
cascade of hazard impacts 18–20

channel improvement 281
Chemical Emergency Preparedness Program
    335
'Chicago School' 4
choice in hazard adjustment(s) 81–83
cholera 247–248
chunam 198
climatic variability 4, 13–14
cold stress, windchill 240–241
compensation after technological disaster
    332
CRED 30–31, 36, 37, 40
cut and fill 144, 145, 189–191

dams: floods from failures of 186–187,
    264, 337; safety of 186–187, 319, 320,
    337
death: probability of 9–11
deaths in disaster 10–11, 32–37. from
    avalanches 182–183; at Bhopal 322,
    328; bias in reporting of 28–29; in
    Chernobyl 330; from civil strife 10–11;
    from cold stress 241; from debris flows
    188; in disaster definition 29–31; from
    drought 10, 290, 297; from earthquakes
    125–126, 131, 139; from epidemics
    239, 244, 245; from famine 10, 238,
    290, 297; from floods 259, 261,
    263–264, 265–266, 268, 269; from heat
    stress 241, 242; from ice and snow
    storms 227; from industrial disasters 17;
    from lahars 164, 165, 169, 170; from
    landslides 125, 134–135, 180–181, 188,
    198–199; from lava flows 161–162;
    from lightning 225; from malnutrition
    342; from mass movements 180–183;
    from road accidents 10, 71, 325–326;
    from smoking 9; from technological
    hazards 316, 319, 320, 321, 322, 323,

Geographical index